AF701056

Please note that the previous printing included a CD-ROM.

The material is now only available on the companion website:

http://booksite.elsevier.com/9781933762418.

Natural Gas Engineering Handbook
2nd Edition

Dr. Boyun Guo

and

Dr. Ali Ghalambor

University of Louisiana at Lafayette

Houston, TX

Natural Gas Engineering Handbook, 2nd Edition

HOUSTON, TX:
Gulf Publishing Company
2 Greenway Plaza, Suite 1020
Houston, TX 77046

ISBN-13: 978-0-12-810339-5

Transferred to Digital Printing in 2014

Library of Congress Cataloging-in-Publication Data

Guo, Boyun.

Natural gas engineering handbook / Boyun Guo, Ali Ghalambor. -- 2nd ed.

p. cm.

Includes bibliographical references and index.

ISBN 1-933762-41-1 (alk. paper)

1. Natural gas–Handbooks, manuals, etc. 2. Petroleum engineering–Handbooks, manuals, etc. 3. CD-ROMs I. Ghalambor, Ali. II. Title.

TN880.G86 2012

665.7–dc23

2012005312

Printed in the United States of America

Text design and composition by TIPS Technical Publishing, Inc.

This book is dedicated to the families of the authors for their understanding and encouragement that were as responsible as the experience and knowledge that have been inscribed herein.

Contents

Preface

It is well recognized that the nineteenth century was a century of coal that supported the initiation of industrial revolution in Europe. The twentieth century was the century of oil that was the primary energy source to support the growth of global economy. The demand of the world's economy for energy is ever increasing. The energy disruptions should be a genuine concern. It will likely cause chronic energy shortage as early as 2010. It will eventually evolve into a serious energy crunch. The way to avoid such a crunch is to expand energy supply and move from oil to natural gas and, eventually, to hydrogen. Natural gas is the only fuel that is superior to other energy sources in economic attractiveness and environmental concerns. At the end of the last century, natural gas took over the position of coal as the number two energy source behind oil. In 2000, total world energy consumption was slightly below 400 quadrillion (10^{15}) Btu. Of this, oil accounted for 39 percent, while natural gas and coal provided 23 percent and 22 percent, respectively. It is a historical imperative that the transition from oil to natural gas must be made in the early twenty-first century. This is not only motivated by environmental considerations but also technological innovations and refinements.

The consumption of natural gas in all end-use classifications (residential, commercial, industrial, and power generation) has increased rapidly since World War II. Natural gas is one of the major fossil energy sources. It provided close to 24 percent of U.S. energy sources over the three-year period of 2000 to 2002. There has been a huge disparity between "proven" reserves and potential reserves of natural gases. Even in the case of the highly mature and exploited United States, depending upon information sources, the potential remaining gas reserve estimates varies from 650 Tcf to 5,000 Tcf. Proved natural gas reserves in 2000 are about 1,050 Tcf in the Unites States and 170 Tcf in Canada. On the global scale, it is more difficult to give a good estimate of natural gas reserves. Major natural gas reserves are found in the former Soviet Union, Middle East, Asia Pacific, Africa, North America, Southern and Central America, and Europe.

Natural gas engineering has supported the natural gas industry since the birth of the industry. Although the principles of natural gas engineering have been documented in numerous books, most of them do not reflect the current practice in the natural gas industry where computer applications play a crucial role in engineering design and analyses. This book fills the gap.

This book is written primarily for natural gas production and processing engineers and college students of senior level. It is not the authors' intention to simply duplicate general information that can be found in other books. This book also gathers the authors' experiences gained through years of teaching the course of natural gas engineering at universities. The mission of the book is to provide engineers a handy guideline to designing, analyzing, and optimizing natural gas production and processing systems. This book can also be used as a reference for college students of undergraduate and graduate levels in petroleum engineering.

This book was intended to cover the full scope of natural gas production engineering. Following the sequence of natural gas production, this book presents its *contents* in twelve chapters. Chapter 1 presents a brief introduction to the natural gas industry. Chapter 2 documents properties of natural gases that are essential for designing and analyzing natural gas production and processing systems. Chapters 3 through 6 cover in detail the performance of gas wells. Chapter 7 focuses on the liquid separation process of natural gases. Chapter 8 describes dehydration processes of natural gases. Chapter 9 presents principles of gas compression and cooling. Chapter 10 describes gas-metering techniques. Chapter 11 presents principles of gas transportation in pipelines. Chapter 12 deals with special problems in natural gas production operations. Appendix A presents real gas pseudopressure charts for sweet natural gases. Appendix B provides charts for determining normalized pressures of sweet natural gases. Appendix C presents orifice meter tables for natural gases. Appendix D presents charts for the minimum gas production rates for water removal in gas wells and Appendix E presents charts for the minimum gas production rates for condensate removal in gas wells.

Because the substance of this book is virtually boundless, knowing what to omit was the greatest difficulty with its editing. The authors believe that it requires many books to describe the foundation of knowledge in natural gas engineering. To counter any deficiency that might arise from the limitations of space, we provide a reference list of books and papers at the end of each chapter so that readers should experience little difficulty in pursuing each topic beyond the presented scope.

As regards *presentation*, this book focuses on presenting principles, criteria, basic data, and spreadsheet programs necessary to quickly perform engineering analyses. Derivation of mathematical models is beyond the scope of this book. Most example calculations are presented with computer spreadsheets. All the spreadsheet programs are included on the following website: http://booksite.elsevier.com/9781933762418/.

This book is based on numerous documents including reports and papers accumulated through many years of work at the University of Louisiana at Lafayette. The authors are grateful to the university for permission to publish the materials. Special thanks go to the ChevronTexaco and American Petroleum Institute (API) for providing ChevronTexaco Professorship and API Professorship in Petroleum Engineering throughout the editing of this book. Last but not least, our thanks are due to friends and colleagues too numerous to mention, who encouraged, assessed, and made possible our editing this book. On the basis of their collective experience, we expect this book to be of value to engineers in the natural gas industry.

Dr. Boyun Guo

ChevronTexaco Endowed Professor
in Petroleum Engineering

University of Louisiana at Lafayette

Dr. Ali Ghalambor

American Petroleum Institute
Endowed Professor

University of Louisiana at Lafayette

List of Spreadsheet Programs

Spreadsheet Programs and Functions

Spreadsheet Name[a]	Function
4-PhaseFlow.xls	Four-phase flow in tubing
AverageTZ.xls	Gas flow in tubing
BHP-Gas Well.xls	Bottom hole pressure in gas wells
BottomHoleNodal.xls	Nodal analysis with bottom hole node
Brill-Beggs-z.xls	Z-factor with Brill-Beggs correlation
Carr-Kobayashi-Burrows Viscosity.xls	Viscosity by Carr-Kobayashi-Burrows correlation
Cullender-Smith.xls	Tubing performance by Cullender-Smith method
DryGasDownChoke.xls	Downstream choke pressure
DryGasUpChoke.xls	Upstream choke pressure
Empirical Deliverability.xls	Empirical reservoir deliverability
Empirical IPR.xls	Empirical IPR curve
GasWellLoading.xls	Gas well loading with Guo method
GlycolInjection.xls	Glycol requirement for hydrate prevention
Hall-Yarborough-Z.xls	Z-factor with Hall-Yarborough correlation
LoopedLine.xls	Capacities of series, parallel, and looped pipelines
LP-Flash.xls	Low-pressure flash
MethanolInjection.xls	Methanol requirement for hydrate prevention
MistFlow.xls	Mist flow in tubing
MixingRule.xls	Mixing rule for gas critical properties
Nodal.xls	Nodal analysis with average T and Z method
NormPCharts.xls	Normalized gas pressure
PipelineCapacity.xls	Single-pipeline capacity
PseudoP.xls	Real gas pseudopressure
PseudoPCharts.xls	Real gas pseudopressure chart
Theoretical Deliverability.xls	Theoretical reservoir deliverability
Theoretical IPR.xls	Theoretical IPR curve
TurnerLoading.xls	Critical gas rate by Turner's method
WellheadNodal.xls	Nodal analysis with wellhead node

a. All spreadsheet programs are on the following website: http://booksite.elsevier.com/9781933762418/.

List of Nomenclature

A = cross-sectional area of flow path, in^2

A_{fb} = the total firebox surface area, ft^2

B_g = gas formation volume factor, rb/scf

C = parameter in IPR model in Mscf/d-psi^{2n}; choke flow coefficient

C_d = drag coefficient

C_l = clearance, fraction

C_g = gas compressibility, psi^{-1}

C_w = water content of natural gas, lb$_m$/MMscf

C' = orifice flow constant

d = choke diameter, in

D = pipe diameter, non-Darcy coefficient in d/Mscf

D_i = diameter of the flow path, ft

d_i = diameter of the flow path, in

E = gas expansion factor, scf/ft^3

E_k = gas-specific kinetic energy, lb$_f$-ft/ft^3

E_{ksl} = kinetic energy required to hold liquid drops stationary, lb$_f$-ft/ft^3

E_v = volumetric efficiency, fraction

f = Moody friction factor

F_a = orifice thermal expansion factor

F_b = basic orifice factor, cfh

F_g = specific gravity factor

F_l = gauge location factor

F_m = manometer factor for mercury meter

F_{pb} = pressure base factor

F_{pv} = supercompressibility factor

F_r = Reynolds number factor

F_{tb} = temperature base factor

F_{tf} = flowing temperature factor

g = 32.2 ft/sec^2
g_c = gravitational conversion factor = 32.17 lb_m ft/lb_fsec^2
H = elevation above sea level, ft
Hp_{MM} = required theoretical compression power, hp/MMcfd
H_t = total heat load on reboiler, Btu/h
h_w = differential pressure in inches of water at 60 °F
I_{100} = pure inhibitor required, lb_m/MMscf
K = unit conversion coefficient
k = permeability in md; specific heat ratio
K_H = Hammerschmidt constant for inhibitor
k = liquid-vapor equilibrium ratio of compound i
L = measured depth in ft; latitude in degrees
lw = mechanical energy (loss of work) converted to heat, ft-lb_f/lb_m
$m(p)$ = real gas pseudopressure, psi^2/cp
MW_a = apparent molecular weight
MW_i = molecular weight of component i
n = parameter in IPR model, polytropic exponent
N_c = number of components
n_L = number of moles of fluid in the liquid phase
N_{Re} = Reynolds number
N_s = number of stages required
N_{sl} = slip speed, rpm
N_t = total operating speed, rpm
n_V = number of moles of fluid in the vapor phase
$n(p)$ = real gas normalized pressure
P = pressure at depth, lb/ft^2
p_{cf} = casing pressure, psia
p_{ci} = critical pressure of component i, psia
p_f = absolute static pressure, psia
P_{hf} = pressure at surface, lb/ft^2
p_{hf} = pressure at surface, psia
p_{mf} = pressure at middepth, psia

p_{pc} = pseudocritical pressure, psia
p_{pr} = pseudoreduced pressure
p_{tf} = tubing pressure, psia
p_{wf} = pressure at bottom hole, psia
q = gas production rate, Mscf/d
q_d = desired gas capacity, cfm
Q_{gc} = critical gas production rate, MMscf/day
q_G = glycol circulation rate, gal/hr
Q_{gm} = minimum required gas flow for liquid removal, MMscf/day
q_{msc} = gas production rate, MMscf/day
q_s = sand production rate, ft^3/day
q_o = oil production rate, stb/day
Q_{sc} = gas production rate, scf/day
q_{sc} = gas production rate, Mscf/day
q_w = water production rate, bbl/day
R_e = Reynolds number
r = cylinder compression ratio
R_h = differential pressure range, in
R_p = static pressure range, psi
r_w = wellbore radius, ft
s = skin factor
S_g = specific gravity of gas, air = 1
t = temperature, °F
T = temperature at depth, °R
T_{av} = average temperature at depth, °R
T_{ci} = critical temperature of component i, °R
T_{hf} = temperature at wellhead, °R
t_m = temperature during orifice boring, °F
T_{mf} = temperature at middepth, psia
T_{pc} = pseudocritical temperature, R
T_{pr} = pseudoreduced temperature
T_{wf} = temperature at bottom hole, psia

V = volume, ft^3
v = velocity, ft/s
V_{dis} = displacement, ft^3/revolution
v_{gc} = critical gas velocity, ft/s
V_s = required settling volume in separator, gal
w = theoretical shaft work required to compress the gas, ft-lb_f/lb_m
v_{gm} = minimum gas velocity required to transport liquid drops, ft/s
v_{sl} = terminal settling velocity, ft/s
v_{tr} = liquid transport velocity, ft/s
W_G = water removed from gas stream, lb_m/MMscf
W_h = weight of pure inhibitor in liquid water phase, %
W_{in} = concentration of inhibitor in inlet inhibitor stream, wt %
W_{out} = concentration of inhibitor in outlet inhibitor stream, wt %
W = water to be removed, lb_m/hr
x_i = mole fraction of compound i in the liquid phase
y_i = mole fraction of compound i in the vapor phase
z = gas compressibility factor
Z = vertical depth, ft
z_{av} = average gas compressibility factor
z_i = mole fraction of compound i in the system
ε = tubing wall roughness, in
γ_g = gas-specific gravity, air = 1
γ_o = oil-specific gravity, fresh water = 1
γ_s = sand-specific gravity, fresh water = 1
γ_w = water-specific gravity, fresh water = 1
ν_g = gas dynamic viscosity, cp
θ = inclination angle, degrees
ρ = density, lbm/ft^3
ρ_g = gas density, lbm/ft^3
ρ_l = liquid density, lbm/ft^3
σ = interfacial tension, dynes/cm
ν_g = gas kinetic viscosity, cs

Chapter 1

Introduction

1.1 What Is Natural Gas?

Natural gas is a subcategory of petroleum that is a naturally occurring, complex mixture of hydrocarbons, with a minor amount of inorganic compounds. Geologists and chemists agree that petroleum originates from plants and animal remains that accumulate on the sea/lake floor along with the sediments that form sedimentary rocks. The processes by which the parent organic material is converted into petroleum are not understood. The contributing factors are thought to be bacterial action; shearing pressure during compaction, heat, and natural distillation at depth; possible addition of hydrogen from deep-seated sources; presence of catalysts; and time (Allison and Palmer 1980).

Table 1–1 shows composition of a typical natural gas. It indicates that methane is a major component of the gas mixture. The inorganic compounds nitrogen, carbon dioxide, and hydrogen sulfide are not desirable because they are not combustible and cause corrosion and other problems in gas production and processing systems. Depending upon gas composition, especially the content of inorganic compounds, the heating value of natural gas usually varies from 700 Btu/scf to 1,600 Btu/scf.

Natural gas accumulations in geological traps can be classified as reservoir, field, or pool. A reservoir is a porous and permeable underground formation containing an individual bank of hydrocarbons confined by impermeable rock or water barriers and is characterized by a single natural pressure system. A field is an area that consists of one or more reservoirs all related to the same structural feature. A pool contains one or more reservoirs in isolated structures. Wells in the same field can be classified as gas wells, condensate wells, and oil wells. Gas wells are wells with producing gas-oil-ratio (GOR) being greater than 100,000 scf/stb;

Table 1–1 Composition of a Typical Natural Gas

Compound	Mole Fraction
Methane	0.8407
Ethane	0.0586
Propane	0.0220
i-Butane	0.0035
n-Butane	0.0058
i-Pentane	0.0027
n-Pentane	0.0025
Hexane	0.0028
Heptanes and Heavier	0.0076
Carbon Dioxide	0.0130
Hydrogen Sulfide	0.0063
Nitrogen	0.0345
Total	1.0000

condensate wells are those with producing GOR being less than 100,000 scf/stb but greater than 5,000 scf/stb; and wells with producing GOR being less than 5,000 scf/stb are classified as oil wells.

Because natural gas is petroleum in a gaseous state, it is always accompanied by oil that is liquid petroleum. There are three types of natural gases: nonassociated gas, associated gas, and gas condensate. Nonassociated gas is from reservoirs with minimal oil. Associated gas is the gas dissolved in oil under natural conditions in the oil reservoir. Gas condensate refers to gas with high content of liquid hydrocarbon at reduced pressures and temperatures.

1.2 Utilization of Natural Gas

Natural gas is one of the major fossil energy sources. When one standard cubic feet of natural gas is combusted, it generates 700 Btu to 1,600 Btu of heat, depending upon gas composition. Natural gas provided close to 24 percent of U.S. energy sources over the three-year period 2000–02. Natural gas is used as a source of energy in all sectors of the economy. Figure 1–1 shows that during the three-year period 2000–02, natural gas consumption was equitably distributed across all sectors of the U.S. economy (except transportation).

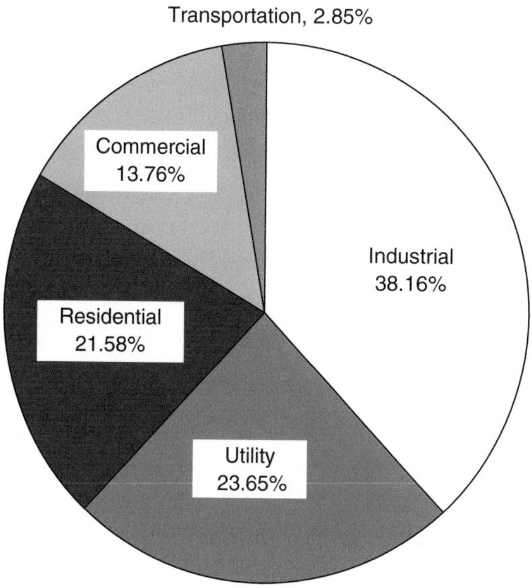

Figure 1–1 Natural gas is used as a source of energy in all sectors of the U.S. economy (Louisiana Department of Natural Resources 2004).

Example Problem 1.1

Natural gas from the Schleicher County, Texas, Straw Reef has a heating value of 1,598 Btu/scf. If this gas is combusted to generate power of 1,000 kW, what is the required gas flow rate in Mscf/day? Assume that the overall efficiency is 50 percent (1 kW = 3,412 Btu/h).

Solution

Output power of the generator:

$$1{,}000 \text{ kW} = (1{,}000 \text{ kW})(3{,}412 \text{ Btu/h per kW})$$

$$= 3.412 \times 10^6 \text{ Btu/h} = 8.19 \times 10^7 \text{ Btu/day}$$

Fuel gas requirement:

$$(8.19 \times 10^7 \text{ Btu/day})/(1{,}598 \text{ Btu/scf})/(0.5)$$
$$= 1.025 \times 10^5 \text{ scf/day} = 102.5 \text{ Mscf/day}$$

1.3 Natural Gas Industry

Natural gas was once a by-product of crude oil production. Since its discovery in the United States in Fredonia, New York, in 1821, natural gas has been used as fuel in areas immediately surrounding the gas fields. In the early years of the natural gas industry, when gas accompanied crude oil, it had to find a market or be flared; in the absence of effective conservation practices, oil-well gas was often flared in huge quantities. Consequently, gas production at that time was often short-lived, and gas could be purchased as low as 1 or 2 cents per 1,000 cu ft in the field (Ikoku 1984).

The consumption of natural gas in all end-use classifications (residential, commercial, industrial, and power generation) has increased rapidly since World War II. This growth has resulted from several factors, including development of new markets, replacement of coal as fuel for providing space and industrial process heat, use of natural gas in making petrochemicals and fertilizers, and strong demand for low-sulfur fuels.

The rapidly growing energy demands of Western Europe, Japan, and the United States could not be satisfied without importing gas from far fields. Natural gas, liquefied by a refrigeration cycle, can now be transported efficiently and rapidly across the oceans of the world by insulated tankers. The use of refrigeration to liquefy natural gas, and hence reduce its volume to the point where it becomes economically attractive to transport across oceans by tanker, was first attempted on a small scale in Hungary in 1934 and later used in the United States for moving gas in liquid form from the gas fields in Louisiana up the Mississippi River to Chicago in 1951 (Ikoku 1984).

The first use of a similar process on a large scale outside the United States was the liquefaction by a refrigerative cycle of some of the gas from the Hassi R'Mel gas field in Algeria and the export from 1964 onward of the resultant liquefied natural gas (LNG) by specially designed insulated tankers to Britain and France. Natural gas is in this way reduced to about one six-hundredth of its original volume and the nonmethane components are largely eliminated. At the receiving terminals, the LNG is reconverted to a gaseous state by passage through a regasifying plant, whence it can be fed as required into the normal gas distribution grid of the importing country. Alternatively, it can be stored for future use in insulated tanks or subsurface storages. Apart from its obvious applications as a storable and transportable form of natural gas, LNG has many applications in its own right, particularly as a nonpolluting fuel for aircraft and ground vehicles. Current production from conventional sources is not sufficient to satisfy all demands for natural gas.

1.4 Natural Gas Reserves

Two terms are frequently used to express natural gas reserves: proved reserves and potential resources. *Proved reserves* are those quantities of gas that have been found by the drill. They can be proved by known reservoir characteristics such as production data, pressure relationships, and other data, so that volumes of gas can be determined with reasonable accuracy. *Potential resources* constitute those quantities of natural gas that are believed to exist in various rocks of the Earth's crust but have not yet been found by the drill. They are future supplies beyond the proved reserves.

Different methodologies have been used in arriving at estimates of the future potential of natural gas. Some estimates were based on growth curves, extrapolations of past production, exploratory footage drilled, and discovery rates. Empirical models of gas discoveries and production have also been developed and converted to mathematical models. Future gas supplies as a ratio of the amount of oil to be discovered is a method that has been used also. Another approach is a volumetric appraisal of the potential undrilled areas. Different limiting assumptions have been made, such as drilling depths, water depths in offshore areas, economics, and technological factors.

There has been a huge disparity between "proven" reserves and potential reserves. Even in the case of the highly mature and exploited United States, depending upon information sources, the potential remaining gas reserve estimates vary from 650 Tcf to 5,000 Tcf (Economides et al. 2001). Proved natural gas reserves in 2000 were about 1,050 Tcf in the United States and 170 Tcf in Canada. On the global scale, it is more difficult to give a good estimate of natural gas reserves. Unlike oil reserves that are mostly (80 percent) found in Organization of Petroleum Exporting Countries (OPEC), major natural gas reserves are found in the former Soviet Union, Middle East, Asia Pacific, Africa, North America, Southern and Central America, and Europe.

1.5 Types of Natural Gas Resources

The natural gases can be classified as conventional natural gas, gas in tight sands, gas in tight shales, coal-bed methane, gas in geopressured reservoirs, and gas in gas hydrates.

Conventional natural gas is either associated or nonassociated gas. Associated or dissolved gas is found with crude oil. Dissolved gas is that portion of the gas dissolved in the crude oil and associated gas (sometimes called gas-cap gas) is free gas in contact with the crude oil. All crude oil reservoirs contain dissolved gas and may or may not contain associated gas. Nonassociated gas is found in a reservoir that contains a minimal quantity of crude oil. Some gases are called gas condensates or simply condensates. Although they occur as gases in underground reservoirs, they have a high content of hydrocarbon liquids. On production, they may yield considerable quantities of hydrocarbon liquids.

Gases in tight sands are found in many areas that contain formations generally having porosities of 0.001 to 1 millidarcy (md). Within the United States, the largest portion of the gas resource is found in the Green River Basin of Wyoming, the Piceance Basin of Colorado, and the Unita Basin of Utah (Ikoku 1984). At higher gas permeabilities, the formations are generally amenable to conventional fracturing and completion methods.

Gases in tight shales are found in the eastern United States (Kentucky, Ohio, Virginia, and West Virginia). Of these, eastern Kentucky and western West Virginia are considered the most important. The shale is generally fissile, finely laminated, and varicolored but predominantly black, brown, or greenish-gray. Core analysis has determined that the shale itself may have up to 12 percent porosity, however, permeability values are commonly less than 1 md. It is thought, therefore, that the majority of production is controlled by naturally occurring fractures and is further influenced by bedding planes and jointing (Ikoku 1984).

Coal-bed methane is the methane gas in minable coal beds with depths less than 3,000 ft. Although the estimated size of the resource base seems significant, the recovery of this type of gas may be limited owing to practical constraints.

In a rapidly subsiding basin area, clays often seal underlying formations and trap their contained fluids. After further subsidence, the pressure and temperature of the trapped fluids exceed those normally anticipated at reservoir depth. These reservoirs, commonly called geopressured reservoirs, have been found in many parts of the world during the search for oil and gas. In the United States they are located predominantly both onshore and offshore in a band along the Gulf of Mexico (Ikoku 1984). In length, the band extends from Florida to Texas; in width, it extends from about 100 miles inland to the edge of the continental shelf.

Gas hydrates, discovered in 1810, are snow-like solids in which each water molecule forms hydrogen bonds with the four nearest water molecules to build a crystalline lattice structure that traps gas molecules in its cavities (Sloan 1990). Gas hydrates contain about 170 times the natural gas by volume under standard conditions. Because gas hydrate is a highly concentrated form of natural gas and extensive deposits of naturally occurring gas hydrates have been found in various regions of the world, they are considered as a future, unconventional resource of natural gas.

1.6 Future of the Natural Gas Industry

It is well recognized that the nineteenth century was a century of coal that supported the initiation of industrial revolution in Europe. The twentieth century was the century of oil that was the primary energy source to support the growth of global economy. Figure 1–2 shows world energy consumption in the past three decades and forecast for the next two decades (DOE/EIA 2001). It indicates that the demand of the world's economy for energy is ever increasing. Simmons (2000) concluded that energy disruptions should be a "genuine concern." Simmons suggests that it will likely cause chronic energy shortage as early as 2010. It will eventually evolve into a serious energy crunch.

The way to avoid such a crunch is to expand energy supply and move from oil to natural gas and, eventually, to hydrogen. Natural gas is the fuel that is superior to other energy sources not only in economic attractiveness but also in environmental concerns. At the end of the last century, natural gas took over the position of coal as the number two energy source behind oil. In 2000, total world energy consumption was slightly below 400 quadrillion

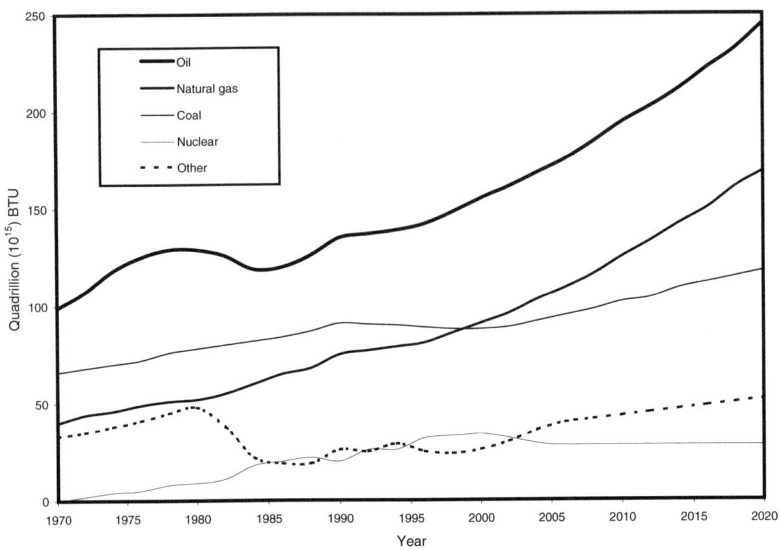

Figure 1–2 World energy consumption and forecast from 1970 to 2020 (DOE/EIA 2001).

(10^{15}) Btu. Of this, oil accounted for 39 percent, while natural gas and coal provided 23 percent and 22 percent, respectively (DOE/EIA 2001). It is a historical imperative that the transition from oil to natural must be made in the early twenty-first century. This is not only motivated by environmental considerations but also by technological innovations and refinements (Economides, Demarchos, Saputelli 2002).

The United States has the world's largest economy and is by far the most voracious user of energy. Figure 1–3 presents U.S. natural gas production history. The U.S. demand for natural gas can also be seen from the increase in gas price in the last three decades (Figure 1–4). The very conservative estimates (DOE/EIA 2001) suggest that while the total annual energy demand between 2000 and 2020 will increase by 30 percent from 98 to 127 quadrillion Btu, natural gas will increase by over 60 percent from 22.5 to 35.6 quadrillion Btu, or about 35 Tcf. This means that the natural gas share of the energy mix will increase from 23 percent to over 28 percent. It is clear that natural gas is now becoming the premier fuel of choice for the world economy. Other so-called alternative energy sources have little chance to compete with natural gas.

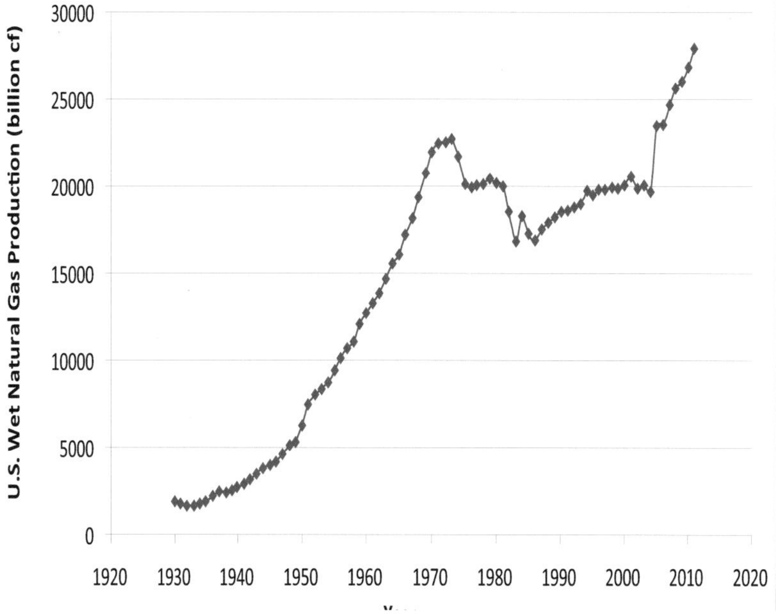

Figure 1–3 U.S. natural gas production history (Louisiana Department of Natural Resources 2011).

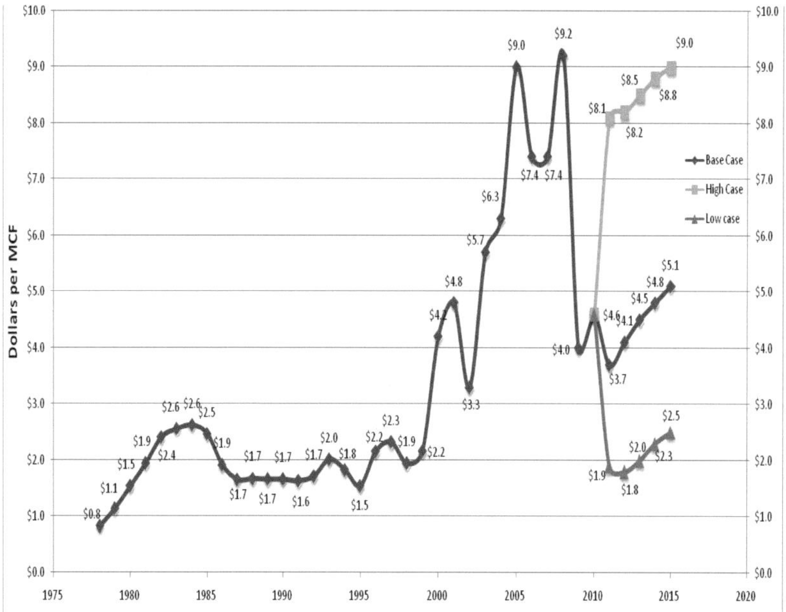

Figure 1–4 U.S. natural gas price from 1978 to 2010 (Louisiana Department of Natural Resources 2011).

1.7 References

Allison, I. S. and D. F. Palmer. *Geology*. 7th ed. New York: McGraw-Hill, 1980.

DOE/EIA. "International Energy Outlook." Energy Information Administration, Department of Energy, Washington DC, 2001.

Economides, M. J., A. S. Demarchos, and L. Saputelli. "Energy Sources and Energy Intensity for the Twenty-First Century." Paper SPE 75504 presented at the SPE Gas Technology Symposium, Calgary, Alberta, Canada, April 30–May 2, 2002.

Economides, M. J., R. E. Oligney, A. S. Demarchos, and P.E. Lewis. "Natural Gas: Beyond All Expectations." Paper SPE 71512 presented at the 2001 SPE Annual Technical Conference and Exhibition, New Orleans, Louisiana, September 30–October 3, 2001.

Ikoku, C. U. *Natural Gas Production Engineering*. New York: John Wiley & Sons, 1984.

Louisiana Department of Natural Resources. *Louisiana Energy Facts*. Baton Rouge, January 2011.

Louisiana Department of Natural Resources. *Louisiana Energy Facts*. Baton Rouge, March 2011.

Simmons, M. R. *An Energy White Paper*. Houston: Simmons and Company International, October 2000.

Sloan, E. D. *Clathrate Hydrates of Natural Gases*. New York: Marcel Dekker Inc., 1990.

1.8 Problems

1-1 Natural gas from the Morgan County, Colorado, D-Sand, has a heating value of 1,228 Btu/scf. If this gas is combusted to drive a gas turbine for a gas compressor of 1,000 hp, what is the required gas flow rate in MMscf/day? Assume that the overall efficiency is 30% (1 hp = 2,544 Btu/h).

1-2 Natural gas from the William County, North Dakota, Red River formation, has a heating value of 1,032 Btu/scf. If this gas is used to generate electricity at a rate of 1 MMscf/day, how many watts of electricity would the generator produce if the overall efficiency is 50% (1 hp = 745 W)?

Chapter 2

Properties of Natural Gas

2.1 Introduction

Properties of natural gas include gas-specific gravity, pseudocritical pressure and temperature, viscosity, compressibility factor, gas density, and gas compressibility. Knowledge of these property values is essential for designing and analyzing natural gas production and processing systems. Because natural gas is a complex mixture of light hydrocarbons with a minor amount of inorganic compounds, it is always desirable to find the composition of the gas through measurements. Once the gas composition is known, gas properties can usually be estimated using established correlations with confidence. This chapter focuses on determination of gas properties with correlations developed from various lab measurements. Example problems are presented and solved using computer programs provided with this book.

2.2 Specific Gravity

Gas-specific gravity (γ_g) is defined as the ratio of the apparent molecular weight of a natural gas to that of air, itself a mixture of gases. The molecular weight of air is usually taken as equal to 28.97 (approximately 79% nitrogen and 21% oxygen). Therefore the gas gravity is

$$\gamma_g = \frac{MW_a}{28.97} \tag{2.1}$$

where the apparent molecular weight of gas can be calculated on the basis of gas composition. Gas composition is usually determined in a laboratory and reported in mole fractions of components in the gas. Let y_i be the mole fraction of component i, the apparent molecular weight of the gas can be formulated using mixing rule as

$$MW_a = \sum_{i=1}^{Nc} y_i MW_i \tag{2.2}$$

where MW_i is the molecular weight of component i, and N_c is the number of components. The molecular weights of compounds (MW_i) can be found in textbooks on organic chemistry or petroleum fluids such as that by McCain (1973). A light gas reservoir is one that contains primarily methane with some ethane. Pure methane would have a gravity equal to (16.04/28.97) = 0.55. A rich or heavy gas reservoir may have a gravity equal to 0.75 or, in some rare cases, higher than 0.9.

2.3 Pseudocritical Properties

Similar to gas apparent molecular weight, the critical properties of a gas can be determined on the basis of the critical properties of compounds in the gas using the mixing rule. The gas critical properties determined in such a way are called pseudocritical properties. Gas pseudocritical pressure (p_{pc}) and pseudocritical temperature (T_{pc}) are, respectively, expressed as

$$p_{pc} = \sum_{i=1}^{N_c} y_i p_{ci} \tag{2.3}$$

and

$$T_{pc} = \sum_{i=1}^{N_c} y_i T_{ci} \tag{2.4}$$

where p_{ci} and T_{ci} are critical pressure and critical temperature of component i, respectively.

Example Problem 2.1

For the gas composition given in the following text, determine apparent molecular weight, pseudocritical pressure, and pseudocritical temperature of the gas.

Component	Mole Fraction
C_1	0.775
C_2	0.083
C_3	0.021
i-C_4	0.006
n-C_4	0.002
i-C_5	0.003
n-C_5	0.008
C_6	0.001
C_{7+}	0.001
N_2	0.050
CO_2	0.030
H_2S	0.020

Solution

This problem is solved with the spreadsheet program MixingRule.xls. Results are shown in Table 2–1.

If the gas composition is not known but gas-specific gravity is given, the pseudocritical pressure and temperature can be determined from various charts or correlations developed based on the charts. One set of simple correlations is

$$p_{pc} = 709.604 - 58.718\gamma_g \quad \textbf{(2.5)}$$

$$T_{pc} = 170.491 + 307.344\gamma_g \quad \textbf{(2.6)}$$

Table 2–1 Results Given by MixingRule.xls[a]

Compound	y_i	MW_i	y_iMW_i	p_{ci} (psia)	y_ip_{ci} (psia)	T_{ci} (°R)	y_iT_{ci} (°R)
C_1	0.775	16.04	12.43	673	521.58	344	266.60
C_2	0.083	30.07	2.50	709	58.85	550	45.65
C_3	0.021	44.10	0.93	618	12.98	666	13.99
i-C_4	0.006	58.12	0.35	530	3.18	733	4.40
n-C_4	0.002	58.12	0.12	551	1.10	766	1.53
i-C_5	0.003	72.15	0.22	482	1.45	830	2.49
n-C_5	0.008	72.15	0.58	485	3.88	847	6.78
C_6	0.001	86.18	0.09	434	0.43	915	0.92
C_{7+}	0.001	114.23	0.11	361	0.36	1024	1.02
N_2	0.050	28.02	1.40	227	11.35	492	24.60
CO_2	0.030	44.01	1.32	1073	32.19	548	16.44
H_2S	0.020	34.08	0.68	672	13.45	1306	26.12
	1.000	MW_a =	20.71	p_{pc} =	661	T_{pc} =	411
		γ_g =	0.71				

a. This spreadsheet calculates gas apparent molecular weight, specific gravity, pseudocritical pressure, and pseudocritical temperature.

which are valid for $H_2S < 3\%$, $N_2 < 5\%$, and total content of inorganic compounds less than 7%.

Corrections for impurities in sour gases are always necessary. The corrections can be made using either charts or correlations such as the Wichert-Aziz (1972) correction expressed as follows:

$$A = y_{H_2S} + y_{CO_2} \tag{2.7}$$

$$B = y_{H_2S} \tag{2.8}$$

$$\varepsilon_3 = 120\left(A^{0.9} - A^{1.6}\right) + 15\left(B^{0.5} - B^{4.0}\right) \tag{2.9}$$

$$T_{pc}' = T_{pc} - \varepsilon_3 \quad \text{(corrected } T_{pc}\text{)} \tag{2.10}$$

$$P_{pc}' = \frac{P_{pc} T_{pc}'}{T_{pc} + B(1 - B)\varepsilon_3} \quad \text{(corrected } p_{pc}\text{)} \tag{2.11}$$

Correlations with impurity corrections for mixture pseudocriticals are also available (Ahmed 1989):

$$p_{pc} = 678 - 50(\gamma_g - 0.5) - 206.7 y_{N_2} + 440 y_{CO_2} + 606.7 y_{H_2S} \tag{2.12}$$

$$T_{pc} = 326 + 315.7(\gamma_g - 0.5) - 240 y_{N_2} - 83.3 y_{CO_2} + 133.3 y_{H_2S} \tag{2.13}$$

Applications of the pseudocritical pressure and temperature are normally found in natural gas engineering through pseudoreduced pressure and temperature defined as:

$$p_{pr} = \frac{p}{p_{pc}} \tag{2.14}$$

$$T_{pr} = \frac{T}{T_{pc}} \tag{2.15}$$

2.4 Viscosity

Gas viscosity is a measure of the resistance to flow exerted by the gas. Dynamic viscosity (μ_g) in centipoises (cp) is usually used in the natural engineering:

$$1 \text{ cp} = 6.72 \times 10^{-4} \text{ lbm/ft-sec}$$

Kinematic viscosity (ν_g) is related to the dynamic viscosity through density (ρ_g)

$$\nu_g = \frac{\mu_g}{\rho_g} \quad (2.16)$$

Kinematic viscosity is not normally used in natural gas engineering.

Direct measurements of gas viscosity are preferred for a new gas. If gas composition and viscosities of gas components are known, the mixing rule can be used for determining the viscosity of the gas mixture:

$$\mu_g = \frac{\sum\left(\mu_{gi} y_i \sqrt{MW_i}\right)}{\sum\left(y_i \sqrt{MW_i}\right)} \quad (2.17)$$

Gas viscosity is very often estimated with charts or correlations developed based on the charts. The gas viscosity correlation of Carr, Kobayashi, and Burrows (1954) involves a two-step procedure: the gas viscosity at temperature and atmospheric pressure is estimated first from gas-specific gravity and inorganic compound content. The atmospheric value is then adjusted to pressure conditions by means of a correction factor on the basis of reduced temperature and pressure state of the gas. The atmospheric pressure viscosity (μ_1) can be expressed as:

$$\mu_1 = \mu_{1HC} + \mu_{1N_2} + \mu_{1CO_2} + \mu_{1H_2S} \quad (2.18)$$

where

$$\mu_{1HC} = 8.188 \times 10^{-3} - 6.15 \times 10^{-3}\log(\gamma_g) + (1.709 \times 10^{-5} - 2.062 \times 10^{-6}\gamma_g)T \quad (2.19)$$

$$\mu_{1N_2} = [9.59 \times 10^{-3} + 8.48 \times 10^{-3}\log(\gamma_g)]y_{N_2} \quad (2.20)$$

$$\mu_{1CO_2} = [6.24 \times 10^{-3} + 9.08 \times 10^{-3}\log(\gamma_g)]y_{CO_2} \quad (2.21)$$

$$\mu_{1H_2S} = [3.73 \times 10^{-3} + 8.49 \times 10^{-3}\log(\gamma_g)]y_{H_2S} \quad (2.22)$$

Dempsey (1965) developed the following relation:

$$\mu_r = \ln\left(\frac{\mu_g}{\mu_1}T_{pr}\right) = a_0 + a_1 p_{pr} + a_2 p_{pr}^2 + a_3 p_{pr}^3$$

$$+T_{pr}(a_4 + a_5 p_{pr} + a_6 p_{pr}^2 + a_7 p_{pr}^3)$$

$$+T_{pr}^2(a_8 + a_9 p_{pr} + a_{10} p_{pr}^2 + a_{11} p_{pr}^3)$$

$$+T_{pr}^3(a_{12} + a_{13} p_{pr} + a_{14} p_{pr}^2 + a_{15} p_{pr}^3) \quad \textbf{(2.23)}$$

where

$a_0 = -2.46211820$

$a_1 = 2.97054714$

$a_2 = -0.28626405$

$a_3 = 0.00805420$

$a_4 = 2.80860949$

$a_5 = -3.49803305$

$a_6 = 0.36037302$

$a_7 = -0.01044324$

$a_8 = -0.79338568$

$a_9 = 1.39643306$

$a_{10} = -0.14914493$

$a_{11} = 0.00441016$

$a_{12} = 0.08393872$

$a_{13} = -0.18640885$

$a_{14} = 0.02033679$

$a_{15} = -0.00060958$

Thus, once the value of μ_r is determined from the right-hand side of this equation, gas viscosity at elevated pressure can be readily calculated using the following relation:

$$\mu_g = \frac{\mu_1}{T_{pr}} e^{\mu_r} \tag{2.24}$$

Other correlations for gas viscosity include Dean-Stiel (1958) and Lee-Gonzalez-Eakin (1966).

Example Problem 2.2

A 0.65 specific gravity natural gas contains 10% nitrogen, 8% carbon dioxide, and 2% hydrogen sulfide. Estimate viscosity of the gas at 10,000 psia and 180 °F.

Solution

This problem is solved with the spreadsheet Carr-Kobayashi-Burrows Viscosity.xls that is attached to this book. The result is shown in Table 2–2.

2.5 Compressibility Factor

Gas compressibility factor is also called deviation factor, or z-factor. Its value reflects how much the real gas deviates from the ideal gas at given pressure and temperature. Definition of the compressibility factor is expressed as:

$$z = \frac{V_{actual}}{V_{ideal\ gas}} \tag{2.25}$$

Introducing the z-factor to the gas law for ideal gas results in the gas law for real gas as:

$$pV = nzRT \tag{2.26}$$

Table 2–2 Results Given by Carr-Kobayashi-Burrows Viscosity.xls[a]

Input Data	
Pressure:	10,000 psia
Temperature:	180 °F
Gas-specific gravity:	0.65 air =1
Mole fraction of N_2:	0.1
Mole fraction of CO_2:	0.08
Mole fraction of H_2S:	0.02
Calculated Parameter Values	
Pseudocritical pressure:	697.164 psia
Pseudocritical temperature:	345.357 °R
Uncorrected gas viscosity at 14.7 psia:	0.012174 cp
N_2 correction for gas viscosity at 14.7 psia:	0.000800 cp
CO_2 correction for gas viscosity at 14.7 psia:	0.000363 cp
H_2S correction for gas viscosity at 14.7 psia:	0.000043 cp
Corrected gas viscosity at 14.7 psia (μ_1):	0.013380 cp
Pseudoreduced pressure:	14.34
Pseudoreduced temperature:	1.85
ln ($\mu_g/\mu_1 {*} T_{pr}$):	1.602274
Gas viscosity:	0.035843 cp

a. This spreadsheet calculates gas viscosity with correlation of Carr, Kobayashi, and Burrows.

where n is the number of moles of gas. When pressure p is entered in psia, volume V in ft^3, and temperature in °R, the gas constant R is equal to

$$10.73 \frac{psia - ft^3}{mole - {}^\circ R}.$$

The gas compressibility factor can be determined on the basis of measurements in PVT laboratories. For a given amount of gas, if temperature is kept constant and volume is measured at 14.7 psia and an elevated pressure p_1, z-factor can then be determined with the following formula:

$$z = \frac{p_1}{14.7} \frac{V_1}{V_0} \tag{2.27}$$

where V_0 and V_1 are gas volumes measured at 14.7 psia and p_1, respectively.

Very often the z-factor is estimated with the chart developed by Standing and Katz (1942). This chart has been set up for computer solution by a number of individuals. Brill and Beggs (1974) yield z-factor values accurate enough for many engineering calculations. Brill and Beggs' z-factor correlation is expressed as follows:

$$A = 1.39(T_{pr} - 0.92)^{0.5} - 0.36T_{pr} - 0.10 \tag{2.28}$$

$$B = (0.62 - 0.23T_{pr})p_{pr} + \left(\frac{0.066}{T_{pr} - 0.86} - 0.037\right)p_{pr}^2 + \frac{0.32p_{pr}^6}{10^E} \tag{2.29}$$

$$C = 0.132 - 0.32\log(T_{pr}) \tag{2.30}$$

$$D = 10^F \tag{2.31}$$

$$E = 9(T_{pr} - 1) \tag{2.32}$$

$$F = 0.3106 - 0.49T_{pr} + 0.1824T_{pr}^2 \tag{2.33}$$

and

$$z = A + \frac{1 - A}{e^B} + Cp_{pr}^D \tag{2.34}$$

Example Problem 2.3

For the natural gas described in Example Problem 2.2, estimate z-factor at 5,000 psia and 180 °F.

Solution

This problem is solved with the spreadsheet program Brill-Beggs-Z.xls. The result is shown in Table 2–3.

Table 2–3 Results Given by Brill-Beggs-Z.xls[a]

Input Data	
Pressure:	5,000 psia
Temperature:	180 °F
Gas-specific gravity:	0.65 1 for air
Mole fraction of N_2:	0.1
Mole fraction of CO_2:	0.08
Mole fraction of H_2S:	0.02
Calculated Parameter Values	
Pseudocritical pressure:	697 psia
Pseudocritical temperature:	345 °R
Pseudo-reduced pressure:	7.17
Pseudo-reduced temperature:	1.85
A =	0.5746
B =	2.9057
C =	0.0463
D =	1.0689
Gas compressibility factor z:	0.9780

a. This spreadsheet calculates gas compressibility factor based on Brill and Beggs correlation.

Hall and Yarborough (1973) presented more accurate correlation to estimate z-factor of natural gas. This correlation is summarized as follows:

$$t = \frac{1}{T_{pr}} \tag{2.35}$$

$$A = 0.06125\ t\ e^{-1.2(1-t)^2} \tag{2.36}$$

$$B = t\left(14.76 - 9.76t + 4.58t^2\right) \tag{2.37}$$

$$C = t\left(90.7 - 242.2t + 42.4t^2\right) \tag{2.38}$$

$$D = 2.18 + 2.82t \tag{2.39}$$

and

$$z = \frac{Ap_{pr}}{Y} \tag{2.40}$$

where Y is the reduced density to be solved from

$$f(Y) = \frac{Y + Y^2 + Y^3 - Y^4}{(1-Y)^3} - Ap_{pr} - BY^2 + CY^D = 0 \tag{2.41}$$

If Newton-Raphson's iteration method is used to solve Equation (2.41) for Y, the following derivative is needed:

$$\frac{df(Y)}{dY} = \frac{1 + 4Y + 4Y^2 - 4Y^3 + Y^4}{(1-Y)^4} - 2BY + CDY^{D-1} \tag{2.42}$$

Example Problem 2.4

For a natural gas with a specific gravity of 0.71, estimate z-factor at 5,000 psia and 180 °F.

Solution

This problem is solved with the spreadsheet program Hall-Yarborogh-z.xls. The result is shown in Table 2–4.

Table 2–4 Results Given by Hall-Yarborogh-z.xls[a]

Instructions: 1) Input data; 2) Run Macro Solution; 3) View result.	
Input Data	
T:	180 °F
p:	5,000 psia
SGFG:	0.71 air = 1
Calculate Critical and Reduced Temperature and Pressure	
Tpc = 169.0 + 314.0*SGFG:	391.94 °R
Ppc = 708.75 – 57.5*SGFG:	667.783 psia
Tpr = (T + 460.0)/Tpc:	1.632902995
t = 1/Tpr:	0.61240625
Ppr = p/Ppc:	7.487462244
Calculate Temperature-dependent Terms	
A = 0.06125*t*EXP(–1.2*(1.–t**2):	0.031322282
B = t*(14.76 – 9.76*t + 4.58*t*t):	6.430635935
C = t*(90.7 – 242.2*t + 42.4*t*t):	–25.55144909
D = 2.18 + 2.82*t:	3.906985625
Calculate Reduced Density (use Macro Solution)	
Y = ASSUMED:	0.239916681
F = –A*Ppr + (Y + Y*Y + Y**3 – Y**4)/(1.-Y)**3 – B*Y*Y + C*Y**D:	–7.30123E–06
Calculate z-Factor	
Z = A*Ppr/Y:	0.97752439

a. This spreadsheet computes gas compressibility factor with the Hall-Yarborough method.

2.6 Gas Density

Because natural gas is compressible, its density depends upon pressure and temperature. Gas density can be calculated from gas law for real gas with good accuracy:

$$\rho = \frac{m}{V} = \frac{MW_a p}{zRT} \tag{2.43}$$

where m is mass of gas and ρ is gas density. Taking air molecular weight 29 and $R = 10.73 \frac{\text{psia} - \text{ft}^3}{\text{mole} - {}^\circ\text{R}}$, Equation (2.43) is rearranged to yield:

$$\rho = \frac{2.7\gamma_g p}{zT} \tag{2.44}$$

where the gas density is in lbm/ft^3. This equation is also coded in the spreadsheet program Hall-Yarborogh-z.xls.

2.7 Formation Volume Factor and Expansion Factor

Formation volume factor is defined as the ratio of gas volume at reservoir condition to the gas volume at standard condition, that is,

$$B_g = \frac{V}{V_{sc}} = \frac{p_{sc}}{p} \frac{T}{T_{sc}} \frac{z}{z_{sc}} = 0.0283 \frac{zT}{p} \tag{2.45}$$

where the unit of formation volume factor is ft^3/scf. If expressed in rb/scf, it takes the form of

$$B_g = 0.00504 \frac{zT}{p} \tag{2.46}$$

Gas formation volume factor is frequently used in mathematical modeling of gas well inflow performance relationship (IPR).

Gas expansion factor is defined, in scf/ft^3, as:

$$E = \frac{1}{B_g} = 35.3\frac{P}{ZT} \tag{2.47}$$

or

$$E = 198.32\frac{p}{zT} \tag{2.48}$$

in scf/rb. It is normally used for estimating gas reserves.

2.8 Compressibility of Natural Gas

Gas compressibility is defined as:

$$C_g = -\frac{1}{V}\left(\frac{\partial V}{\partial p}\right)_T \tag{2.49}$$

Because the gas law for real gas gives $V = \frac{nzRT}{p}$,

$$\left(\frac{\partial V}{\partial p}\right) = nRT\left(\frac{1}{p}\frac{\partial z}{\partial p} - \frac{z}{p^2}\right) \tag{2.50}$$

Substituting Equation (2.50) into Equation (2.49) yields:

$$C_g = \frac{1}{p} - \frac{1}{z}\frac{\partial z}{\partial p} \tag{2.51}$$

2.9 Real Gas Pseudopressure

Real gas pseudopressure $m(p)$ is defined as

$$m(p) = \int_{p_b}^{p} \frac{2p}{\mu z} dp \tag{2.52}$$

where p_b is the base pressure (14.7 psia in most states in the U.S.). The pseudopressure is considered to be a "pseudoproperty" of gas because it depends on gas viscosity and compressibility factor, which are properties of the gas. The pseudopressure is widely used for mathematical modeling of IPR of gas wells. Determination of the pseudopressure at a given pressure requires knowledge of gas viscosity and z-factor as functions of pressure and temperature. As these functions are complicated and not explicit, a numerical integration technique is frequently used.

Example Problem 2.5

Natural gas from a gas reservoir has a specific gravity of 0.71. It also contains the following compounds:

Mole fraction of N_2:	0.10
Mole fraction of CO_2:	0.08
Mole fraction of H_2S:	0.02

Calculate pseudopressure at 10,000 psia and 180 °F. Plot pressure against pseudopressure in the pressure range from 14.7 psia to 10,000 psia.

Solution

The spreadsheet PseudoP.xls can be used for calculating real gas pseudopressure. In the spreadsheet, gas viscosity is calculated with the correlation of Carr, Kobayashi, and Burrows. Gas deviation factor is calculated with the correlation of Brill and Beggs. Numerical integration is performed with a trapezoidal method. Table 2–5 shows data input and calculated parameters.

Table 2–5 Input Data and Calculated Parameters Given by PseudoP.xls[a]

Input Data	
Base pressure:	14.7 psia
Maximum pressure:	10,000 psia
Temperature:	60 °F
Gas-specific gravity:	0.6 1 for air
Mole fraction of N_2:	0
Mole fraction of CO_2:	0
Mole fraction of H_2S:	0
Calculated Parameter Values	
Pseudocritical pressure:	673 psia
Pseudocritical temperature:	357.57 °R
Uncorrected gas viscosity at 14.7 psia:	0.010504 cp
N_2 correction for gas viscosity at 14.7 psia:	0.000000 cp
CO_2 correction for gas viscosity at 14.7 psia:	0.000000 cp
H_2S correction for gas viscosity at 14.7 psia:	0.000000 cp
Corrected gas viscosity at 14.7 psia (μ_1):	0.010504 cp
Pseudoreduced temperature:	1.45

a. This spreadsheet computes real gas pseudopressures.

Calculated gas viscosities, z-factors, and pseudopressures at pressures between 9,950 psia and 10,000 psia are presented in Table 2–6. Pseudopressure values in the whole range of pressure are plotted in Figure 2–1.

For the convenience of engineering applications, pseudopressures of sweet natural gases at various pressures and temperatures have been generated with PseudoP.xls. The results are presented in Appendix A.

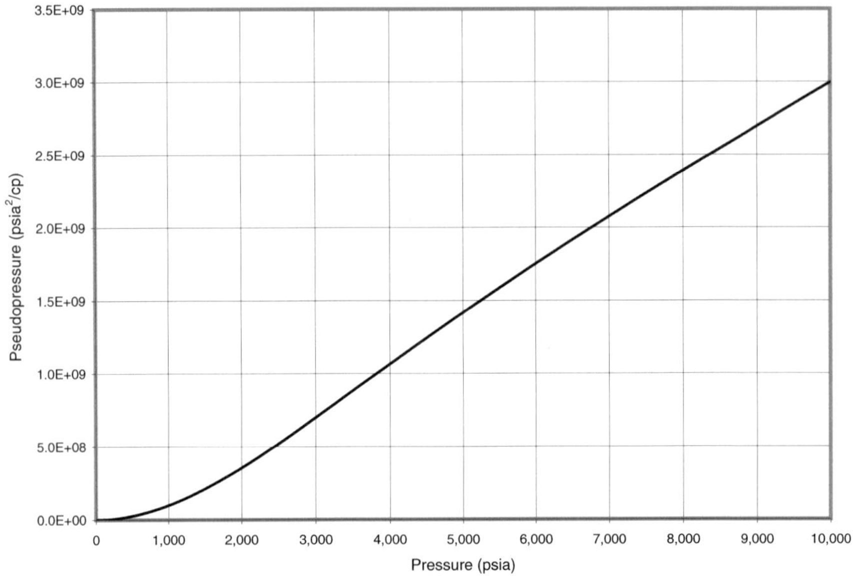

Figure 2–1 Plot of pseudopressures calculated by PseudoP.xls.

2.10 Real Gas Normalized Pressure

Real gas normalized gas pressure *n(p)* is defined as

$$n(p) = \int_0^{p_r} \frac{p_r}{z} dp_r \tag{2.53}$$

where p_r is the pseudoreduced pressure. For the convenience of engineering applications, the normalized gas pressures of sweet natural gases at various pressures and temperatures have been generated with the spreadsheet program NormP.xls. The results are presented in Appendix B.

Table 2-6 Partial Output Given by PseudoP.xls

p (psia)	μ (cp)	z	2p/(μz)	m(p)
9,950	0.045325	1.462318	300,244	2,981,316,921
9,952	0.045329	1.462525	300,235	2,981,916,517
9,954	0.045333	1.462732	300,226	2,982,516,096
9,956	0.045337	1.462939	300,218	2,983,115,657
9,958	0.045341	1.463146	300,209	2,983,715,201
9,960	0.045345	1.463353	300,200	2,984,314,727
9,962	0.045349	1.463560	300,191	2,984,914,236
9,964	0.045353	1.463767	300,182	2,985,513,727
9,966	0.045357	1.463974	300,174	2,986,113,200
9,968	0.045361	1.464182	300,165	2,986,712,656
9,970	0.045365	1.464389	300,156	2,987,312,094
9,972	0.045369	1.464596	300,147	2,987,911,515
9,974	0.045373	1.464803	300,138	2,988,510,918
9,976	0.045377	1.465010	300,130	2,989,110,304
9,978	0.045381	1.465217	300,121	2,989,709,672
9,980	0.045385	1.465424	300,112	2,990,309,022
9,982	0.045389	1.465631	300,103	2,990,908,355
9,984	0.045393	1.465838	300,094	2,991,507,670
9,986	0.045397	1.466045	300,086	2,992,106,968
9,988	0.045401	1.466252	300,077	2,992,706,248
9,990	0.045405	1.466459	300,068	2,993,305,510
9,992	0.045409	1.466666	300,059	2,993,904,755
9,994	0.045413	1.466873	300,050	2,994,503,982
9,996	0.045417	1.467080	300,041	2,995,103,191
9,998	0.045421	1.467287	300,033	2,995,702,383
10,000	0.045425	1.467494	300,024	2,996,301,557

2.11 References

Ahmed, T. *Hydrocarbon Phase Behavior*. Houston: Gulf Publishing Company, 1989.

Brill, J. P., and H. D. Beggs. "Two-Phase Flow in Pipes." INTERCOMP Course, The Hague, 1974.

Carr, N.L., R. Kobayashi, and D. B. Burrows. "Viscosity of Hydrocarbon Gases under Pressure." *Trans. AIME* **201** (1954): 264–72.

Dempsey, J. R. "Computer Routine Treats Gas Viscosity as a Variable." *Oil & Gas Journal* (Aug. 16, 1965): 141.

Dean, D. E. and L. I. Stiel. "The Viscosity of Non-polar Gas Mixtures at Moderate and High Pressures." *AIChE Journal* **4** (1958): 430–6.

Hall, K. R. and L. Yarborough. "A New Equation of State for Z-Factor Calculations." *Oil & Gas Journal* (June 18, 1973): 82.

Lee, A. L., M. H. Gonzalez, and B. E. Eakin. "The Viscosity of Natural Gases." *Journal of Petroleum Technology* (Aug. 1966): 997–1000.

McCain, W. D., Jr. *The Properties of Petroleum Fluids*, Tulsa: PennWell Books, 1973.

Standing, M. B. and D. L. Katz. "Density of Natural Gases." *Trans. AIME* **146**: (1954) 140–9.

Standing, M. B.: *Volumetric and Phase Behavior of Oil Field Hydrocarbon Systems*. Society of Petroleum Engineers of AIME, Dallas, 1977.

Wichert, E. and K. Aziz. "Calculate Zs for Sour Gases." *Hydrocarbon Processing* **51** (May 1972): 119.

2.12 Problems

2-1 Estimate gas viscosities of a 0.70 specific gravity gas at 200 °F and 100 psia, 1,000 psia, 5,000 psia, and 10,000 psia.

2-2 Calculate gas compressibility factors of a 0.65 specific gravity gas at 150 °F and 50 psia, 500 psia, and 5,000 psia with Hall-

Yarborough method. Compare the results with that given by the Brill and Beggs' correlation. What is your conclusion?

2-3 For a 0.65 specific gravity gas at 250 °F, calculate and plot pseudopressures in a pressure range from 14.7 psia and 8,000 psia. Under what condition is the pseudopressure linearly proportional to pressure?

2-4 Prove that the compressibility of an ideal gas is equal to inverse of pressure, that is, $C_g = \frac{1}{p}$.

Chapter 3

Gas Reservoir Deliverability

3.1 Introduction

Gas reservoir deliverability is evaluated using well inflow performance relationship (IPR). Gas well IPR determines gas production rate as a nonlinear function of pressure drawdown (reservoir pressure minus bottom hole pressure). Gas well IPR also depends on flow conditions, that is, transient, steady state, or pseudosteady state flow, which are determined by reservoir boundary conditions. This chapter presents methods that can be used for establishing gas well IPR under different flow conditions. Both analytical methods and empirical methods are discussed. Example problems are illustrated and solved using computer programs provided with this book.

3.2 Analytical Methods

A general solution to pseudosteady state flow in a radial-flow gas reservoir is expressed as (Economides 1994):

$$q = \frac{kh\left[m(\overline{p}) - m(p_{wf})\right]}{1424T\left[\ln\left(\frac{0.472r_e}{r_w}\right) + s + Dq\right]} \qquad \textbf{(3.1)}$$

where q is the gas production rate in Mscf/d, k is the effective permeability to gas in md, h is the thickness of pay zone in ft, $m(\overline{p})$ is the real gas pseudopressure in psi^2/cp at the reservoir pressure $\overline{p}$ in psi, $m(p_{wf})$ is the real gas pseudopressure in psi^2/cp at the flowing bottom hole pressure

p_{wf}, T is the reservoir temperature in R, γ_w is the radius of drainage area in ft, γ_w is wellbore radius in ft, s is skin factor, and D is the non-Darcy coefficient in d/Mscf. The skin factor and non-Darcy coefficient can be estimated on the basis of pressure transient analyses.

As the real gas pseudopressure is difficult to evaluate without a computer program, approximations to Equation (3.1) are usually used in the natural gas industry. At pressures lower than 2,000 psia,

$$m(p) = \int_{p_b}^{p} \frac{2p}{\mu z}\, dp \approx \frac{p^2 - p_b^2}{\bar{\mu}\ \bar{z}} \tag{3.2}$$

where p_b is the base pressure, $\bar{\mu}$ is the average gas viscosity, and $\bar{z}$ is the average gas compressibility factor. Equation (3.1) can then be simplified using a pressure-squared approach such as:

$$q = \frac{kh\left(\bar{p}^2 - p_{wf}^2\right)}{1424\ \bar{\mu}\ \bar{z}\ T\left[\ln\left(\frac{0.472 r_e}{r_w}\right) + s + Dq\right]} \tag{3.3}$$

At pressures higher than 3,000 psia, highly compressed gases behave like liquids. Equation (3.1) can be approximated using pressure approach as:

$$q = \frac{kh\left(\bar{p} - p_{wf}\right)}{141.2\times 10^3\, \bar{B}_g \bar{\mu}\left[\ln\left(\frac{0.472 r_e}{r_w}\right) + s + Dq\right]} \tag{3.4}$$

where $\bar{B}_g$ is the average formation volume factor of gas in rb/scf.

Example Problem 3.1

A gas well produces 0.65 specific gravity natural gas with N_2, CO_2, and H_2S of mole fractions 0.1, 0.08, and 0.02, respectively. The well diameter is 7-7/8 inches. It drains gas from a 78-ft thick pay zone in an area of 160 acres. The average reservoir pressure is 4,613 psia. Reservoir temperature is 180 °F. Assuming a Darcy skin factor of 5 and a non-Darcy coefficient of 0.001 day/Mscf, estimate the deliverability of the gas reservoir under pseudosteady state flow condition at a flowing bottom hole pressure of 3,000 psia.

Solution

This problem is solved with the spreadsheet program Theoretical Deliverability.xls. The appearance of the first section of the spreadsheet is shown in Table 3–1. Results are shown in Table 3–2.

Table 3–1 The First Section of Theoretical Deliverability.xls[a]

Instructions: 1) Go to the Solution section and enter a value for real gas pseudopressure at the flowing bottom hole pressure; 2) Run Macro Solution and view results.

Input Data	
Effective permeability to gas:	0.17 md
Pay zone thickness:	78 ft
Equivalent drainage radius:	1,490 ft
Wellbore radius:	0.328 ft
Darcy skin factor:	5
Non-Darcy coefficient:	0.001d/Mscf
Reservoir pressure:	4,613 psia
Flowing bottom hole pressure:	3,000 psia
Temperature:	180 °F
Gas-specific gravity:	0.65 1 for air
Mole fraction of N_2:	0.1
Mole fraction of CO_2:	0.08
Mole fraction of H_2S:	0.02

a. This spreadsheet calculates theoretical gas reservoir deliverability.

Table 3–2 Results Given by Theoretical Deliverability.xls

Solution 1) Based on the property table, enter a value for the real gas pseudopressure at pressure 3,000 psia 604,608,770 psi^2/cp; 2) Run Macro Solution to get result.	
$q=\dfrac{kh\left[m(\overline{p})-m(p_{wp})\right]}{1424T\left[\ln\left(\dfrac{0.472r_e}{r_w}\right)+s+Dq\right]}$	= 709 Mscf/d
$q=\dfrac{kh\left(\overline{p}^2-p_{wf}^2\right)}{1424\ \overline{\mu}\ \overline{z}\ T\left[\ln\left(\dfrac{0.472r_e}{r_w}\right)+s+Dq\right]}$	= 632 Mscf/d
$q=\dfrac{kh\left(\overline{p}-p_{wf}\right)}{141.2\times10^3\,\overline{B}_g\overline{\mu}\left[\ln\left(\dfrac{0.472r_e}{r_w}\right)+s+Dq\right]}$	= 759 Mscf/d

3.3 Empirical Methods

Very often it is difficult and costly to obtain values of all the parameters in equations (3.1), (3.3), and (3.4). Empirical models are therefore more attractive and widely employed in field applications. Two commonly used empirical models are the Forchheimer model and backpressure model. They take the following forms, respectively:

$$m(\overline{p})-m(p_{wf})=Aq+Bq^2 \tag{3.5}$$

and

$$q=C[m(\overline{p})-m(p_{wf})]^n \tag{3.6}$$

where A, B, C, and n are empirical constants that can be determined based on test points. The value of n is usually between 0.5 and 1. It is obvious that a multirate test is required to estimate values of these constants. If two test points are (q_1, p_{wf1}) and (q_2, p_{wf2}), expressions of these constants are:

$$B = \frac{[m(\bar{p}) - m(p_{wf1})]q_2 - [m(\bar{p}) - m(p_{wf2})]q_1}{q_1^2 q_2 - q_2^2 q_1} \tag{3.7}$$

$$A = \frac{[m(\bar{p}) - m(p_{wf1})] - Bq_1^2}{q_1} \tag{3.8}$$

$$n = \frac{\log\left(\frac{q_1}{q_2}\right)}{\log\left(\frac{m(\bar{p}) - m(p_{wf1})}{m(\bar{p}) - m(p_{wf2})}\right)} \tag{3.9}$$

$$C = \frac{q_1}{\left[m(\bar{p}) - m(p_{wf1})\right]^n} \tag{3.10}$$

Similar to Equation (3.1), Equation (3.5) and Equation (3.6) can be simplified using the pressure-squared approach as follows:

$$\bar{p}^2 - p_{wf}^2 = Aq + Bq^2 \tag{3.11}$$

and

$$q = C(\bar{p}^2 - p_{wf}^2)^n \tag{3.12}$$

Similarly, the constants can be determined with test points as follows:

$$B = \frac{(\bar{p}^2 - p_{wf1}^2)q_2 - (\bar{p}^2 - p_{wf2}^2)q_1}{q_1^2 q_2 - q_2^2 q_1} \tag{3.13}$$

$$A = \frac{(\bar{p}^2 - p_{wf1}^2) - Bq_1^2}{q_1} \quad \textbf{(3.14)}$$

$$n = \frac{\log\left(\frac{q_1}{q_2}\right)}{\log\left(\frac{\bar{p}^2 - p_{wf1}^2}{\bar{p}^2 - p_{wf2}^2}\right)} \quad \textbf{(3.15)}$$

$$C = \frac{q_1}{\left[\bar{p}^2 - p_{wf1}^2\right]^n} \quad \textbf{(3.16)}$$

Example Problem 3.2

A gas well produces 0.65 specific gravity natural gas with N_2, CO_2, and H_2S of mole fractions 0.1, 0.08, and 0.02, respectively. The average reservoir pressure is 4,505 psia. Reservoir temperature is 180 °F. The well was tested at two flow rates:

Test point 1	
Flow rate:	1,152 Mscf/d
Bottom hole pressure:	3,025 psia
Test point 2	
Flow rate:	1,548 Mscf/d
Bottom hole pressure:	1,685 psia

Estimate the deliverability of the gas reservoir under a pseudosteady state flow condition at a flowing bottom hole pressure of 1,050 psia.

Solution

This problem is solved with the spreadsheet program Empirical Deliverability.xls. The appearance of the first section of the spreadsheet is shown in Table 3–3. Results are shown in Table 3–4.

Table 3–3 The First Section of Empirical Deliverability.xls[a]

Instructions: 1) Go to the Solution section and enter values for real gas pseudopressure at the tested and desired flowing bottom hole pressures; 2) Run Macro Solution, and view results.

Input Data	
Reservoir pressure:	4,505 psia
Test point 1, flow rate:	1,152 Mscf/d
bottom hole pressure:	3,025 psia
Test point 2, flow rate:	1,548 Mscf/d
bottom hole pressure:	1,685 psia
Flowing bottom hole pressure:	1,050 psia
Temperature:	180 °F
Gas-specific gravity:	0.65 1 for air
Mole fraction of N_2:	0.1
Mole fraction of CO_2:	0.08
Mole fraction of H_2S:	0.02

a. This spreadsheet calculates gas reservoir deliverability with empirical models.

Table 3–4 Results Given by Empirical Deliverability.xls

Solution	
1) Use Forchheimer equation with real gas pseudopressure:	
Enter real gas pseudopressure at pressure	3,025 psia: 588,157,460 psi²/cp
Enter real gas pseudopressure at pressure	1,685 psia: 198,040,416 psi²/cp
Enter real gas pseudopressure at the desired pressure	1,050 psia: 79,585,534 psi²/cp
$B=\dfrac{[m(\overline{p})-m(p_{wf1})]q_2-[m(\overline{p})-m(p_{wf2})]q_1}{q_1^2q_2-q_2^2q_1}$	= 288

Table 3–4 Results Given by Empirical Deliverability.xls (Continued)

$$A=\frac{[m(\bar{p})-m(p_{wf1})]-Bq_1^2}{q_1}$$	= 208,000
Run Macro Solution to get result.	
$$m(\bar{p})-m(p_{wf})=Aq+Bq^2$$	gives q = 1,653 Mscf/d
2) Use Forchheimer equation with pressure-squared approach:	
$$B=\frac{(\bar{p}^2-p_{wf1}^2)q_2-(\bar{p}^2-p_{wf2}^2)q_1}{q_1^2q_2-q_2^2q_1}$$	= 4.05
$$A=\frac{(\bar{p}^2-p_{wf1}^2)-Bq_1^2}{q_1}$$	= 5,012
Run Macro Solution to get result.	
$$\bar{p}^2-p_{wf}^2=Aq+Bq^2$$	gives q = 1,645 Mscf/d
3) Use backpressure model with real gas pseudopressure:	
$$n=\frac{\log\left(\frac{q_1}{q_2}\right)}{\log\left(\frac{m(\bar{p})-m(p_{wf1})}{m(\bar{p})-m(p_{wf2})}\right)}$$	= 0.61
$$C=\frac{q_1}{\left[m(\bar{p})-m(p_{wf1})\right]^n}$$	= 0.0053
$$q=C[m(\bar{p})-m(p_{wf})]^n$$	gives q = 1,656 Mscf/d

Table 3–4 Results Given by Empirical Deliverability.xls (Continued)

4) Use backpressure model with pressure-squared approach:	
$n = \dfrac{\log\left(\dfrac{q_1}{q_2}\right)}{\log\left(\dfrac{\bar{p}^2 - p_{wf1}^2}{\bar{p}^2 - p_{wf2}^2}\right)}$	= 0.66
$C = \dfrac{q_1}{\left[\bar{p}^2 - p_{wf1}^2\right]^n}$	= 0.0264
$q = C(\bar{p}^2 - p_{wf}^2)^n$	gives q = 1,648 Mscf/d

3.4 Construction of Inflow Performance Relationship Curve

Once a deliverability equation is established using either a theoretical or an empirical equation, it can be used to construct well IPR curves.

Example Problem 3.3

Construct IPR curve for the well specified in Example Problem 3.1 with both pressure and pressure-squared approaches.

Solution

This problem is solved with the spreadsheet program Theoretical IPR.xls. The appearance of the spreadsheet is shown in Table 3–5 and Table 3–6. IPR curves are shown in Figure 3–1.

Table 3–5 Input Data Given by Theoretical IPR.xls[a]

Instructions: 1) Update input data; 2) Run Macro Solution and view result and plot.	
Input Data	
Effective permeability to gas:	0.17 md
Pay zone thickness:	78 ft
Equivalent drainage radius:	1,490 ft
Wellbore radius:	0.328 ft
Darcy skin factor:	5
Non-Darcy coefficient:	0.001 d/Mscf
Reservoir pressure:	4,613 psia
Temperature:	180 °F
The average gas viscosity:	0.022 cp
The average gas compressibility factor:	0.96
Effective permeability to gas:	78 ft

a. This spreadsheet calculates and plots theoretical gas well IPR curves.

Table 3–6 Solution Given by Theoretical IPR.xls

p_{wf} (psia)	q (Mscf/d)	
	p Approach	p^2 Approach
Solution		
B_g = 0.000671281 rb/SCF		
15	1,994	1,067
245	1,906	1,064
475	1,817	1,057
704	1,726	1,044
934	1,635	1,026
1,164	1,543	1,004

Table 3–6 Solution Given by Theoretical IPR.xls (Continued)

p_{wf} (psia)	q (Mscf/d)	
	p Approach	p^2 Approach
1,394	1,450	976
1,624	1,355	943
1,854	1,259	905
2,084	1,163	862
2,314	1,064	814
2,544	965	760
2,774	864	700
3,004	762	635
3,234	658	563
3,463	553	486
3,693	446	403
3,923	337	312
4,153	227	216
4,383	114	111
4,613	0	0

Example Problem 3.4

Construct IPR curve for the well specified in Example Problem 3.2 with both Forchheimer and backpressure equations.

Solution

This problem is solved with the spreadsheet program Empirical IPR.xls. The appearance of the spreadsheet is shown in Table 3–7 and Table 3–8. IPR curves are shown in Figure 3–2.

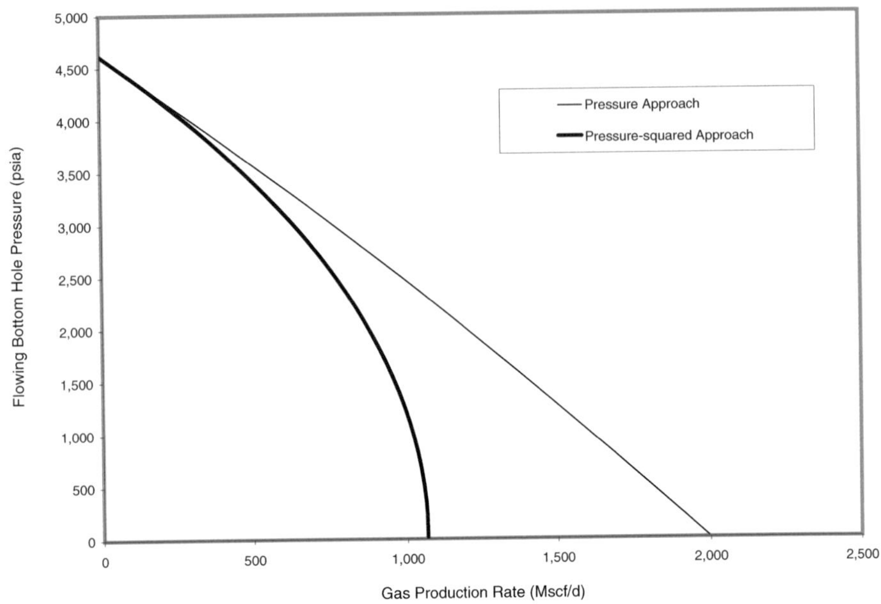

Figure 3–1 IPR curves given by the spreadsheet program Theoretical IPR.xls.

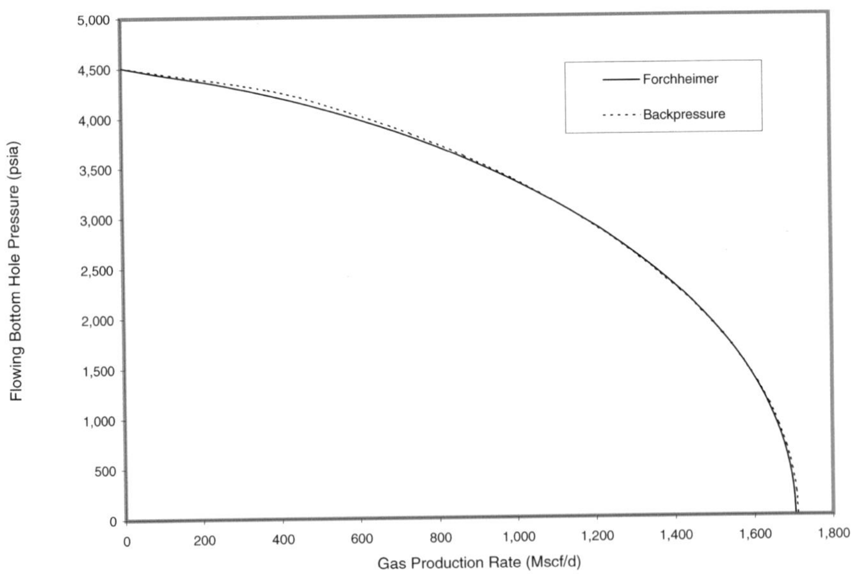

Figure 3–2 IPR curves given by the spreadsheet program Empirical IPR.xls.

Table 3–7 Input Data and Solution Given by Empirical IPR.xls[a]

Instructions: 1) Update data; 2) Run Macro Solution and view results.	
Input Data	
Reservoir pressure:	4,505 psia
Test point 1, flow rate:	1,152 Mscf/d
bottom hole pressure	3,025 psia
Test point 2, flow rate:	1,548 Mscf/d
bottom hole pressure	1,685 psia
Solution	
$B = \dfrac{(\bar{p}^2 - p_{wf1}^2)q_2 - (\bar{p}^2 - p_{wf2}^2)q_1}{q_1^2 q_2 - q_2^2 q_1}$	= 4.05
$A = \dfrac{(\bar{p}^2 - p_{wf1}^2) - Bq_1^2}{q_1}$	= 5,012
$n = \dfrac{\log\left(\dfrac{q_1}{q_2}\right)}{\log\left(\dfrac{\bar{p}^2 - p_{wf1}^2}{\bar{p}^2 - p_{wf2}^2}\right)}$	= 0.66
$C = \dfrac{q_1}{\left[\bar{p}^2 - p_{wf1}^2\right]^n}$	= 0.0264

a. This spreadsheet calculates and plots well IPR curve with empirical models.

Table 3–8 Results Given by Empirical IPR.xls

p_{wf} (psia)	q (Mscf/d)	
	Forchheimer	Backpressure
15	1,704	1,709
239	1,701	1,706
464	1,693	1,698
688	1,679	1,683
913	1,660	1,663
1,137	1,634	1,637
1,362	1,603	1,605
1,586	1,566	1,567
1,811	1,523	1,522
2,035	1,472	1,471
2,260	1,415	1,412
2,484	1,349	1,346
2,709	1,274	1,272
2,933	1,190	1,189
3,158	1,094	1,095
3,382	984	990
3,607	858	871
3,831	712	734
4,056	535	572
4,280	314	368
4,505	0	0

3.5 Horizontal Wells

Joshi (1988) presented a mathematical model considering steady-state flow of oil in the horizontal plane and pseudosteady-state flow in the vertical plane. Joshi's equation was modified by Economides et al. (1991) to include the effect of reservoir anisotropy. Guo et al. (2007) pointed out that Joshi's equation is optimistic for high-productivity reservoirs due to neglecting the effect of frictional pressure in the horizontal wellbore. Guo et al. (2008) suggests that the following modified Joshi equation be applied to gas wells:

$$q_g = \frac{k_H h\left(p_e^2 - p_{wf}^2\right)}{1424\bar{\mu}_g \bar{z} T\left\{\ln\left[\dfrac{a+\sqrt{a^2-(L/2)^2}}{L/2}\right]+\dfrac{I_{ani}h}{L}\ln\left[\dfrac{I_{ani}h}{r_w(I_{ani}+1)}\right]+s+Dq_g\right\}} F_g \quad \textbf{(3.17)}$$

where

$$a = \frac{L}{2}\sqrt{\frac{1}{2}+\sqrt{\left[\frac{1}{4}+\left(\frac{r_{eH}}{L/2}\right)^4\right]}} \quad \textbf{(3.18)}$$

and

q_g= gas production rate (Mscf/day)

$$I_{ani} = \sqrt{\frac{k_H}{k_V}}$$

k_H = the average horizontal permeability (md)

k_V = vertical permeability (md)

r_{eH} = radius of drainage area of horizontal well (ft)

L = length of horizontal wellbore ($L/2 < 0.9r_{eH}$) (ft)

F_g = correction factor for drain hole friction, (1 for no-friction hole)

T = reservoir temperature (^{o}R)

$\bar{\mu}_g$ = average gas viscosity (cp)

$\bar{z}$ = average gas compressibility factor (dimensionless)

s = skin factor (dimensionless)

D = NON-Darcy flow coefficient (day/Mscf)

The method for obtaining the correction factor F_g was given by Guo et al. (2008) and is presented in Appendix F.

3.6 Multi-Fractured Horizontal Wells

Some low-productivity horizontal wells, such as those in shale gas reservoirs, are purposely drilled in parallel to the direction of the minimum horizontal stress in the formation. This specific wellbore orientation allows multiple transverse fractures to be hydraulically created for enhancing productivity. Linear flow may exist initially before fractures begin to influence each other. Radial flow may prevail later if the drainage area is sufficiently large compared to the fractured region of the reservoir.

Raghavan and Joshi (1993) presented a mathematical model that can predict the productivities of horizontal wells with multiple transverse fractures. The model uses the effective wellbore radius (in radial flow) to simulate fluid flow toward the fractured well. Flow within the fracture itself was not considered. Li et al. (1996) presented an analytical model for predicting productivities of horizontal wells with multiple transverse fractures. The model incorporates linear flow from the fractured reservoir region to the fractures, linear flow within the fractures, radial flow within the fractures to the horizontal wellbore, and flow from the fractured region directly to the horizontal wellbore.

Most fractured horizontal wells are drilled in low-permeability reservoirs, in which fluid flow from the unfractured regions directly to the horizontal wellbore is often negligible. As demonstrated by Guo and Yu (2008), predictions of the long-term productivity of multi-fractured horizontal wells must consider the following sequence:

1 Reservoir radial flow within the drainage boundary to the fractured region of reservoir.

2 Reservoir linear flow between fractures in the reservoir to the fracture faces.

3 Fracture linear flow in the fracture to the near-wellbore region.

4 Wellbore radial flow in the fracture to the wellbore, where a "choking" effect occurs.

Figure 3-3 shows two regions of the reservoir. The inner region is the fractured region, and the outer region is the non-fractured region (Guo et al., 2008). Figure 3-4 illustrates flow in the fracture.

For the outer region, the following well productivity model may apply:

$$q = \frac{k_H h\left(\bar{p}^2 - p_L^2\right)}{1424\bar{z}\bar{\mu}_g T\left(\frac{1}{2}\ln\frac{4A}{\gamma C_A r_L^2}\right)} \quad \textbf{(3.19)}$$

where

$$r_L = \sqrt{\frac{4n\bar{z}_e\bar{x}_f}{\pi}} \quad \textbf{(3.20)}$$

where $\bar{z}_e$ and $\bar{x}_f$ are the average half-distances between fractures and fracture half-length, respectively. The aspect ratio (length to width) of the drainage area may be taken as

$$R_A = \frac{n\bar{z}_e}{\bar{x}_f}$$

and the shape factor may be estimated as C_A =39.51-8.5214R_A.

The reservoir-fracture cross-flow model of Guo and Schechter (1997) gives:

$$q = \sum_{i=1}^{n} \frac{4.475\times10^{-4} h}{\bar{\mu}_g \bar{z} T\sqrt{c_i}\left[\frac{\left(z_{ei} - z_{si}\right)}{k_H} + \frac{z_{si}}{k_{si}}\right]}\left(1 - e^{-\sqrt{c_i}x_{fi}}\right)\left(p_L^2 - p_r^2\right) \quad \textbf{(3.21)}$$

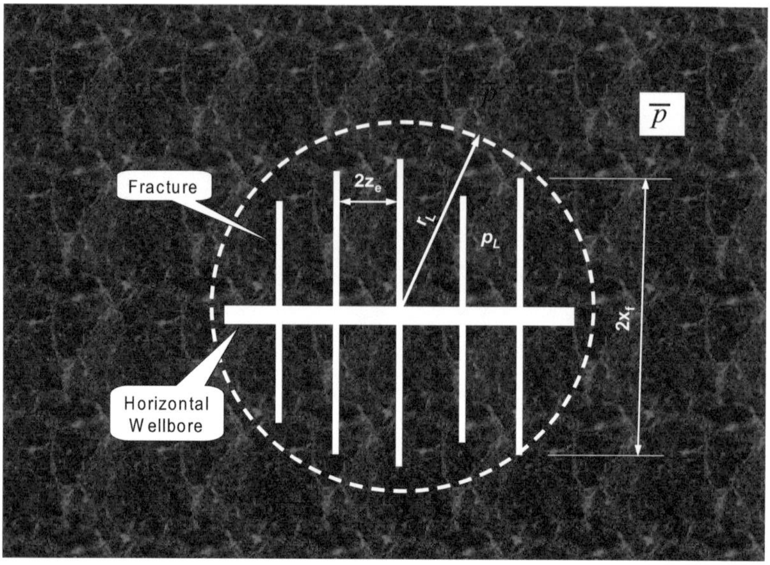

Figure 3–3 A reservoir section drained by a multi-fractured horizontal wellbore

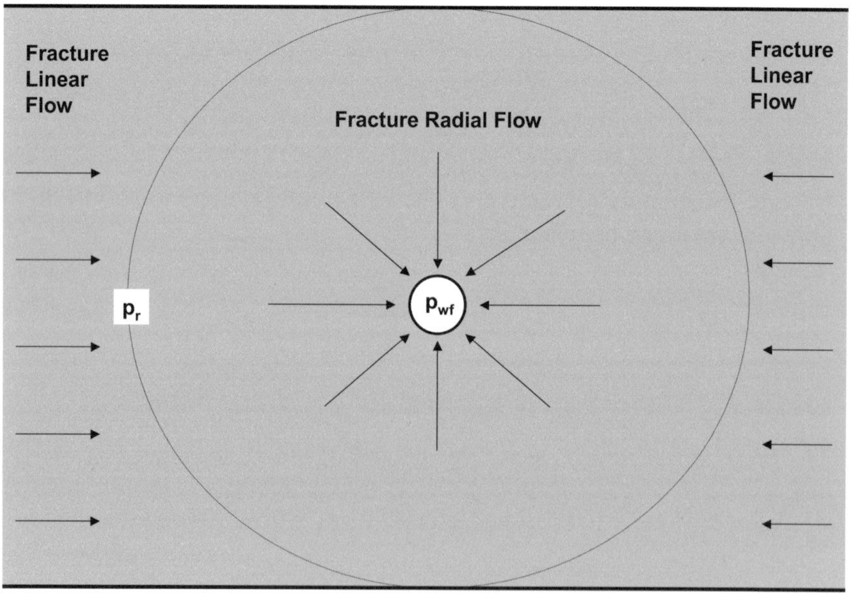

Figure 3–4 Fluid low in a fracture to a horizontal wellbore

where

$$c_i = \frac{24}{k_{fi} w_i \left[\frac{(z_{ei} - z_{si})}{k_H} + \frac{z_{si}}{k_{si}} \right]}$$

z_{ei} is half the distance between the ith and $(i+1)^{th}$ fractures, z_{si} is the depth of the altered zone near the surface of fracture i, k_{si} is the permeability of the altered zone near the surface of fracture i, and p_r represents the pressure in the fracture before the onset of flow convergence to wellbore (Figure 3-4).

When the linear-radial flow model of Furui et al. (2003) is used, the well deliverability through n uniformly distributed fractures can be expressed as:

$$q = \sum_{i=1}^{n} \frac{5.85 \times 10^{-5} k_{fwi} w_{wi} \left(p_r^2 - p_{wf}^2 \right)}{\bar{z} \bar{\mu}_g T \left\{ \ln \left[\frac{h}{2 r_{w_i}} \right] + \pi - \left(1.224 - s_i - Dq \right) \right\}} \tag{3.22}$$

where p_{wf} is the flowing bottom-hole pressure. The k_{fwi} is fracture permeability in the near-wellbore region, and w_{wi} is the width of the ith fracture in the near-wellbore r-gion. These two parameters, plus the non-Darcy flow coefficient D, can be used to simulate choked fracture.

Combining Equations (3.19) through (3.22) yields a reservoir deliverability equation, expressed as:

$$q = \frac{1}{\left(\frac{1}{J_R} + \frac{1}{J_L} + \frac{1}{J_r} \right)} \left(\bar{p}^2 - p_{wf}^2 \right) \tag{3.23}$$

where

$$J_R = \frac{k_H h}{1424 \bar{z} \bar{\mu}_g T \left(\frac{1}{2} \ln \frac{4A}{\gamma C_A r_L^2} \right)} \tag{3.24}$$

$$J_L = \sum_{i=1}^{n} \frac{4.475 \times 10^{-4} h}{\bar{\mu}_g \bar{z} T \sqrt{c_i} \left[\frac{\left(z_{ei} - z_{si}\right)}{k_H} + \frac{z_{si}}{k_{si}} \right]} \left(1 - e^{-\sqrt{c_i} x_{fi}}\right) \tag{3.25}$$

$$J_r = \sum_{i=1}^{n} \frac{5.85 \times 10^{-5} k_{fwi} w_{wi}}{\bar{z} \bar{\mu}_g T \left\{ \ln\left[\frac{h}{2 r_{w_i}}\right] + \pi - \left(1.224 - s_i - Dq\right) \right\}} \tag{3.26}$$

3.7 Shale Gas Wells

Shale gas is natural gas produced from shale sequences. Shale gas production, as an unconventional source of natural gas, has had a long history in the United States, dating back about 80 years. Shale gas has become an increasingly important source of natural gas in several regions of the world over the past decade. An analyst expects shale gas to supply as much as half the natural gas production in North America by 2020 (Polczer, 2009).

Gas shales are essentially lithified clays with organic matter present in varying amounts. The organic matter is believed to be an integral constituent of productive gas shale. Quantities of gas can be stored either as a dissolved phase in liquid hydrocarbons, or as an adsorbed phase on other materials within the shales of the kerogen, i.e., certain forms of illite. The phenomena of gas storage and flow in shale gas sediments are believed to be a combination of different controlling processes. Gas is stored as compressed gas in pores, as adsorbed gas to the pore walls and as soluble gas in solid organic materials and clays. According to Katsube (2000), gas flows through a network of pores with different diameters ranging from nanometres (nm = 10^{-9}m) to micrometres (μm = 10^{-6}m). More specifically, the diameter of pores in shale gas sediments ranges from a few nanometres to a few micrometres.

The fine-grained rocks in the shale are micro-porous with extremely low permeabilities. Javadpour et al. (2007) presented experimentally obtained permeabilities of 152 samples from nine shale gas reservoirs with pore-size distribution of several shale samples at 60,000 psi mercury injection pressure. They described the fractal-like sequence of gas production at different length scales. Gas production from a newly drilled wellbore is through the large pores and then the smaller pores. During reservoir depletion the thermodynamic equilibrium between kerogen/clays and the gas phase in the pore spaces changes. Gas desorbs from the surface of the kerogen/clays. This non-equilibrium process further drives the gas molecules to diffuse from the bulk of the kerogen to the surface of the kerogen exposed to the pore network.

Traditionally, flow of gas in shale gas reservoirs was described using dual-porosity models that were originally developed for naturally fractured conventional reservoirs (Warren and Root, 1963; Streltsova, 1983). The problem with the term "dual-porosity" in shales, however, is that shales have very little open porosity. Unlike conventional gas reservoirs, where gas is stored in the open pore space of the rock, shales store a very large amount of gas in an adsorbed state (Shettler et al., 1987). Carlson and Mercer (1991) pointed out that the behavior of gas flow in the shale matrix is different from that in conventional gas reservoirs due to desorption and diffusion effects. They showed that, while neglecting the special gas storage and flow properties in the short term is appropriate, such neglect in the long term will result in an underestimation of gas production. Carlson and Mercer (1994) found that for Devonian shales, the matrix is so tight that it takes a long time (many years) for the effects of a pressure drawdown in the fracture network to be felt deep in the interiors of the matrix. They referred to this effect as semi-infinite behavior because the rock matrix has infinite-acting characteristics. He concluded that when the semi-infinite flow prevails, the performance of the single-well system depends on fracture permeability, the external drainage radius, and a group of parameters including matrix properties, fluid viscosity, and fracture spacing. Accurate modeling of long-term well performance requires application of unconventional properties to simulate the semi-infinite flow behavior.

Jenkins et al. (2008) presented cumulative gas production data for 23 wells in a 1-sq mile area. All of the wells were drilled and completed in essentially the same way, but there was still significant variation in gas well productivity. The high variability was believed to be due to the local changes in permeability as a result of fracture intensity and fracture aperture width (Weida et al., 2005). Jenkins et al.'s (2008) plot also shows a dual-slope behavior of gas wells. Data from three typical wells plotted in

Figure 3-5. It is believed that the first slope is determined by the permeability of the local fractures and the second slope is controlled by the strength of feeding by the matrix.

Due to the low-permeability nature of shales, hydraulic fracturing is frequently needed to improve well productivity. King (2010) summarized the evolution of the fracturing technique for shale gas formations. The recent engineering achievements in multi-stage fractures in long horizontal well laterals have inevitably increased the interest in exploitation of shale gas reservoirs.

It should be very difficult, if not impossible, to couple the Darcy gas flow in fractures with the nanoscale gas flow in the nanopores, not mentioning that there are several levels of fractal-like scales (Javadpour et al., 2007) in addition to the hydraulic fractures which is the largest scale of flow network in shale gas reservoirs. The gas flow mechanism in the shale gas reservoirs is still not fully understood. Guo et al. (2011) pointed out that the role of formation damage in hydraulic fracturing shale gas wells is minimal, which is consistent with field observations that show that wells with high-water flowback volumes are normally low-productivity wells while those with low-water flowback volumes are unexpectedly high-productivity wells. Perhaps the best way to evaluate shale gas well productivity is to use the mathematical model presented in section 3.6 with validation by data from well deliverability testing.

3.8 Well Deliverability Testing

Well deliverability testing provides a direct means of estimating productivity of gas wells. Testing procedures include

- Flow-after-flow test (stabilized flow test),
- Isochronal test, and
- Modified isochronal test.

The objective of these tests is to deliver values of C and n or A and B that are used for defining well inflow performance relationship (IPR).

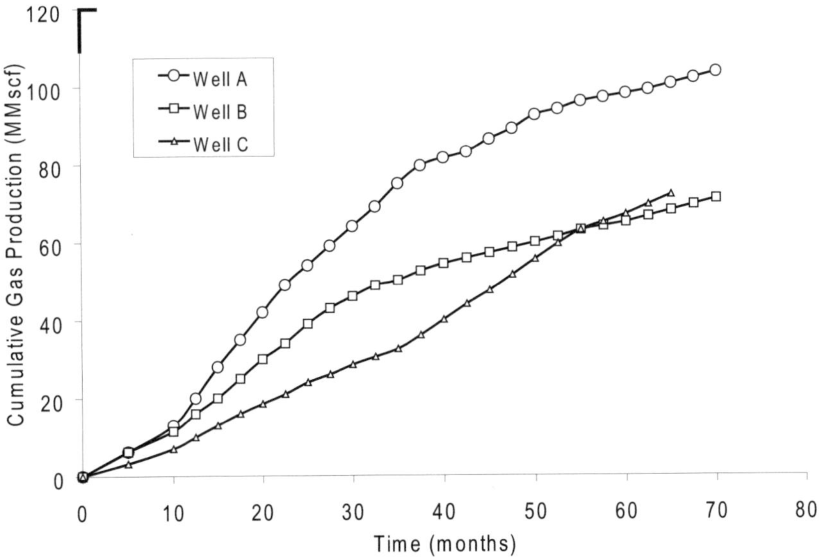

Figure 3–5 Gas production curves of 3 wells in a 1-sq. mile (Jenkins et al., 2008)

3.8.1 Flow-After-Flow Test

In this testing procedure, a well flows at a selected constant rate until pressure stabilizes to reach the pseudosteady state flow condition. The rate is then changed and the well flows until the pressure stabilizes again at the new rate. The process is repeated for a total of 3 to 4 rates. Flow rates and pressures follow the pattern depicted in Figure 3-6.

Both the back pressure model and the Forchheimer model can be used to analyze the test data. Figure 3-7 shows the data plot and technique for deriving the *C* and *n* values from the flow-after-flow test.

Figure 3-8 shows the data plot and technique for deriving the *A* and *B* values from the flow-after-flow test.

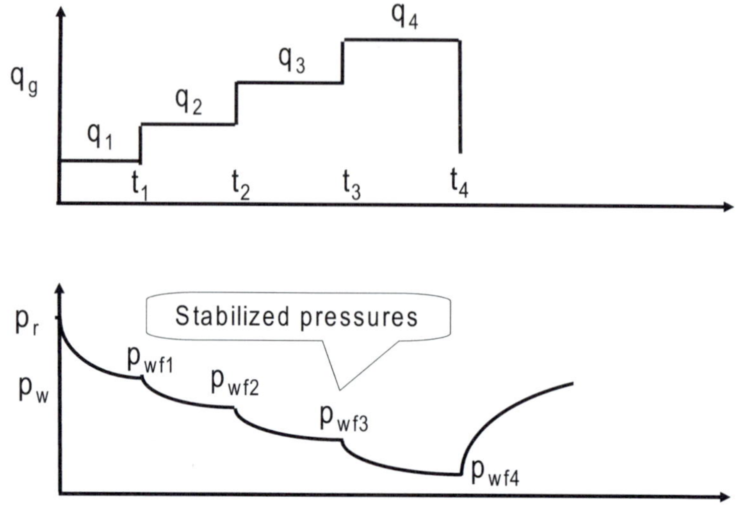

Figure 3–6 Sequence of flow-after-flow test

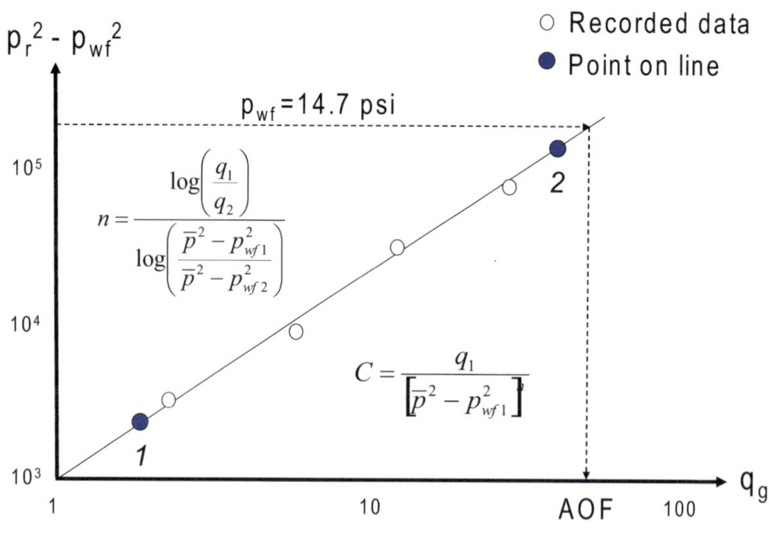

Figure 3–7 Technique for deriving the *C* and *n* values from the flow-after-flow test

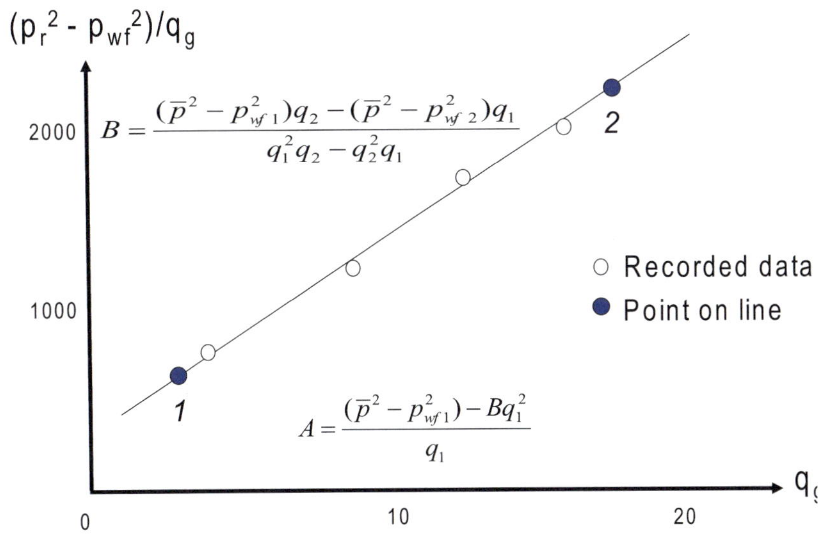

Figure 3–8 Technique for deriving the *A* and *B* values from flow-after-flow test

3.8.2 Isochronal Test

The isochronal testing procedure was developed to obtain data with reduced test time. An isochronal test is conducted by flowing a well at a fixed rate, then shutting it in until the pressure builds up to an unchanging value. The well is then flowed at a second rate for the same length of time, followed by another shut-in, The process is repeated for a total of 3 to 4 rates. If possible, the final flow period should be long enough to achieve stabilized flow condition. Isochronal testing sequence is shown in Figure 3-9.

Both the back pressure model and the Forchheimer model can be used to analyze the test data. Figure 3-10 illustrates the technique for obtaining the C and n values from the isochronal test.

Figure 3-11 depict the technique for obtaining the A and B values from the isochronal test.

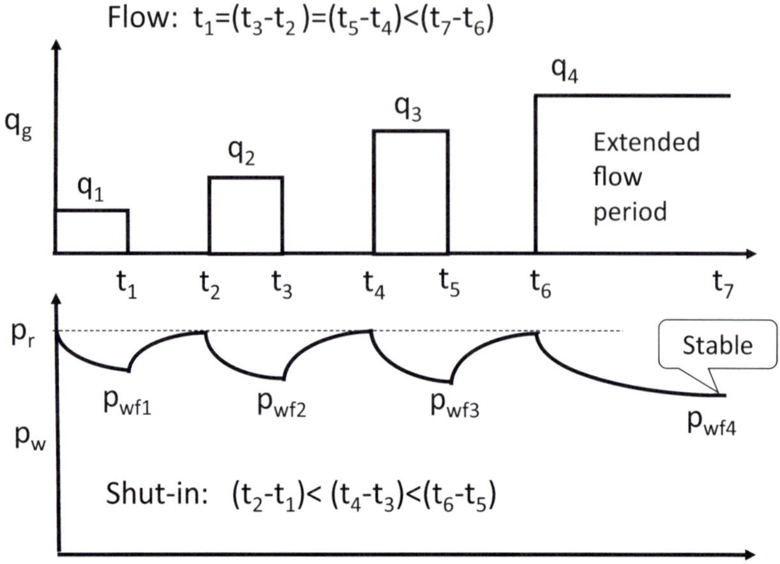

Figure 3–9 Sequence of isochronal test

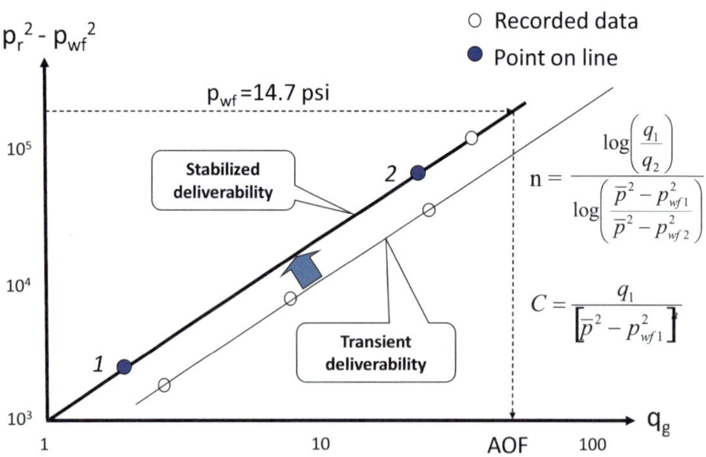

$$n = \frac{\log\left(\frac{q_1}{q_2}\right)}{\log\left(\frac{\bar{p}^2 - p_{wf1}^2}{\bar{p}^2 - p_{wf2}^2}\right)}$$

$$C = \frac{q_1}{\left[\bar{p}^2 - p_{wf1}^2\right]^n}$$

Figure 3–10 Technique for deriving the *C* and *n* values from isochronal test

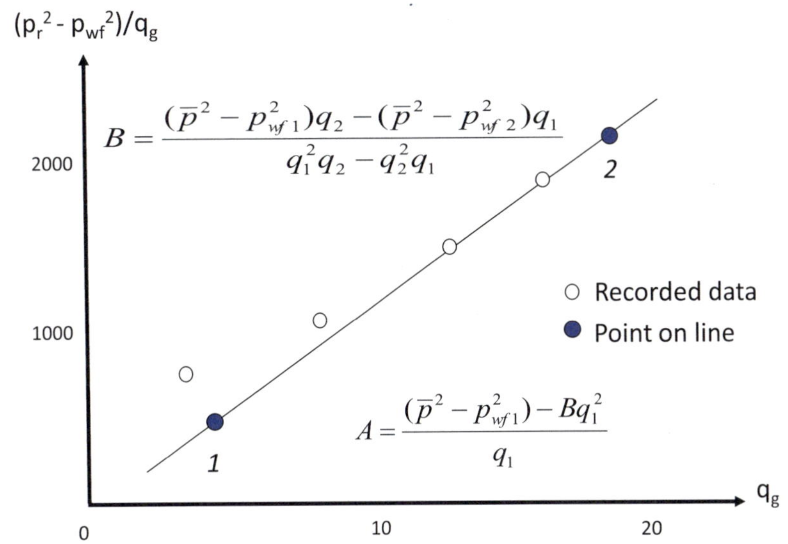

Figure 3–11 Technique for deriving the *A* and *B* values from isochronal test

3.8.3 Modified Isochronal Test

The modified isochronal testing procedure was developed to obtain the same data as in the isochronal test without spending the lengthy shut-in time required for pressure to stabilize completely before each flow test is run. The modified isochronal testing sequence is shown in Figure 3-12.

While the isochronal tests are modeled exactly by rigorous theory, the modified isochronal tests are not. It uses approximations. Both the back pressure model and the Forchheimer model can be used to analyze the test data. Figure 3-13 illustrates the technique for obtaining the approximate values of C and n from the modified isochronal test.

Figure 3-14 demonstrates the technique for obtaining the approximate values of A and B from the modified isochronal test.

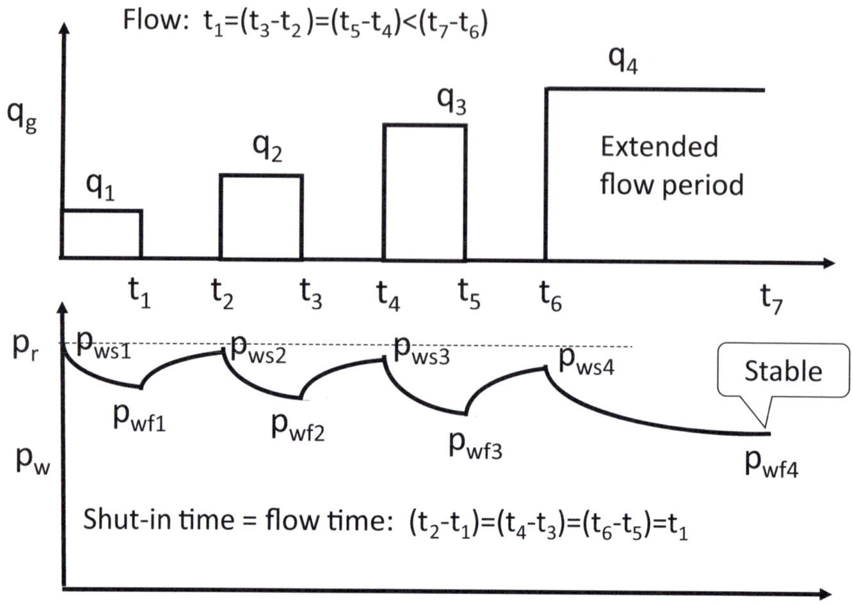

Figure 3–12 Sequence of modified isochronal test

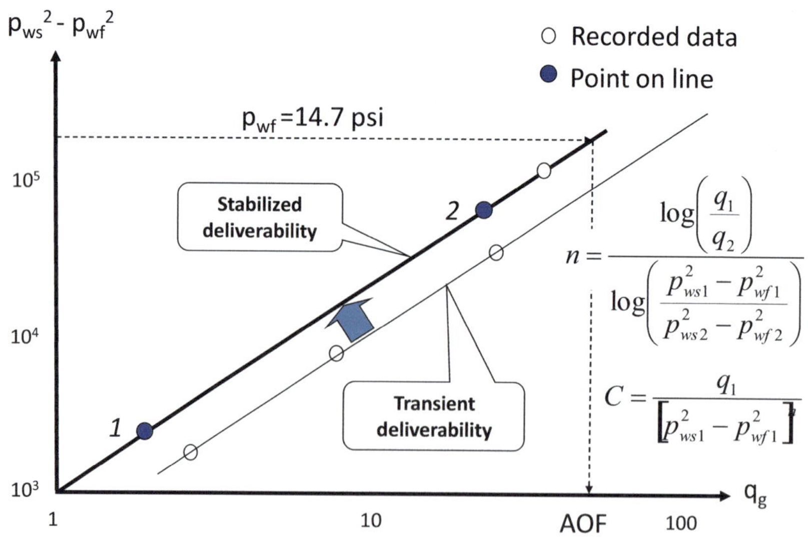

Figure 3–13 Technique for deriving the approximate values of *C* and *n* from the modified isochronal test

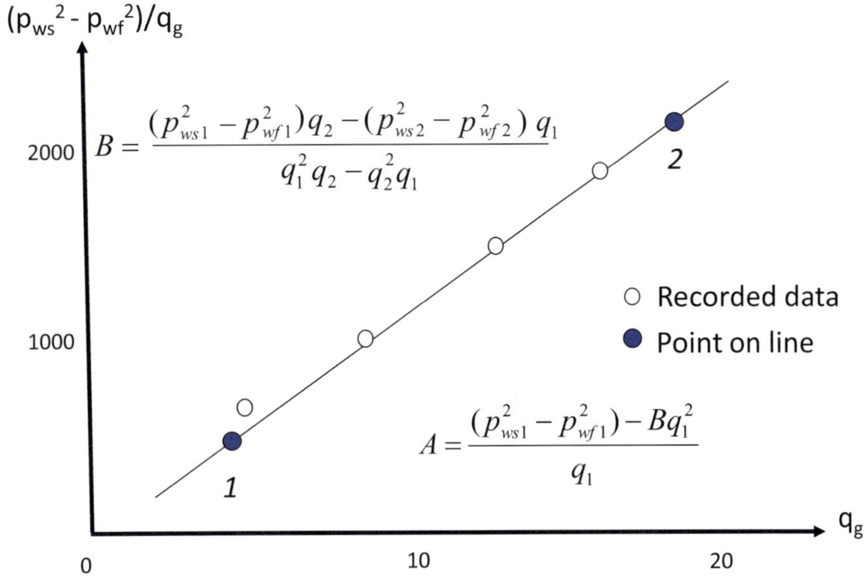

Figure 3–14 Technique for deriving the approximate values of *A* and *B* from the modified isochronal test

3.9 References

Carlson, E.S.: "Characterization of Devonian Shale Gas Reservoirs Using Coordinated Single Well Analytical Models," paper SPE 29199 presented at the SPE Eastern Regional Conference and Exhibition held in Charleston, WV, November 8-10, 1994.

Carlson, E.S. and Mercer, J.C.: "Devonian Shale Gas Production: Mechanisms and Simple Models," JPT (April 1991), 476-82.

Economides, M. J., A. D. Hill, and C. Ehlig-Economides. *Petroleum Production Systems*, New Jersey: Prentice Hall PTR, 1994, 74–5.

Furui, K., Zhu, D., and Hill, A.D.: "A Rigorous Formation Damage Skin Factor and Reservoir Inflow Model for a Horizontal Well," *SPERE* (Aug. 2003): 151.

Guo, B. and Schechter, D.S.: "A Simple and Rigorous IPR Equation for Vertical and Horizontal Wells Intersecting Long Fractures," *Journal of Canadian Petroleum Technology* (July 1999).

Guo, B., Gao, D., and Wang, Q.: "The Role of Formation Damage in Hydraulic Fracturing Shale Gas Wells," paper SPE 148778, proceedings of the SPE Eastern Regional Meeting held in Columbus, Ohio, 17-19 August 2011.

Guo, B., Sun, K. and Ghalambor, A.: *Well Productivity Handbook,* Gulf Publishing Company, Houston (2008).

Guo, B. and Yu, X.: "A Simple and Accurate Mathematical Model for Predicting Productivity of Multifractured Horizontal Wells," paper SPE 114452 presented at the CIPC/SPE Gas Technology Symposium 2008 Joint Conference held 16–19 June 2008 in Calgary, Canada.

Javadpour, F., Fisher, D., and Unsworth, M.: "Nanoscale Gas Flow in Shale Gas Sediments," JCPT (Oct. 2007), Vol. 46, No. 10. , 55-61.

Jenkins, C.D., MacNoughton, D. , and Boyer, C.M.: "Coalbed- and Shale-Gas Reservoirs," JPT (Feb. 2008), 92-99.

Joshi, S.D.: "Augmentation of Well Productivity with Slant and Horizontal Wells," Journal of Petroleum Technologies (June 1988): 729-739.

Katsube, T.J.: "Shale Permeability and Pore-Structure Evolution Characteristics," *Research 2000 E15,Geological Survey of Canada, Ottawa, 2000.*

King, G.E.: "Thirty Years of Gas Shale Fracturing: What Have We Learned," paper SPE 133456 presented at the SPE ATCE held in Florence, Italy, Sept. 19-22, 2010.

Lee, J. *Well Testing*. Dallas: SPE, 1982, 76–85.

Li, H., Jia, Z., and Wei, Z.: "A New Method to Predict Performance of Fractured Horizontal Wells," paper SPE 37051 presented at the SPE International Conference on Horizontal Well Technology held 18-20 November 1996 in Calgary, Canada.

Polczer, S.: "Shale Expected to Supply Half of North America's Gas," *Calgary Herald*, 9 April 2009, accessed 27 August 2009.

Raghavan, R. and Joshi, S.D.: "Productivity of Multiple Drainholes or Fractured Horizontal Wells," SPE Formation Evaluation (March 1993): 11-16.

Shettler, P.D., Parmely, C.R., and Lee, W.J.: "Gas Storage and Transport in Devonian Shales," SPEFE (Sept. 1989), Trans., AIME, Vol. 287, 371-76.

Streltsavo, T.D.: "Well Pressure Behavior of a Naturally Fractured Reservoir," SPEJ (Oct. 1983), 769-80.

Warren, J.E., and Root, P.J.: "The Behavior of Naturally Fractured Reservoirs," SPEJ (Sept. 1963), Trans., AIME, Vol. 228, 245-255.

Weida, S.D., Lambert, S.W., and Boyer, C.M.: "Challenging the Traditional Coalbed Methane Exploration and Evaluation Model," paper SPE 98069 presented at the SPE Eastern Regional Meeting held in Morgantown, WV, September 14-16, 2005.

3.10 Problems

3-1 Run spreadsheet program Theoretical Deliverability.xls for a typical gas reservoir at various bottom hole pressures. Make a conclusion regarding the accuracies of the p- and p^2-approaches.

3-2 Run spreadsheet program Empirical Deliverability.xls for a typical gas reservoir at various bottom hole pressures. Make a conclusion regarding the accuracies of the Forchheimer and backpressure models.

3-3 Run spreadsheet program Theoretical IPR.xls for a typical gas reservoir. Under what condition is the IPR sensitive to the Non-Darcy coefficient?

3-4 On the basis of Forchheimer model, derive an expression for gas flow rate as an explicit function of flowing bottom hole pseudopressure.

Chapter 4

Wellbore Performance

4.1 Introduction

Chapter 3 stated that reservoir properties control the inflow performance of wells, or the potential of gas production rate from the well. However, the achievable gas production rate from the well is determined by wellhead pressure and the flow performance of production string, that is, tubing, casing, or both. The flow performance of production string depends on geometries of the production string and properties of fluids being produced. The fluids in gas wells are mainly gases with small fractions of water, condensate, and sand from the productive zones. Wellbore performance analysis involves establishing a relationship between tubular size, wellhead and bottom hole pressure, gas flow rate, and fluid properties. Understanding wellbore flow performance is vitally important to gas engineers for designing gas well equipment and optimizing well production conditions.

Gas can be produced through tubing, casing, or both in a gas well, depending on which flow path has a better performance. Producing gas through tubing is a better option in most cases to prevent liquid loading. The term Tubing Performance Relationship (TPR) is used in this chapter. However the mathematical models are also valid for casing flow and casing-tubing annular flow as long as hydraulic diameter is used. This chapter focuses on determination of TPR and its applications to system analysis (Nodal analysis) for prediction of gas production rate from the well. Both single-phase and multi-phase fluids are considered. Flow regime transition and flow stability are analyzed. Liquid loading problems are described and their solutions are discussed. Calculation examples are illustrated with the computer spreadsheets that are provided with this book.

4.2 Single-Phase Gas Well

Tubing Performance Relationship is defined as a relation between tubing size, fluid properties, fluid flow rate, wellhead pressure, and bottom hole pressure. In most engineering analyses, it is desired to know the bottom hole pressure at a given wellhead pressure and gas flow rate in a gas well. The first law of thermodynamics (conservation of energy) governs gas flow in tubing. The effect of kinetic energy change is negligible due to the fact that the variation in tubing diameter is insignificant in most gas wells. With no shaft work device installed along the tubing string, the first law of thermodynamics yields the following mechanical balance equation:

$$\frac{dP}{\rho}+\frac{g}{g_c}dZ+\frac{fv^2dL}{g_cD_i}=0 \tag{4.1}$$

Because $dZ=\cos\theta dL$, $\rho=\frac{29\gamma_g P}{ZRT}$, and $v=\frac{4q_{sc}zP_{sc}T}{\pi D_i^2T_{sc}P}$, Equation (4.1) can be rewritten as:

$$\frac{zRT}{29\gamma_g}\frac{dP}{P}+\left\{\frac{g}{g_c}\cos\theta+\frac{8fQ_{sc}^2P_{sc}^2}{\pi^2g_cD_i^5T_{sc}^2}\left[\frac{zT}{P}\right]^2\right\}dL=0 \tag{4.2}$$

which is an ordinary differential equation governing gas flow in tubing. Although the temperature T can be approximately expressed as a linear function of length L through geothermal gradient, the compressibility factor z is a function of pressure P and temperature T. This makes it difficult to solve the equation analytically. Fortunately, the pressure P at length L is not a strong function of temperature and compressibility factor. Approximate solutions to Equation (4.2) have been sought and used in the natural gas industry.

4.2.1 The Average Temperature and Compressibility Factor Method

If single, average values of temperature and compressibility factor over the entire tubing length can be assumed, Equation (4.2) becomes:

$$\frac{\bar{z}R\bar{T}}{29\gamma_g}\frac{dP}{P}+\left\{\frac{g}{g_c}\cos\theta+\frac{8f\,Q_{cs}^2P_{sc}^2\bar{z}^2\bar{T}^2}{\pi^2g_cD_i^5T_{sc}^2P^2}\right\}dL=0 \tag{4.3}$$

By separation of variables, Equation (4.3) can be integrated over the full length of tubing to yield:

$$P_{wf}^2=Exp(s)P_{hf}^2+\frac{8f\,[Exp(s)-1]Q_{sc}^2P_{sc}^2\bar{z}^2\bar{T}^2}{\pi^2g_cD_i^5T_{sc}^2\cos\theta} \tag{4.4}$$

where

$$s=\frac{58\gamma_g gL\cos\theta}{g_cR\bar{z}\bar{T}} \tag{4.5}$$

Equation (4.4) and Equation (4.5) take the following forms when U.S. field units (q_{sc} in Mscf/d), are used (Katz et al. 1959):

$$p_{wf}^2=Exp(s)p_{hf}^2+\frac{6.67\times10^{-4}[Exp(s)-1]fq_{sc}^2\bar{z}^2\bar{T}^2}{d_i^5\cos\theta} \tag{4.6}$$

and

$$s=\frac{0.0375\gamma_g L\cos\theta}{\bar{z}\bar{T}} \tag{4.7}$$

The Moody friction factor f_m can be found in the conventional manner for a given tubing diameter, wall roughness, and Reynolds number. However, if one assumes fully turbulent flow, which is the case for most gas wells, then a simple empirical relation may be used for typical tubing strings (Katz and Lee 1990):

$$f=\frac{0.01750}{d_i^{0.224}}\quad \text{for } d_i\le 4.277 \text{ in} \tag{4.8}$$

$$f = \frac{0.01603}{d_i^{0.164}} \qquad \text{for } d_i > 4.277 \text{ in} \tag{4.9}$$

Guo (2001) used the following Nikuradse friction factor correlation for fully turbulent flow in rough pipes:

$$f = \left[\frac{1}{1.74 - 2\log\left(\frac{2\varepsilon}{d_i}\right)} \right]^2 \tag{4.10}$$

Because the average compressibility factor is a function of pressure itself, a numerical technique such as the Newton-Raphson iteration is required to solve Equation (4.6) for bottom hole pressure. This computation can be performed automatically with the spreadsheet program AverageTZ.xls. Users need to input parameter values in the *Input Data* section and run Macro Solution to get results.

Example Problem 4.1

Suppose that a vertical well produces 2 MMscf/d of 0.71 gas-specific gravity gas through a 2 7/8-in tubing set to the top of a gas reservoir at a depth of 10,000 ft. At tubing head, the pressure is 800 psia and the temperature is 150 °F; the bottom hole temperature is 200 °F. The relative roughness of tubing is about 0.0006. Calculate the pressure profile along the tubing length and plot the results.

Solution

This example problem is solved with the spreadsheet program AverageTZ.xls. Table 4–1 shows the appearance of the spreadsheet for the data input and result sections. Calculated pressure profile is plotted in Figure 4–1.

Table 4–1 Input Data and Results Given by AverageTZ.xls[a]

Instructions: 1) Input your data in the Input Data section; 2) Run Macro Solution to get results; 3) View results in table and in the Profile graph sheet.

Input Data			
γ_g =	0.71		
d =	2.259 in		
ε/d =	0.0006		
L =	10,000 ft		
θ =	0°		
p_{hf} =	800 psia		
T_{hf} =	150 °F		
T_{wf} =	200 °F		
q_{sc} =	2,000 Mscf/d		
Solution			
f =	0.017397		
Depth (ft)	**T (°R)**	**p (psia)**	**Z_{av}**
0	610	800	0.9028
1,000	615	827	0.9028
2,000	620	854	0.9027
3,000	625	881	0.9027
4,000	630	909	0.9026
5,000	635	937	0.9026
6,000	640	965	0.9026
7,000	645	994	0.9026
8,000	650	1,023	0.9027
9,000	655	1,053	0.9027
10,000	660	1,082	0.9028

a. Tubing Performance Relationship (TPR)

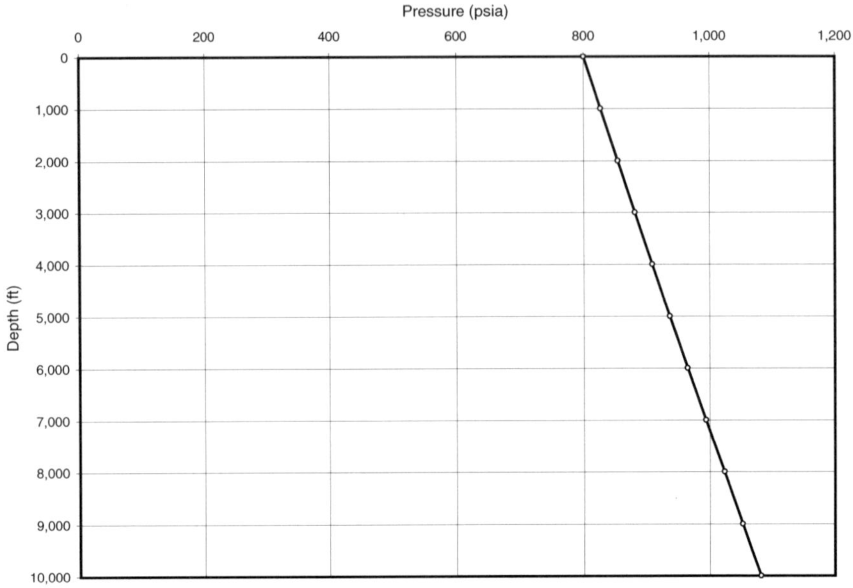

Figure 4–1 Calculated tubing pressure profile for Example Problem 4.1.

4.2.2 The Cullender and Smith Method

Equation (4.2) can be solved for bottom hole pressure using a fast numerical algorithm originally developed my Cullender and Smith (Katz et al. 1959). Equation (4.2) can be rearranged as:

$$\frac{\frac{P}{zT}dp}{\frac{g}{g_c}\cos\theta\left(\frac{P}{zT}\right)^2+\frac{8f\,Q_{sc}{}^2P_{sc}{}^2}{\pi^2 g_c D_i^5 T_{sc}{}^2}}=-\frac{29\gamma_g}{R}dL \tag{4.11}$$

that takes an integration form of

$$\int_{P_{hf}}^{P_{wf}}\left[\frac{\frac{P}{zT}}{\frac{g}{g_c}\cos\theta\left(\frac{P}{zT}\right)^2+\frac{8f\,Q_{sc}{}^2P_{sc}{}^2}{\pi^2 g_c D_i^5 T_{sc}{}^2}}\right]dp=\frac{29\gamma_g L}{R} \tag{4.12}$$

In U.S. field units (q_{msc} in MMscf/d), Equation (4.12) has the following form:

$$\int_{p_{hf}}^{p_{wf}} \left[\frac{\frac{p}{zT}}{0.001\cos\theta \left(\frac{p}{zT}\right)^2 + 0.6666\frac{f\,q_{msc}^2}{d_i^5}} \right] dp = 18.75\gamma_g L \tag{4.13}$$

If the integrant is denoted with symbol I, that is,

$$I = \frac{\frac{p}{zT}}{0.001\cos\theta \left(\frac{p}{zT}\right)^2 + 0.6666\frac{fq_{sc}{}^2}{d_i^5}} \tag{4.14}$$

Equation (4.13) becomes:

$$\int_{p_{hf}}^{p_{wf}} I dp = 18.75\gamma_g L \tag{4.15}$$

In the form of numerical integration, Equation (4.15) can be expressed as:

$$\frac{(p_{mf} - p_{hf})(I_{mf} + I_{hf})}{2} + \frac{(p_{wf} - p_{mf})(I_{wf} + I_{mf})}{2} = 18.75\gamma_g L \tag{4.16}$$

where p_{mf} is the pressure at the mid-depth. The I_{hf}, I_{mf}, and I_{wf} are integrant Is evaluated at p_{hf}, p_{mf}, and p_{wf}, respectively. Assuming the first and second terms in the right-hand side of Equation (4.16) each represents half of the integration, that is,

$$\frac{(p_{mf} - p_{hf})(I_{mf} + I_{hf})}{2} = \frac{18.75\gamma_g L}{2} \tag{4.17}$$

$$\frac{(p_{wf} - p_{mf})(I_{wf} + I_{mf})}{2} = \frac{18.75\gamma_g L}{2} \tag{4.18}$$

the following expressions are obtained:

$$p_{mf} = p_{hf} + \frac{18.75\gamma_g L}{I_{mf} + I_{hf}} \tag{4.19}$$

$$p_{wf} = p_{mf} + \frac{18.75\gamma_g L}{I_{wf} + I_{mf}} \tag{4.20}$$

Because I_{mf} is a function of pressure p_{mf} itself, a numerical technique such as the Newton-Raphson iteration is required to solve Equation (4.19) for p_{mf}. Once p_{mf} is computed, p_{wf} can be solved numerically from Equation (4.20). These computations can be performed automatically with the spreadsheet program Cullender-Smith.xls. Users need to input parameter values in the *Input Data* section and run Macro Solution to get results.

Example Problem 4.2

Solve the problem in Example Problem 4.1 with the Cullender and Smith Method.

Solution

This example problem is solved with the spreadsheet program Cullender-Smith.xls. Figure 4–2 shows the appearance of the spreadsheet for the data input and result sections. The pressures at depths of 5,000 ft and 10,000 ft are 937 psia and 1,082 psia, respectively. These results are exactly the same as those given by the Average Temperature and Compressibility Factor Method.

4.3 Mist Flow in Gas Wells

In addition to gas, almost all gas wells produce a certain amount of liquids. These liquids are formation water and/or gas condensate (light oil). Depending on pressure and temperature, in some wells gas condensate is

Table 4–2 Input Data and Results Given by Cullender-Smith.xls[a]

Instructions: 1) Input your data in the Input Data section; 2) Run Macro Solution to get results.

Input Data

γ_g =	0.71
d =	2.259 in
ε/d =	0.0006
L =	10,000 ft
θ =	0°
p_{hf} =	800 psia
T_{hf} =	150 °F
T_{wf} =	200 °F
q_{msc} =	2 MMscf/d

Solution

f = 0.017397

Depth (ft)	T (°R)	p (psia)	Z	p/ZT	I
0	610	800	0.9028	1.45263	501.137
5,000	635	937	0.9032	1.63324	472.581
10,000	660	1,082	0.9057	1.80971	445.349

a. Tubing Performance Relationship

not seen at surface, but it exists in the wellbore. Some gas wells produce sand and coal particles. These wells are called multiphase-gas wells.

The TPR equations presented in the previous section are not valid for multiphase gas wells. To analyze TPR of multiphase gas wells, a gas-oil-water-solid four-phase flow model is presented in this section. It is warned that the four-phase flow model is valid only for gas wells producing multiphase fluid with gas being the main component. Specifically, the model can be used with good accuracy when mist flow exists in the wellbore. When the flow velocity drops to below a critical velocity at which the liquid droplets cannot be carried up to surface by gas, annular flow or even slug flow may develop in the well. TPR equations for annular flow and slug flow are available from literature of oil well performance.

The gas-oil-water-solid four-phase flow model was first presented by Guo (2001) for coal-bed methane production wells. Guo formulated the governing equation assuming homogeneous mixture of the four phases, which may exist in misting flow. Guo also presented an approximate solution to the governing equation by breaking down the solution into three terms. Guo, Sun, and Ghalambor (2004) presented an exact solution to the governing equation and applied the solution to aerated fluid hydraulics. The exact solution is summarized as follows.

According to Guo, Sun, and Ghalambor (2004) the following equation can be used for calculating pressure P (in lbf/ft^2) at depth L:

$$b\left(P-P_{hf}\right)+\frac{1-2bM}{2}\ln\left|\frac{\left(P+M\right)^2+N}{\left(P_{hf}+M\right)^2+N}\right|$$
$$-\frac{M+\frac{b}{c}N-bM^2}{\sqrt{N}}\left[\tan^{-1}\left(\frac{P+M}{\sqrt{N}}\right)-\tan^{-1}\left(\frac{P_{hf}+M}{\sqrt{N}}\right)\right]$$
$$=a\left(\cos\theta+d^2e\right)L \quad \mathbf{(4.21)}$$

where the group parameters are defined as

$$a=\frac{0.0765\gamma_g Q_{sc}+350\gamma_o q_o+350\gamma_w q_w+62.4\gamma_s q_s}{4.07T_{av}Q_{sc}} \quad \mathbf{(4.22)}$$

$$b=\frac{5.615q_o+5.615q_w+q_s}{4.07T_{av}Q_{sc}} \quad \mathbf{(4.23)}$$

$$c=0.00678\frac{T_{av}Q_{sc}}{A} \quad \mathbf{(4.24)}$$

$$d=\frac{0.00166}{A}\left(5.615q_o+5.615q_w+q_s\right) \quad \mathbf{(4.25)}$$

$$e=\frac{f}{2gD_i} \quad \mathbf{(4.26)}$$

$$M = \frac{cde}{\cos\theta + d^2 e} \quad \textbf{(4.27)}$$

$$N = \frac{c^2 e \cos\theta}{\left(\cos\theta + d^2 e\right)^2} \quad \textbf{(4.28)}$$

It has been found that the bottom hole pressures given by Equation (4.21) are about 1.5% lower than those given by Equation (4.6) in single-phase wells because ideal gas was assumed in formulation of the four-phase flow model.

Example Problem 4.3

Solve the problem in Example Problem 4.1 for bottom hole pressure with the following additional data:

Condensate Gas Ratio (CGR): 0.02 bbl/Mscf

Water Cut (WC): 50%

Oil gravity: 60 °API

Water-specific gravity: 1.03

Sand production: 0.1 ft^3/d

Sand-specific gravity: 2.65

Solution

This example problem is solved with the spreadsheet program MistFlow.xls. Table 4–3 shows the appearance of the spreadsheet for the data input and results sections. It indicates a flowing bottom hole pressure of 1,103 psia.

Table 4–3 Input Data and Results Sections for MistFlow.xls[a]

Instructions: 1) Input your data in the Input Data section; 2) Run Macro Solution to get results.	
Input Data	
$L =$	10,000 ft
$\theta =$	0°
$d_i =$	2.259 in
$Q_{sc} =$	2,000,000 scfd
$\gamma_g =$	0.71
$q_0 =$	40 stb/d
$\gamma_0 =$	0.74
$q_w =$	20 bbl/d
$\gamma_w =$	1.03
$q_s =$	0.1 ft^3/d
$\gamma_s =$	2.65
$T_{hf} =$	150 °F
$T_{wf} =$	200 °F
$p_{hf} =$	800 psia
$\varepsilon =$	0.000113 ft
Solution	
$A =$	4.0059186 in^2
$D_i =$	0.18825 ft
$T_{av} =$	635 °R
$\cos(\theta) =$	1
$f_m =$	0.017397
$a =$	2.442E–05
$b =$	6.52E–08
$c =$	2149469.5
$d =$	0.1402036
$e =$	0.0014363
$M =$	432.8487
$N =$	6.636E+09
$p_{wf} =$	1,103 psia

a. Tubing Performance Relationship

4.4 References

Guo, B. "An Analytical Model for Gas-Water-Coal Particle Flow in Coalbed-Methane Production Wells." Paper SPE 72369, presented at the SPE Eastern Regional Meeting, Canton, Ohio, October 17–19, 2001.

Guo, B., K. Sun, and A. Ghalambor. "A Closed Form Hydraulics Equation for Aerated Mud Drilling in Inclined Wells." *SPE Drilling & Completion Journal* (June 2004).

Katz, D. L., D. Cornell, R. Kobayashi, F. H. Poettmann, J. A. Vary, J. R. Elenbaas, and C. F. Weinaug. *Handbook of Natural Gas Engineering*. New York: McGraw-Hill Publishing Company, 1959.

Katz, D. L. and R. L. Lee. *Natural Gas Engineering—Production and Storage*. New York: McGraw-Hill Publishing Company, 1990.

4.5 Problems

4-1 Suppose 3 MMscf/d of 0.75 specific gravity gas are produced through a 3 1/2-in tubing string set to the top of a gas reservoir at a depth of 8,000 ft. At tubing head, the pressure is 1,000 psia and the temperature is 120 °F; the bottom hole temperature is 180 °F. The relative roughness of the tubing is about 0.0006. Calculate the flowing bottom hole pressure with three methods: a) the average temperature and compressibility factor method; b) the Cullender and Smith method; and c) the four-phase flow method. Make comments on your results.

4-2 Solve Problem 4-1 for gas production through a K-55, 17 lb/ft, 5 1/2-in casing.

4-3 Suppose 2 MMscf/d of 0.65 specific gravity gas are produced through a 2 7/8-in (2.259-in ID) tubing string set to the top of a gas reservoir at a depth of 5,000 ft. Tubing head pressure is 300 psia and the temperature is 100 °F; the bottom hole temperature is 150 °F. The relative roughness of the tubing is about 0.0006. Calculate the flowing bottom pressure with the average temperature and compressibility factor method.

4-4 Plot TPR for the well data given in Problem 4-3.

4-5 Suppose the data for the well described in Problem 4-3 also produces 20 stb/d of 0.85 specific gravity oil, 10 bbl/d of 1.02 specific gravity water, and 0.05 ft^3/d of 2.65 specific gravity sand. Calculate the flowing bottom pressure.

4-6 Plot TPR for the well data given in Problem 4-5. Assume the liquid and sand production rates will remain constant.

4-7 Develop a spreadsheet program to calculate tubing pressure profile using the data given in Problem 4-5.

Chapter 5

Choke Performance

5.1 Introduction

Gas production rates from individual wells are controlled for preventing water coning and/or sand production, meeting limitations of rate or pressure imposed by production facilities, and satisfying production limits set by regulatory authorities. Choke is a device installed at wellhead or down hole to cause a restriction to flow of fluids, and thus control gas production rate. Traditionally chokes are classified as nozzle-type and orifice-type with fixed diameters. Modern chokes of adjustable diameters are also available. This chapter presents the performance of chokes under different flow conditions. Both pure gas and gas/ liquid mixtures are considered.

5.2 Sonic and Subsonic Flow

Pressure drop across well chokes is usually very significant. There is no universal equation for predicting pressure drop across the chokes for all types of production fluids. Different choke flow models are available from literature, and they have to be chosen based on the gas fraction in the fluid and flow regimes, that is, subsonic or sonic flow.

Both sound wave and pressure wave are mechanical waves. When the fluid flow velocity in a choke reaches the traveling velocity of sound in the fluid under the in situ condition, the flow is called sonic flow. Under sonic flow conditions, the pressure wave downstream of the choke cannot go upstream through the choke because the medium (fluid) is traveling in the opposite direction at the same velocity. Therefore, a pressure discontinuity exists at the choke, that is, the downstream pressure does not affect the upstream pressure. Because of the pressure discontinuity at the choke, any change in the downstream pressure cannot be detected from the upstream pressure gauge. Of course, any change in the upstream pressure cannot be detected from the downstream pressure gauge either. This sonic flow provides a unique well choke feature that stabilizes well production rate and separation operation conditions.

Whether or not a sonic flow exists at a choke depends on a downstream to upstream pressure ratio. If this pressure ratio is less than a critical pressure ratio, sonic (critical) flow exists. If this pressure ratio is greater or equal to the critical pressure ratio, subsonic (subcritical) flow exists. The critical pressure ratio through chokes is expressed as:

$$\left(\frac{p_{outlet}}{p_{up}}\right)_c = \left(\frac{2}{k+1}\right)^{\frac{k}{k-1}} \tag{5.1}$$

where p_{outlet} is the pressure at choke outlet, p_{up} is the upstream pressure, and $k = C_p/C_v$ is the specific heat ratio. The value of the k is 1.4 for air and 1.28 for natural gas. Thus, the critical pressure ratio is 0.528 for air and 0.549 for natural gas.

5.3 Dry Gas Flow through Chokes

Pressure equations for choke flow are derived based on isentropic process. This is because there is no time for heat to transfer (adiabatic) and the friction loss is negligible (assuming reversible) at chokes. In addition to the concern of pressure drop across the chokes, temperature drop associated with choke flow is also an important issue for gas wells as hydrates may form that may plug flow lines.

5.3.1 Subsonic Flow

Under subsonic flow conditions, gas passage through a choke can be expressed as:

$$Q_{sc} = 1{,}248CAp_{up}\sqrt{\frac{k}{(k-1)\gamma_g T_{up}}\left[\left(\frac{p_{dn}}{p_{up}}\right)^{\frac{2}{k}} - \left(\frac{p_{dn}}{p_{up}}\right)^{\frac{k+1}{k}}\right]} \tag{5.2}$$

where

Q_{sc} = gas flow rate, Mscf/d

p_{up} = upstream pressure at choke, psia

A = cross-sectional area of choke, in^2

T_{up} = upstream temperature, °R

g = acceleration of gravity, 32.2 ft/sec^2

γ_g = gas specific gravity related to air

C = choke flow coefficient

The choke flow coefficient can be determined using charts in Figure 5–1 and Figure 5–2 for nozzle-type and orifice-type chokes, respectively. The following correlation has been found to give reasonable accuracy for Reynolds numbers between 10^4 and 10^6 for nozzle-type chokes:

$$C = \frac{d}{D} + \frac{0.3167}{\left(\frac{d}{D}\right)^{0.6}} + 0.025[\log(N_{\mathrm{Re}}) - 4] \tag{5.3}$$

where

d = choke diameter, in

D = pipe diameter, in

N_{Re} = Reynolds number

Gas velocity under subsonic flow conditions is less than the sound velocity in the gas at the in situ conditions:

$$v = \sqrt{v_{up}{}^2 + 2g_c C_p T_{up}\left[1 - \frac{z_{up}}{z_{dn}}\left(\frac{p_{down}}{p_{up}}\right)^{\frac{k-1}{k}}\right]} \tag{5.4}$$

where C_p = specific heat of gas at constant pressure (187.7 lbf-ft/lbm-R for air).

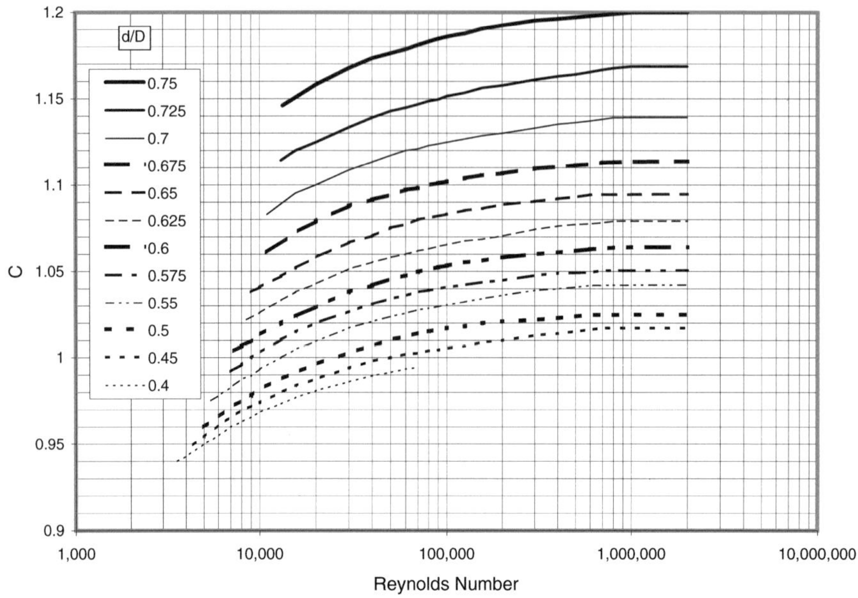

Figure 5–1 Choke flow coefficient for nozzle-type chokes.

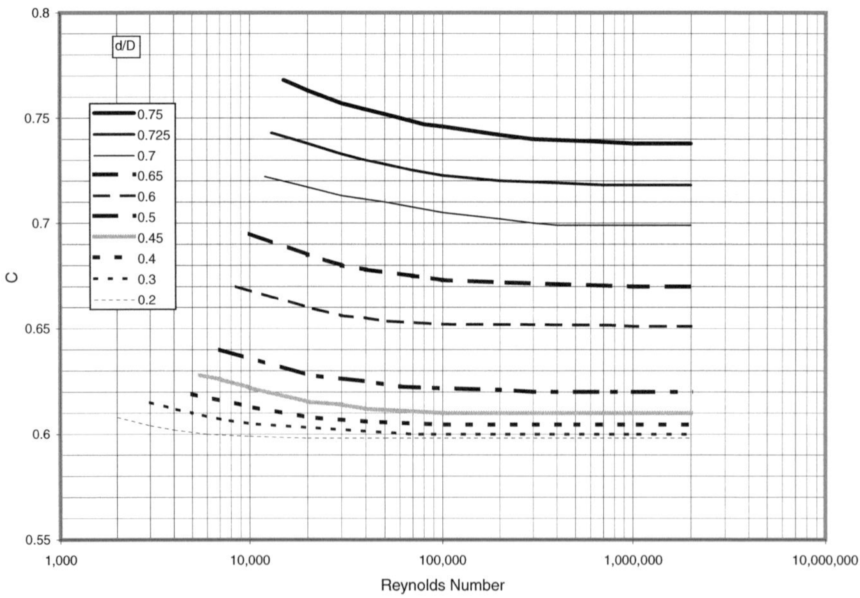

Figure 5–2 Choke flow coefficient for orifice-type chokes.

5.3.2 Sonic Flow

Under sonic flow conditions the gas passage rate reaches its maximum value. Gas passage rate is expressed in the following equation for ideal gases:

$$Q_{sc} = 879CAp_{up}\sqrt{\left(\frac{k}{\gamma_g T_{up}}\right)\left(\frac{2}{k+1}\right)^{\frac{k+1}{k-1}}} \tag{5.5}$$

The choke flow coefficient C is not sensitive to the Reynolds number for Reynolds number values greater than 10^6. Thus the C value at the Reynolds number of 10^6 can be assumed for C values at higher Reynolds numbers.

Gas velocity under sonic flow conditions is equal to sound velocity in the gas under the in situ conditions:

$$v = \sqrt{v_{up}{}^2 + 2g_c C_p T_{up}\left[1 - \frac{z_{up}}{z_{outlet}}\left(\frac{2}{k+1}\right)\right]} \tag{5.6}$$

or

$$v \approx 44.76\sqrt{T_{up}} \tag{5.7}$$

5.3.3 Temperature at Choke

Depending on upstream to downstream pressure ratio, the temperature at choke can be much lower than expected. This low temperature is due to the Joule-Thomson cooling effect, that is, a sudden gas expansion below the nozzle causes a significant temperature drop. The temperature can easily drop to below ice point resulting in ice-plugging if water exists. Even though the temperature still can be above ice point, hydrates can form and cause plugging problems. Assuming an isentropic process for an ideal gas flowing through chokes, the temperature at the choke downstream can be predicted using the following equation:

$$T_{dn} = T_{up}\frac{z_{up}}{z_{outlet}}\left(\frac{p_{outlet}}{p_{up}}\right)^{\frac{k-1}{k}} \tag{5.8}$$

The outlet pressure is equal to the downstream pressure in subsonic flow conditions.

5.3.4 Applications

Equation (5.1) through Equation (5.8) can be used for estimating

- Downstream temperature (Equation (5.8))
- Gas passage rate at given upstream and downstream pressures
- Upstream pressure at given downstream pressure and gas passage
- Downstream pressure at given upstream pressure and gas passage

To estimate gas passage rate at given upstream and downstream pressures, the following procedure can be taken:

1. Calculate the critical pressure ratio with Equation (5.1).
2. Calculate the downstream to upstream pressure ratio.
3. If the downstream to upstream pressure ratio is greater than the critical pressure ratio, use Equation (5.2) to calculate gas passage. Otherwise, use Equation (5.5) to calculate gas passage.

Example Problem 5.1

A 0.6 specific gravity gas flows from a 2-in pipe through a 1-in orifice-type choke. The upstream pressure and temperature are 800 psia and 75 °F, respectively. The downstream pressure is 200 psia (measured 2 ft from the orifice). The gas-specific heat ratio is 1.3. (a) What is the expected daily flow rate? (b) Does heating need to be applied to assure that the frost does not clog the orifice? (c) What is the expected pressure at the orifice outlet?

Solution

(a)

$$\left(\frac{P_{outlet}}{P_{up}}\right)_c = \left(\frac{2}{k+1}\right)^{\frac{k}{k-1}} = \left(\frac{2}{1.3+1}\right)^{\frac{1.3}{1.3-1}} = 0.5459$$

$$\frac{P_{dn}}{P_{up}} = \frac{200}{800} = 0.25 < 0.5459 \text{ Sonic flow exists.}$$

$$\frac{d}{D} = \frac{1''}{2''} = 0.5$$

Assuming $N_{Re} > 10^6$, Figure 5–2 gives $C = 0.62$.

$$Q_{sc} = 879CAP_{up}\sqrt{\left(\frac{k}{\gamma_g T_{up}}\right)\left(\frac{2}{k+1}\right)^{\frac{k+1}{k-1}}}$$

$$Q_{sc} = (879)(0.62)(\pi/4*1^2)(800)\sqrt{\left(\frac{1.3}{(0.6)(75+460)}\right)\left(\frac{2}{1.3+1}\right)^{\frac{1.3+1}{1.3-1}}}$$

$$Q_{sc} = 12{,}743 \text{ Mscf/d}$$

Check N_{Re}:

$\mu = 0.01245$ cp by the Carr-Kobayashi-Burrows correlation.

$$N_{\text{Re}} = \frac{20Q_{sc}\gamma_g}{\mu d} = \frac{(20)(12{,}743)(0.6)}{(0.01245)(1)} = 1.23\times10^7 \;\; > 10^6$$

(b)

$$T_{dn} = T_{up}\frac{z_{up}}{z_{outlet}}\left(\frac{P_{outlet}}{P_{up}}\right)^{\frac{k-1}{k}} = (75+460)(1)(0.5459)^{\frac{1.3-1}{1.3}} = 465 \;\; ^\circ\text{R} = 5 \;\; ^\circ\text{F} < 32 \;\; ^\circ\text{F}$$

Therefore, heating is needed to prevent icing.

(c)

$$P_{outlet} = P_{up}\left(\frac{P_{outlet}}{P_{up}}\right) = (800)(0.5459) = 437 \text{ psia}$$

Example Problem 5.2

A 0.65 specific gravity natural gas flows from a 2-in pipe through a 1.5-in nozzle-type choke. The upstream pressure and temperature are 100 psia and 70 °F, respectively. The downstream pressure is 80 psia (measured 2 ft from the nozzle). The gas specific heat ratio is 1.25. (a) What is the expected daily flow rate? (b) Is icing a potential problem? (c) What is the expected pressure at the nozzle outlet?

Solution

(a)

$$\left(\frac{P_{outlet}}{P_{up}}\right)_c = \left(\frac{2}{k+1}\right)^{\frac{k}{k-1}} = \left(\frac{2}{1.25+1}\right)^{\frac{1.25}{1.25-1}} = 0.5549$$

$\frac{P_{dn}}{P_{up}} = \frac{80}{100} = 0.8$ > 0.5549 Subsonic flow exists.

$$\frac{d}{D} = \frac{1.5''}{2''} = 0.75$$

Assuming $N_{Re} > 10^6$, Figure 5–1 gives $C = 1.2$.

$$Q_{sc} = 1{,}248CAP_{up}\sqrt{\frac{k}{(k-1)\gamma_g T_{up}}\left[\left(\frac{P_{dn}}{P_{up}}\right)^{\frac{2}{k}} - \left(\frac{P_{dn}}{P_{up}}\right)^{\frac{k+1}{k}}\right]}$$

$$Q_{sc} = (1{,}248)(1.2)(\pi/4 * 1.5^2)(100)\sqrt{\frac{1.25}{(1.25-1)(0.65)(530)}\left[\left(\frac{80}{100}\right)^{\frac{2}{1.25}} - \left(\frac{80}{100}\right)^{\frac{1.25+1}{1.25}}\right]}$$

$$Q_{sc} = 5{,}572 \text{ Mscf/d}$$

Check N_{Re}:

$\mu = 0.0108$ cp by the Carr-Kobayashi-Burrows correlation.

$$N_{\text{Re}} = \frac{20Q_{sc}\gamma_g}{\mu d} = \frac{(20)(5{,}572)(0.65)}{(0.0108)(1.5)} = 4.5\times10^6 \;\; >10^6$$

(b)

$$T_{dn} = T_{up}\frac{z_{up}}{z_{outlet}}\left(\frac{P_{outlet}}{P_{up}}\right)^{\frac{k-1}{k}} = (70+460)(1)(0.8)^{\frac{1.25-1}{1.25}} = 507^\circ\text{R} = 47^\circ\text{F} > 32^\circ\text{F}$$

Heating may not be needed. Hydrate curve may need to be checked (see Chapter 12).

(c)

$$P_{outlet} = P_{dn} = 80 \;\; \text{psia for subcritical flow.}$$

To estimate upstream pressure at given downstream pressure and gas passage, the following procedure can be taken:

1 Calculate the critical pressure ratio with Equation (5.1).

2 Calculate the minimum upstream pressure required for sonic flow by dividing the downstream pressure by the critical pressure ratio.

3 Calculate gas flow rate at the minimum sonic flow condition with Equation (5.5).

4 If the given gas passage is less than the calculated gas flow rate at the minimum sonic flow condition, use Equation (5.2) to solve upstream pressure numerically. Otherwise, Equation (5.5) to calculate upstream pressure.

Example Problem 5.3

For the following given data, estimate upstream pressure at choke:

Downstream pressure: 300 psia

Choke size: 32 1/64 in

Flowline ID: 2 in

Gas production rate: 5,000 Mscf/d

Gas-specific gravity: 0.75 1 for air

Gas-specific heat ratio: 1.3

Upstream temperature: 110 °F

Choke discharge coefficient: 0.99

Solution

This problem is solved with the spreadsheet program DryGasUpChoke.xls. The result is shown in Table 5–1.

Downstream pressure cannot be calculated on the basis of given upstream pressure and gas passage under sonic flow conditions. But it can be calculated under subsonic flow conditions. The following procedure can be followed:

1 Calculate the critical pressure ratio with Equation (5.1).

2 Calculate the maximum downstream pressure for minimum sonic flow by multiplying the upstream pressure by the critical pressure ratio.

3 Calculate gas flow rate at the minimum sonic flow condition with Equation (5.5).

4 If the given gas passage is less than the calculated gas flow rate at the minimum sonic flow condition, use Equation (5.2) to solve downstream pressure numerically. Otherwise, the downstream pressure cannot be calculated. The maximum possible downstream pressure for sonic flow can be estimated by multiplying the upstream pressure by the critical pressure ratio.

Example Problem 5.4

For the following given data, estimate downstream pressure at choke:

Upstream pressure: 600 psia

Choke size: 32 1/64 in

Flowline ID: 2 in

Gas production rate: 2500 Mscf/d

Table 5–1 Solution Given by DryGasUpchoke.xls[a]

Instructions: 1) Update parameter values; 2) Run Macro Solution and view results.

Input Data	
Downstream pressure:	300 psia
Choke size:	32 1/64 in
Flowline ID:	2 in
Gas production rate:	5,000 Mscf/d
Gas-specific gravity:	0.75 1 for air
Gas-specific heat ratio (k):	1.3
Upstream temperature:	110 °F
Choke discharge coefficient:	0.99
Calculated Values	
Choke area:	0.19625 in^2
Critical pressure ratio:	0.5457
Minimum upstream pressure required for sonic flow:	549.72 psia
Flow rate at the minimum sonic flow condition:	3,029.76 Mscf/d
Flow regime (1 = sonic flow; –1 = subsonic flow):	1
Upstream pressure given by sonic flow equation:	907.21 psia
Upstream pressure given by subsonic flow equation:	1,088.04 psia
Estimated upstream pressure:	907.21 psia

a. This spreadsheet calculates upstream pressure at choke for dry gases

Gas-specific gravity: 0.75 1 for air

Gas-specific heat ratio: 1.3

Upstream temperature: 110 °F

Choke discharge coefficient: 0.99

Solution

This problem is solved with the spreadsheet program DryGas-DownChoke.xls. The result is shown in Table 5–2.

5.4 Wet Gas Flow through Chokes

Wet gas is referred to as natural gas with condensate (oil). Fortunati (1972) presented a model that can be used to calculate critical and subcritical two-phase flow through chokes. Ashford (1974) also developed a relation for two-phase critical flow based on the work of Ros (1960). Gould (1974) plotted the critical/subcritical boundary defined by Ashford, showing that different values of the polytropic exponents yield different boundaries. Ashford and Pierce (1975) derived an equation to predict the critical pressure ratio. Their model assumes that the derivative of flow rate with respect to the downstream pressure is zero at critical conditions. One set of equations was recommended for both critical and subcritical flow conditions. Pilehvari (1981) also studied choke flow under subcritical conditions. Sachdeva (1986) extended the work of Ashford and Pierce and proposed a relationship to predict critical pressure ratio. He also derived an expression in order to find the boundary between critical and subcritical flow. Sachdeva's model was used in a multi-phase compositional well simulator for production optimization. On the basis of case studies with data from a Southwest Louisiana gas condensate field, Guo, Al-Bemani, and Ghalambor (2002) concluded that Sachdeva's choke model is accurate for predicting liquid rates of oil wells and gas rates of gas condensate wells with a discharge coefficient of $C = 1.08$. A discharge coefficient of $C = 0.78$ should be used for predicting gas rates of oil wells, and $C = 1.53$ should be used for predicting liquid rates of gas condensate wells.

Table 5–2 Solution Given by DryGasDownChoke.xls[a]

Instructions: 1) Update parameter values in blue; 2) Run Macro Solution; 3) View results.

Input Data	
Upstream pressure:	600 psia
Choke size:	32 1/64 in
Flowline ID:	2 in
Gas production rate:	2,500 Mscf/d
Gas-specific gravity:	0.75 1 for air
Gas-specific heat ratio (k):	1.3
Upstream temperature:	110 °F
Choke discharge coefficient:	0.99
Calculated Values	
Choke area:	0.19625 in^2
Critical pressure ratio:	0.5457
Maximum downstream pressure for minimum sonic flow:	327.44 psia
Flow rate at the minimum sonic flow condition:	3,306.84 Mscf/d
Flow regime (1 = sonic flow; –1 = subsonic flow):	–1
The maximum possible downstream pressure in sonic flow:	327.44 psia
Downstream pressure given by subsonic flow equation:	508.15 psia
Estimated downstream pressure:	508.15 psia

a. This spreadsheet calculates downstream pressure at choke for dry gases.

5.5 References

Ashford, F. E. "An Evaluation of Critical Multiphase Flow Performance through Wellhead Chokes." *Journal of Petroleum Technology* **26** (August 1974): 843–8.

Ashford, F. E. and P. E. Pierce. "Determining Multiphase Pressure Drop and Flow Capabilities in Down Hole Safety Valves." *Journal of Petroleum Technology* **27** (September 1975): 1145–52.

Fortunati, F. "Two-Phase Flow through Wellhead Chokes." Paper SPE 3742 presented at the SPE European Spring Meeting in Amsterdam, The Netherlands, May 16–18 1972.

Guo, B., A. S. Al-Bemani, and A. Ghalambor. "Applicability of Sachdeva's Choke Flow Model in Southwest Louisiana Gas Condensate Wells." Paper SPE 75507 presented at the SPE Gas Technology Symposium, Calgary, Alberta, Canada, April 30–May 2, 2002.

Gould, T. L. "Discussion of An Evaluation of Critical Multiphase Flow Performance through Wellhead Chokes." *Journal of Petroleum Technology* **26** (August 1974): 849–50.

Pilehvari, A. A. "Experimental Study of Critical Two-Phase Flow through Wellhead Chokes." University of Tulsa Fluid Flow Projects Report, Tulsa, Oklahoma, June 1981.

Ros, N. C. J. "An Analysis of Critical Simultaneous Gas/Liquid Flow through a Restriction and Its Application to Flow Metering." *Applied Science Research* Sec. A, 9 (1960): 374–89.

Sachdeva, R., Z. Schmidt, J. P. Brill, and R. M. Blais. "Two-Phase Flow Through Chokes." Paper SPE 15657 presented at the SPE 61st Annual Technical Conference and Exhibition, New Orleans, Louisiana, October 5–8, 1986.

5.6 Problems

5-1 A 0.66 specific gravity gas flows from a 2-in pipe through a 1.5-in orifice-type choke. The upstream pressure and temperature are 600 psia and 75 °F, respectively. The downstream pressure is 200 psia (measured 2 ft from the orifice). The gas-specific heat ratio is 1.3. (a) What is the expected daily flow rate? (b) Does heating need to be applied to assure that the frost does not clog the orifice? (c) What is the expected pressure at the orifice outlet?

5-2 A 0.60 specific gravity natural gas flows from a 2-in pipe through a 1-in nozzle-type choke. The upstream pressure and temperature are 120 psia and 70 °F, respectively. The downstream pressure is 90 psia (measured 2 ft from the nozzle). The gas-specific heat ratio is 1.3. (a) What is the expected daily flow rate? (b) Is icing a potential problem? (c) What is the expected pressure at the nozzle outlet?

5-3 For the following given data, estimate upstream pressure at choke:

Downstream pressure: 500 psia

Choke size: 48 1/64 in

Flowline ID: 2 in

Gas production rate: 4,000 Mscf/d

Gas-specific gravity: 0.70 1 for air

Gas-specific heat ratio: 1.3

Upstream temperature: 100 °F

Choke discharge coefficient: 1.05

5-4 For the following given data, estimate downstream pressure at choke:

Upstream pressure: 500 psia

Choke size: 32 1/64 in

Flowline ID: 2 in

Gas production rate: 2,400 Mscf/d

Gas-specific gravity: 0.70 1 for air

Gas-specific heat ratio: 1.3

Upstream temperature: 100 °F

Choke discharge coefficient: .98

Chapter 6

Well Deliverability

6.1 Introduction

Well deliverability is determined by the combination of well inflow performance (see Chapter 3) and wellbore flow performance (see Chapter 4). While the former describes the deliverability of the reservoir, the latter presents the resistance to flow of production string. This chapter focuses on prediction of achievable gas production rates from gas reservoirs with specified production string characteristics. The technique of analysis is called Nodal analysis (a Schlumburger patent). Calculation examples are illustrated with computer spreadsheets that are provided with this book.

6.2 Nodal Analysis

Fluid properties, such as gas z-factor and gas viscosity, change with the location-dependent pressure and temperature in the gas production system. To simulate the fluid flow in the system, it is necessary to "break" the system into discrete nodes that separate system elements (equipment sections). Fluid properties at the elements are evaluated locally. The system analysis for determination of fluid production rate and pressure at a specified node is called Nodal analysis in petroleum engineering.

Nodal analysis is performed on the principle of pressure continuity, that is, there is only one unique pressure value at a given node no matter whether the pressure is evaluated from the performance of upstream equipment or downstream equipment. The performance curve (pressure-rate relation) of upstream equipment is called inflow performance curve; the performance curve of downstream equipment is called outflow performance curve. The intersection of the two performance curves defines the operating point, that is, operating flow rate and pressure, at the specified node. For the convenience of using pressure data measured normally at either bottom hole or wellhead, Nodal analysis is usually conducted using the bottom hole or wellhead as the solution node.

6.2.1 Analysis with the Bottom Hole Node

When the bottom hole is used as a solution node in Nodal analysis, the inflow performance is the well Inflow Performance Relationship (IPR) and the outflow performance is the Tubing Performance Relationship (TPR), if the tubing shoe is set to the top of the pay zone. Well IPR can be established with different methods presented in Chapter 3. TPR can be modeled with various approaches as discussed in Chapter 4.

Traditionally, Nodal analysis at the bottom hole is carried out by plotting the IPR and TPR curves and visually finding the solution at the intersection point of the two curves. With modern computer technologies, the solution can be computed quickly without plotting the curves, although the curves are still plotted for visual verification.

Consider the bottom hole node of a gas well. If the IPR of the well is defined by

$$q_{sc} = C\left(\bar{p}^2 - p_{wf}^2\right)^n \tag{6.1}$$

and if the outflow performance relationship of the node (TPR) is defined by

$$p_{wf}^2 = Exp(s)p_{hf}^2 + \frac{6.67\times10^{-4}[Exp(s)-1]fq_{sc}^2\bar{z}^2\bar{T}^2}{d_i^5\cos\theta} \tag{6.2}$$

then the operating flow rate q_{sc} and pressure p_{wf} at the bottom hole node can be determined graphically by plotting Equation (6.1) and Equation (6.2) and finding the intersection point.

The operating point can also be solved numerically by combining Equation (6.1) and Equation (6.2). In fact, Equation (6.1) can be rearranged as:

$$p_{wf}^2 = \bar{p}^2 - \left(\frac{q_{sc}}{C}\right)^{\frac{1}{n}} \tag{6.3}$$

Substituting Equation (6.3) into Equation (6.2) yields:

$$\bar{p}^2 - \left(\frac{q_{sc}}{C}\right)^{\frac{1}{n}} - Exp(s)p_{hf}^2 - \frac{6.67 \times 10^{-4}[Exp(s)-1]fq_{sc}^2\bar{z}^2\bar{T}^2}{D_i^5 \cos\theta} = 0 \tag{6.4}$$

which can be solved with a numerical technique such as the Newton-Raphson iteration for gas flow rate q_{sc}. This computation can be performed automatically with the spreadsheet program BottomHoleNodal.xls. Users need to input parameter values in the *Input Data* section and run Macro Solution to get results.

Example Problem 6.1

Suppose that a vertical well produces 0.71 specific gravity gas through a 2 7/8-in tubing set to the top of a gas reservoir at a depth of 10,000 ft. At tubing head, the pressure is 800 psia and the temperature is 150 °F, the bottom hole temperature is 200 °F. The relative roughness of tubing is about 0.0006. Calculate the expected gas production rate of the well using the following data for IPR:

Reservoir pressure: 2,000 psia

IPR model parameter *C*: 0.01 Mscf/d-psi^{2n}

IPR model parameter *n*: 0.8

Solution

This example problem is solved with the spreadsheet program BottomHoleNodal.xls. Table 6–1 shows the appearance of the spreadsheet for the data input and result sections. It indicates that the expected gas flow rate is 1,478 Mscf/d at a bottom hole pressure of 1,050 psia. The inflow and outflow performance curves plotted in Figure 6–1 confirm this operating point.

Table 6–1 Input Data and Results Given by BottomHoleNodal.xls[a]

Instructions: 1) Input your data in the Input Data section; 2) Run Macro Solution to get results; 3) View results in table and in the Plot graph sheet.

Input Data		
Gas-specific gravity (γ_g):	0.71	
Tubing inside diameter (D):	2.259 in	
Tubing relative roughness (ε/D):	0.0006	
Measured depth at tubing shoe (L):	10,000 ft	
Inclination angle (θ):	0°	
Wellhead pressure (p_{hf}):	800 psia	
Wellhead temperature (T_{hf}):	150 °F	
Bottom hole temperature (T_{wf}):	200 °F	
Reservoir pressure (p):	2,000 psia	
C-exponent in backpressure IPR model:	0.01Mscf/d-psi^{2n}	
n-constant in backpressure IPR model:	0.8	
Solution		
T_{av} =	635 °R	
Z_{av} =	0.8626	
s =	0.486062358	
e^s =	1.62590138	
f_m =	0.017396984	
AOF =	1,912.70 Mscf/d	
q_{sc} (Mscf/d)	**IPR**	**TPR**
0	2,000	1,020
191	1,943	1,021
383	1,861	1,023
574	1,764	1,026
765	1,652	1,031
956	1,523	1,037
1,148	1,374	1,044
1,339	1,200	1,052
1,530	987	1,062
1,721	703	1,073
1,817	498	1,078
1,865	353	1,081
1,889	250	1,083
1,913	0	1,084
Operating flow rate =	1,478 Mscrf/d	
Operating pressure =	1,050 psia	

a. This spreadsheet calculates well deliverability with bottom hole node

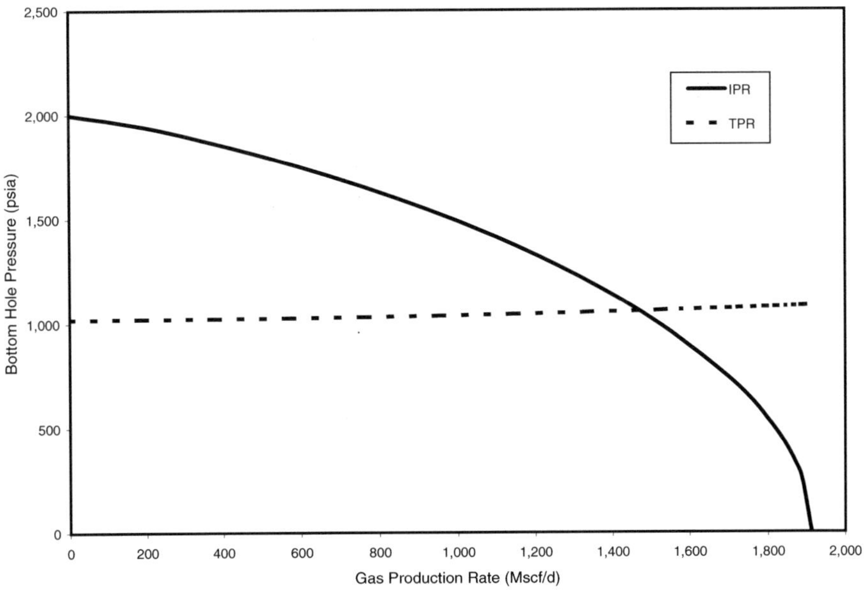

Figure 6–1 Nodal analysis for Example Problem 6.1.

6.2.2 Analysis with Wellhead Node

When the wellhead is used as a solution node in Nodal analysis, the inflow performance curve is the Wellhead Performance Relationship (WPR) that is obtained by transforming the IPR to wellhead through TPR. The outflow performance curve is the wellhead Choke Performance Relationship (CPR). Some TPR models are presented in Chapter 4. CPR models are discussed in Chapter 5.

Nodal analysis with wellhead being a solution node is carried out by plotting the WPR and CPR curves and finding the solution at the intersection point of the two curves. Again, with modern computer technologies, the solution can be computed quickly without plotting the curves, although the curves are still plotted for verification.

If the IPR of the well is defined by Equation (6.1), and TPR is represented by Equation (6.2), substituting Equation (6.2) into Equation (6.1) gives

$$\bar{p}^2 - \left(\frac{q_{sc}}{C}\right)^{\frac{1}{n}} - Exp(s)p_{hf}^2 - \frac{6.67\times10^{-4}[Exp(s)-1]fq_{sc}^2\bar{z}^2\bar{T}^2}{D_i^5\cos\theta} = 0 \quad \textbf{(6.5)}$$

which defines a relationship between wellhead pressure p_{hf} and gas production rate q_{sc}, that is WPR. If the CPR is defined by Equation (5.5), that is,

$$q_{sc} = 879CAp_{hf}\sqrt{\left(\frac{k}{\gamma_g T_{up}}\right)\left(\frac{2}{k+1}\right)^{\frac{k+1}{k-1}}} \quad \textbf{(6.6)}$$

then the operating flow rate q_{sc} and pressure p_{hf} at the wellhead node can be determined graphically by plotting Equation (6.5) and Equation (6.6) and finding the intersection point.

The operating point can also be solved numerically by combining Equation (6.5) and Equation (6.6). In fact, Equation (6.6) can be rearranged as:

$$p_{hf} = \frac{q_{sc}}{879CA\sqrt{\left(\frac{k}{\gamma_g T_{up}}\right)\left(\frac{2}{k+1}\right)^{\frac{k+1}{k-1}}}} \quad \textbf{(6.7)}$$

Substituting Equation (6.7) into Equation (6.6) gives

$$q_{sc} = C\left[\bar{p}^2 - \left(Exp(s)\left(\frac{q_{sc}}{879CA\sqrt{\left(\frac{k}{\gamma_g T_{up}}\right)\left(\frac{2}{k+1}\right)^{\frac{k+1}{k-1}}}}\right)^2 + \frac{6.67\times10^{-4}[Exp(s)-1]fq_{sc}^2\bar{z}^2\bar{T}^2}{d_i^5\cos\theta}\right)\right]^n$$

which can be solved numerically for gas flow rate q_{sc}. This computation can be performed automatically with the spreadsheet program Wellhead-

Nodal.xls. Users need to input parameter values in the *Input Data* section and run Macro Solution to get results.

Example Problem 6.2

Use the following given data to estimate gas production rate of the well:

Gas-specific gravity: 0.71

Tubing inside diameter: 2.259 in

Tubing wall relative roughness: 0.0006

Measured depth at tubing shoe: 10,000 ft

Inclination angle: 0°

Wellhead choke size: 16 1/64 in

Flowline diameter: 2 in

Gas-specific heat ratio: 1.3

Gas viscosity at wellhead: 0.01 cp

Wellhead temperature: 120 °F

Bottom hole temperature: 180 °F

Reservoir pressure: 2,000 psia

C-constant in backpressure IPR model: 0.01 Mscf/dpsi^{2n}

n-exponent in backpressure IPR model: 0.8

Solution:

This example problem is solved with the spreadsheet program WellheadNodal.xls. Table 6–2 and Table 6–3 show the appearance of the spreadsheet for the data input and result sections. It indicates that the expected gas flow rate is 1,470 Mscf/d at a bottom hole pressure of 797 psia. The inflow and outflow performance curves plotted in Figure 6–2 confirm this operating point.

6.3 Production Forecast

Due to the high compressibility of gas and low-permeability of gas reservoir rock, the transient flow period can last significantly long before pseudo-steady state flow is fully established. During the transient flow,

Table 6–2 Input Data and Solution Given by WellheadNodal.xls[a]

Instructions: 1) Input your data in the Input Data section; 2) Run Macro Solution to get results; 3) View results in table and in the Plot graph sheet.	
Input Data	
Gas-specific gravity (γ_g):	0.71
Tubing inside diameter (D):	2.259 in
Tubing relative roughness (ε/D):	0.0006
Measured depth at tubing shoe (L):	10,000 ft
Inclination angle (θ):	0°
Wellhead choke size (D_{ck}):	16 1/64 in
Flowline diameter (D_{fl}):	2 in
Gas-specific heat ratio (k):	1.3
Gas viscosity at wellhead (μ):	0.01 cp
Wellhead temperature (T_{hf}):	120 °F
Bottom hole temperature (T_{wf}):	180 °F
Reservoir pressure ($p\sim$):	2,000 psia
C-constant in backpressure IPR model:	0.01 Mscf/d-psi^{2n}
n-exponent in backpressure IPR model:	0.8
Solution	
T_{av} =	610 °R
Z_{av} =	0.8786
s =	0.4968
e^s =	1.6434
f_m =	0.0174
AOF =	1,913 Mscf/d
D_{ck}/D_{fl} =	0.125
Re =	8,348,517
C_{ck} =	1.3009
A_{ck} =	0.0490625 in^2

a. This spreadsheet calculates well deliverability with wellhead node.

Table 6–3 Results Section of WellheadNodal.xls

q_{sc} (Mscf/d)	WPR	CPR
0	1,600	0
191	1,554	104
383	1,489	207
574	1,411	311
765	1,321	415
956	1,218	518
1,148	1,099	622
1,339	960	726
1,530	789	830
1,721	562	933
1,817	399	985
1,865	282	1,011
1,889	200	1,024
1,913	1	1,037
Operating flow rate =	1,470 Mscf/d	
Operating pressure =	797 psia	

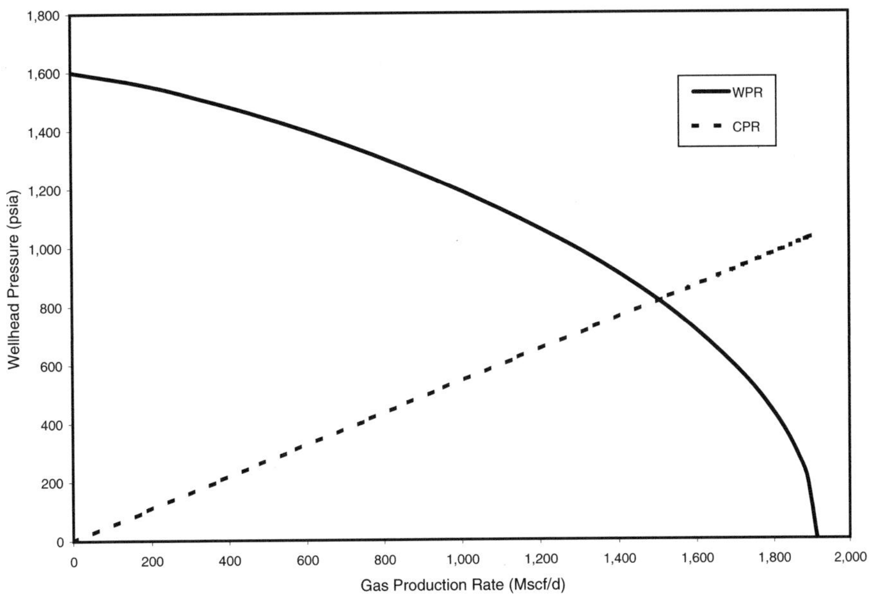

Figure 6–2 Nodal analysis for Example Problem 6.2.

gas production rate can be predicted by Nodal analysis using transient IPR and steady flow TPR. The transient IPR model for gas wells is described in Chapter 3, i.e.,

$$q = \frac{kh\left[m(p_i) - m(p_{wf})\right]}{1638T\left(\log t + \log\frac{k}{\varphi\mu\ c_t r_w^{\ 2}} - 3.23 + 0.87S\right)} \tag{6.8}$$

This equation can be used for generating IPR curves for future time t before any reservoir boundary is "felt". After all reservoir boundaries are reached, a pseudo-steady state flow should prevail for a volumetric gas reservoir. For a circular reservoir, the time required for the pressure wave to reach the reservoir boundary can be estimated by with

$$t_{pss} \approx 1200\frac{\varphi\mu c_t r_e^2}{k}$$

The same TPR is usually used in the transient flow period assuming fluid properties remain the same in the well over the period. The average temperature – average z-factor method can be used for constructing TPR.

Gas production during the pseudo-steady state flow period is due to gas expansion. IPR changes over time due to the change in reservoir pressure. An IPR model is described in Chapter 3, i.e.,

$$q = \frac{kh\left[m(\overline{p}) - m(p_{wf})\right]}{1424T\left(\ln\frac{r_e}{r_w} - \frac{3}{4} + S + Dq\right)} \tag{6.9}$$

Again a constant TPR is usually assumed if liquid loading is not a problem and the wellhead pressure is kept constant over time.

Gas production schedule can be established through material balance equation, i.e.,

$$G_p = G_i\left(1 - \frac{\frac{\bar{p}}{z}}{\frac{p_i}{z_i}}\right) \quad \textbf{(6.10)}$$

where G_p and G_i are the cumulative gas production and initial gas-in-place, respectively.

If gas production rate is predicted by Nodal analysis at a given reservoir pressure level and the cumulative gas production is estimated with Eq (6.10) at the same reservoir pressure level, the corresponding production time can be calculated and thus production forecast can be carried out.

Example Problem 6.3:

Use the following data and develop a forecast of a well production after transient flow until the average reservoir pressure declines to 2,000 psia:

Reservoir depth: 10,000 ft

Initial reservoir pressure: 4,613 psia

Reservoir temperature: 180°F

Pay zone thickness: 78 ft

Formation permeability: 0.17md

Formation porosity: 0.14

Water saturation: 0.27

Gas specific gravity: 0.7air = 1

Total compressibility: $1.5\text{x}10^{-4}\text{psi}^{-1}$

Darcy skin factor: 0

Non-Darcy flow coefficient: 0

Drainage area: 40 acres

Wellbore radius: 0.328 ft

Tubing inner diameter:2.441 in.

Desired flowing bottom hole pressure: 1,500 psia

Solution:

Spreadsheet program Carr-Kobayashi-Burrows-GasViscosity.xls gives a gas viscosity value of 0.0251 cp at the initial reservoir pressure of 4,613 psia and temperature of 180 °F for the 0.7 specific gravity gas. Spreadsheet program Hall-Yarborogh-z.xls gives a z-factor value of 1.079 at the same conditions. Formation volume factor at the initial reservoir pressure is calculated with Eq (2.45):

$$B_{gi} = 0.0283 \frac{(1.079)(180+460)}{4,613} = 0.004236 \text{ ft}^3/\text{scf}$$

The initial gas-in-place within the 40 acres is:

$$G_i = \frac{(43,560)(40)(78)(0.14)(1-0.27)}{0.004236} = 3.28 \times 10^9 \text{ scf}$$

Assuming a circular drainage area, the equivalent radius of the 40 acres is 745 ft. The time required for the pressure wave to reach the reservoir boundary is estimated as:

$$t_{pss} \approx 1200 \frac{(0.14)(0.0251)(1.5\times10^{-4})(745)^2}{0.17} = 2,065 \text{ hours} = 86 \text{ days}$$

Spreadsheet program PseudoPressure.xls gives

$$m(p_i) = m(4613) = 1.27\times10^9 \text{ psi}^2/\text{cp}$$

$$m(p_{wf}) = m(1500) = 1.85\times10^8 \text{ psi}^2/\text{cp}$$

Substituting these and other given parameter values to Eq (6.8) yields:

$$q = \frac{(0.17)(78)\left[1.27\times10^9 - 1.85\times10^8\right]}{1638(180+460)\left(\log(2065) + \log\frac{0.17}{(0.14)(0.0251)(1.5\times10^{-4})(0.328)^2} - 3.23\right)} = 2,092 \text{ Mscf/day}$$

Substituting q = 2,092 Mscf/day into Eq (6.9) gives:

$$2{,}092 = \frac{(0.17)(78)\left[m(\bar{p}) - 1.85\times10^{8}\right]}{1424(180+460)\left(\ln\frac{745}{0.328} - \frac{3}{4} + 0\right)}$$

which results in $m(\bar{p}) = 1.19\times10^{9}$ psi^2/cp. Spreadsheet program PseudoPressure.xls gives $\bar{p}$ = 4,409 psia at the beginning of the pseudo-steady state flow period.

If the flowing bottom hole pressure is maintained at a level of 1,500 psia during the pseudo-steady state flow period (after 86 days of transient production), Eq. (6.8) is simplified as

$$q = \frac{(0.17)(78)\left[m(\bar{p}) - 1.85\times10^{8}\right]}{1424(180+460)\left(\ln\frac{745}{0.328} - \frac{3}{4} + 0\right)}$$

or

$$q = 2.09\times10^{-6}\left[m(\bar{p}) - 1.85\times10^{8}\right]$$

which, combined with Eq (6.10), gives the production forecast shown in Table 6.4 where z-factors and real gas pseudo-pressures were obtained using spreadsheet programs Hall-Yarborogh-z.xls and PseudoPressure.xls, respectively. The production forecast result is also plotted in Figure 6-3.

Table 6-4: Result of Production Forecast for Example Problem 6.3

Reservoir Pressure (psia)	z	Pseudo-pressure (10^8 psi^2/cp)	G_p (MMscf)	ΔG_p (MMscf)	q (Mscf/d)	Δt (day)	t (day)
4,409	1.074	11.90	130				
4,200	1.067	11.14	260	130	1,942	67	67
4,000	1.060	10.28	385	125	1,762	71	138
3,800	1.054	9.50	514	129	1,598	81	218
3,600	1.048	8.73	645	131	1,437	91	309
3,400	1.042	7.96	777	132	1,277	103	413
3,200	1.037	7.20	913	136	1,118	122	534
3,000	1.032	6.47	1,050	137	966	142	676
2,800	1.027	5.75	1,188	139	815	170	846
2,600	1.022	5.06	1,328	140	671	209	1,055
2,400	1.018	4.39	1,471	143	531	269	1,324
2,200	1.014	3.76	1,615	144	399	361	1,686
2,000	1.011	3.16	1,762	147	274	536	2,222

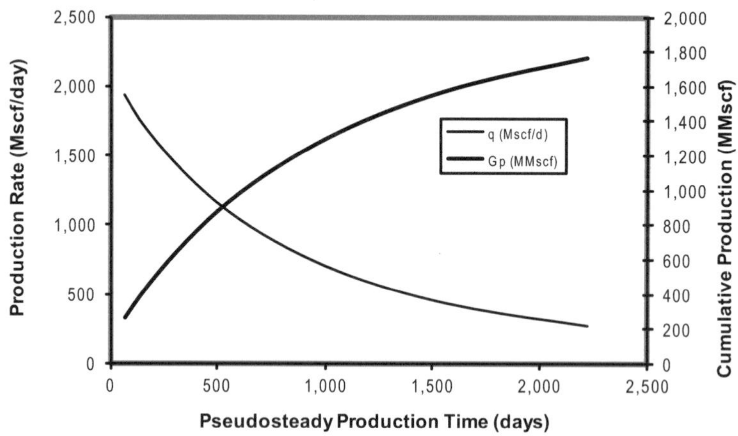

Figure 6–3 Result of production forecast for Example Problem 6.3

6.4 References

Greene, W. R. "Analyzing the Performance of Gas Wells." *Journal of Petroleum Technology* (July 1983): 31–9.

Nind, T. E. W. *Principles of Oil Well Production*. 2nd Ed. New York: McGraw-Hill, 1981.

Russell, D. G., J. H. Goodrich, G. E. Perry, and J. F. Bruskotter. "Methods for Predicting Gas Well Performance." *Journal of Petroleum Technology* (January 1966): 50–7.

6.5 Problems

6-1 A vertical well produces 0.75 specific-gravity gas through a 2 7/8-in (ID 2.441 in) tubing set to the top of a gas reservoir at a depth of 8,000 ft. Tubing head temperature is 90 °F, and bottom hole temperature is 160 °F. The relative roughness of tubing is about 0.0006. Calculate the expected gas production rates of the well at wellhead pressures of 200 psia, 300 psia, 400 psia, 500 psia, and 600 psia using the following data for IPR:

Reservoir pressure: 1,800 psia

IPR model parameter C: 0.15 Mscf/d-psi^{2n}

IPR model parameter n: 0.85

6-2 Calculate the expected gas production rates of the well described in Problem 6-1 for a 2.259-in ID tubing.

6-3 Use the following data to calculate expected gas production rate of the well:

Gas-specific gravity: 0.75

Tubing inside diameter: 2.259 in

Tubing wall relative roughness: 0.0006

Measured depth at tubing shoe: 8,000 ft

Inclination angle: 0°

Wellhead choke size: 24 1/64 in

Flowline diameter: 2 in

Gas-specific heat ratio: 1.3

Gas viscosity at wellhead: 0.01 cp

Wellhead temperature: 120 °F

Bottom hole temperature: 180 °F

Reservoir pressure: 2,000 psia

C-constant in backpressure IPR model: 0.01 Mscf/dpsi^{2n}

n-exponent in backpressure IPR model: 0.8

6-4 Modify spreadsheet program BottomHoleNodal.xls to incorporate the Forchheimer equation for IPR. Solve Problem 6-1 using estimated *A* and *B* values from *C* and *n* values.

6-5 Modify spreadsheet program WellheadNodal.xls to incorporate the subsonic choke flow equation. Solve Problem 6-3 for flow line pressures of 200 psia, 300 psia, 400 psia, 500 psia, and 600 psia.

Chapter 7

Separation

7.1 Introduction

Natural gases produced from gas wells are normally complex mixtures of hundreds of different compounds. A typical gas well stream is a high-velocity, turbulent, constantly expanding mixture of gases and hydrocarbon liquids, intimately mixed with water vapor, free water, and sometimes solids. The well stream should be processed as soon as possible after bringing it to the surface. Field processing consists of four basic processes: (1) separating the gas from free liquids such as crude oil, hydrocarbon condensate, water, and entrained solids; (2) processing the gas to remove condensable and recoverable hydrocarbon vapors; (3) processing the gas to remove condensable water vapor; and (4) processing the gas to remove other undesirable compounds, such as hydrogen sulfide or carbon dioxide. This chapter focuses on the principles of separation and selection of required separators.

7.2 Separation of Gas and Liquids

Separation of well stream gas from free liquids is the first and most critical stage of field-processing operations. Composition of the fluid mixture determines what type and size of separator is required. However, pressure is another key factor affecting selection of separators. Separators are also used in other locations such as upstream and downstream of compressors, dehydration units, and gas sweetening units. At these locations, separators are referred to as scrubbers, knockouts, and free liquid knockouts. All these vessels are used for the same purpose: to separate free liquids from the gas stream.

Separators should be designed to perform the following basic functions:

- cause a primary-phase separation of the mostly liquid hydrocarbons from the gas stream
- refine the primary separation by further removing most of the entrained liquid mist from the gas
- refine the separation by further removing the entrained gas from the liquid stream
- discharge the separated gas and liquid from the vessel and ensure that no reentrainment of one into the other occurs

7.2.1 Principles of Separation

Most separators work based on the principles of gravity segregation and/or centrifugal segregation. A separator is normally constructed in such a way that it has the following features:

- it has a centrifugal inlet device where the primary separation of the liquid and gas is made
- it provides a large settling section of sufficient height or length to allow liquid droplets to settle out of the gas stream with adequate surge room for slugs of liquid
- it is equipped with a mist extractor or eliminator near the gas outlet to coalesce small particles of liquid that do not settle out by gravity
- it allows adequate controls consisting of level control, liquid dump valve, gas backpressure valve, safety relief valve, pressure gauge, gauge glass, instrument gas regulator, and piping

The centrifugal inlet device makes the incoming stream spin around. Depending upon the mixture flow rate, the reaction force from the separator wall can be up to 500 G of centripetal acceleration. This action forces the liquid droplets together where they fall to the bottom of the separator into the settling section.

The settling section allows the turbulence of the fluid stream to subside and the liquid droplets to fall to the bottom of the vessel due to gravity segregation. A large open space in the vessel is required for this purpose.

Use of internal baffling or plates may produce more liquid to be discharged from the separator. However, the product may not be stable due to the light ends entrained in it. Sufficient surge room is essential in the settling section to handle slugs of liquid without carryover to the gas outlet. This can be achieved by placing the liquid level control in the separator, which in turn determines the liquid level. The amount of surge room required depends on the surge level of the production steam and the separator size used for a particular application.

Small liquid droplets that do not settle out of the gas stream due to little gravity difference between them and the gas phase tend to be entrained and pass out of the separator with the gas. A mist eliminator or extractor near the gas outlet allows this to be almost eliminated. The small liquid droplets will hit the eliminator or extractor surfaces, coalesce, and collect to form larger droplets that will then drain back to the liquid section in the bottom of the separator. A stainless steel woven-wire mesh mist eliminator can remove up to 99.9% of the entrained liquids from the gas stream. Cane mist eliminators can be used in areas where there is entrained solid material in the gas phase that may collect and plug a wire mesh mist eliminator.

7.2.2 Types of Separators

Three types of separators are generally available from manufacturers: vertical, horizontal, and spherical separators. Horizontal separators are further classified into two categories: single tube and double tube. Each type of separator has specific advantages and limitations. Selection of separator type is based on several factors including characteristics of production steam to be treated, floor space availability at the facility site, transportation, and cost.

7.2.2.1 Vertical Separators

Vertical separators are often used to treat low to intermediate gas/oil ratio well streams and streams with relatively large slugs of liquid. They handle greater slugs of liquid without carryover to the gas outlet, and the action of the liquid level control is not as critical as in Figure 7–1. Vertical separators occupy less floor space, which is important for facility sites such as those on offshore platforms where space is limited. Owing to

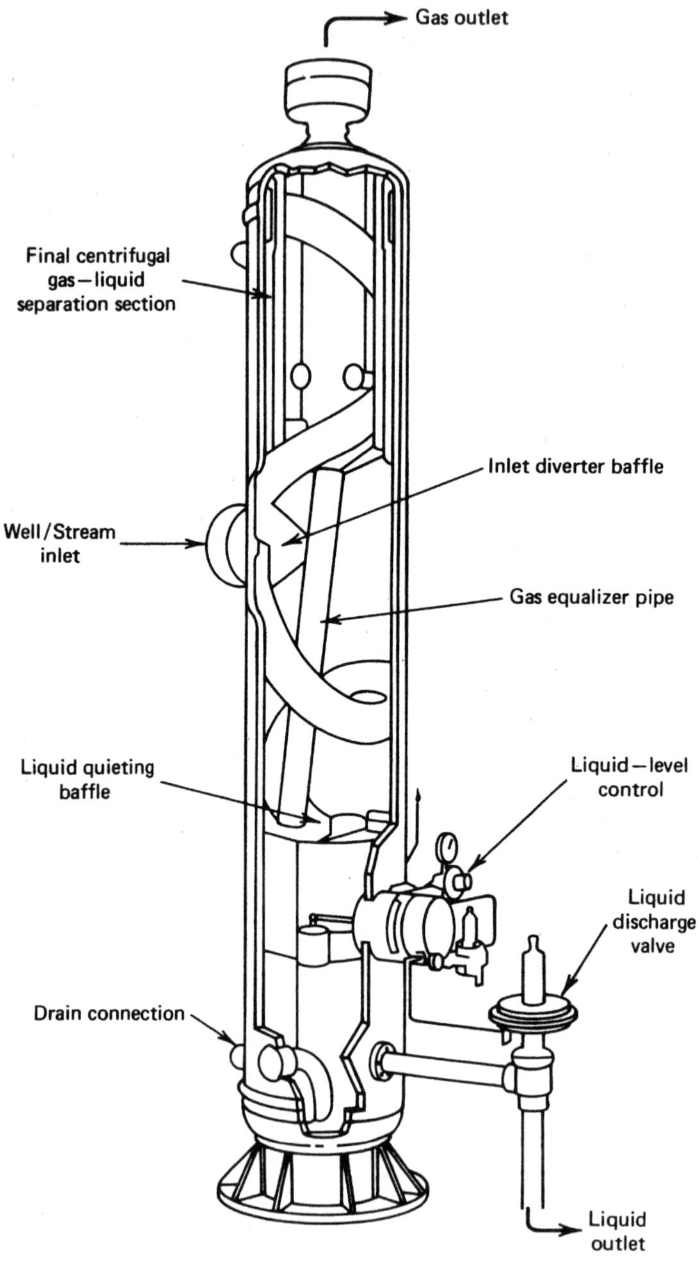

Figure 7–1 Conventional vertical separator (Courtesy Petroleum Extension Services).

the large vertical distance between the liquid level and the gas outlet, the chance for liquid to revaporize into the gas phase is limited. However, due to the natural upward flow of gas in a vertical separator against the falling droplets of liquid, adequate separator diameter is required. Vertical separators are more costly to fabricate and ship in skid-mounted assemblies.

7.2.2.2 Horizontal Separators

Separators (Figure 7–2) are usually the first choice because of their low costs. Horizontal separators are widely used for high gas/oil ratio well streams, foaming well streams, or liquid-from-liquid separation. They have much greater gas/liquid interface due to a large, long, baffled gas-separation section. Horizontal separators are easier to skid-mount and service, and require less piping for field connections. Individual separators can be stacked easily into stage-separation assemblies to minimize space requirements. In horizontal separators, gas flows horizontally and, at the same time, liquid droplets fall toward the liquid surface. The moisture gas flows in the baffle surface and forms a liquid film that is drained away to the liquid section of the separator. The baffles need to be longer than the distance of liquid trajectory travel. The liquid-level control placement is more critical in a horizontal separator than in a vertical separator due to limited surge space.

A horizontal double-tube separator (Figure 7–3) consists of two tube sections. The upper tube section is filled with baffles, and gas flows straight through and at higher velocities, and the incoming free liquid is immediately drained away from the upper tube section into the lower tube section. Horizontal double-tube separators have all the advantages of normal horizontal single-tube separators plus much higher liquid capacities.

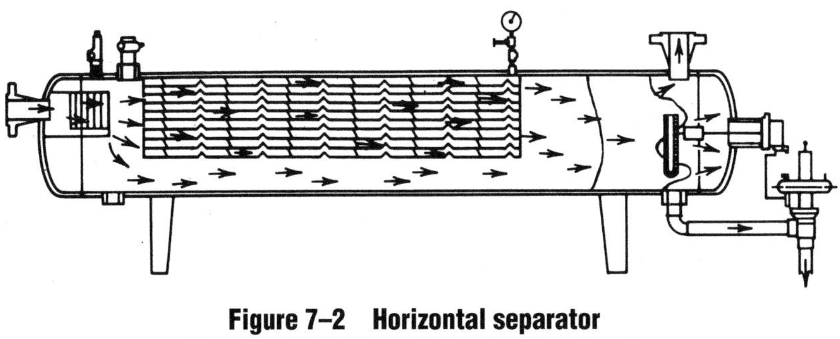

Figure 7–2 Horizontal separator (Courtesy Petroleum Extension Services).

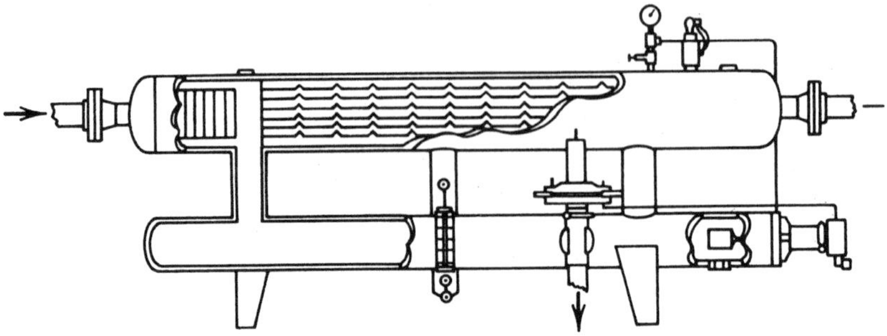

Figure 7–3 Conventional horizontal double-barrel separator (Courtesy Petroleum Extension Services).

7.2.2.3 Spherical Separators

Spherical separators offer an inexpensive and compact means of the separation arrangement shown in Figure 7–4. Owing to their compact configurations, this type of separator has a very limited surge space and liquid settling section. Also, the placement and action of the liquid-level control in this type of separator is more critical.

Oil/gas/water three-phase separators (Figure 7–5) are commonly used for well testing and in instances where free water readily separates from the oil or condensate. Three-phase separation can be accomplished in any type of separator. This can be achieved by installing either special internal baffling to construct a water leg or a water siphon arrangement. It can also be achieved by using an interface liquid-level control. The three-phase separation feature is difficult to install in spherical separators because of their limited available internal space. In three-phase operations, two liquid dump valves are required.

7.2.3 Factors Affecting Separation

For a given separator, factors that affect separation of liquid and gas phases include separator operating pressure, separator operating temperature, and fluid stream composition. For a given fluid wellstream in a specified separator, changes in any one of these factors will change the amount of gas and liquid leaving the separator. An increase in operating pressure or a decrease in operating temperature generally increases the

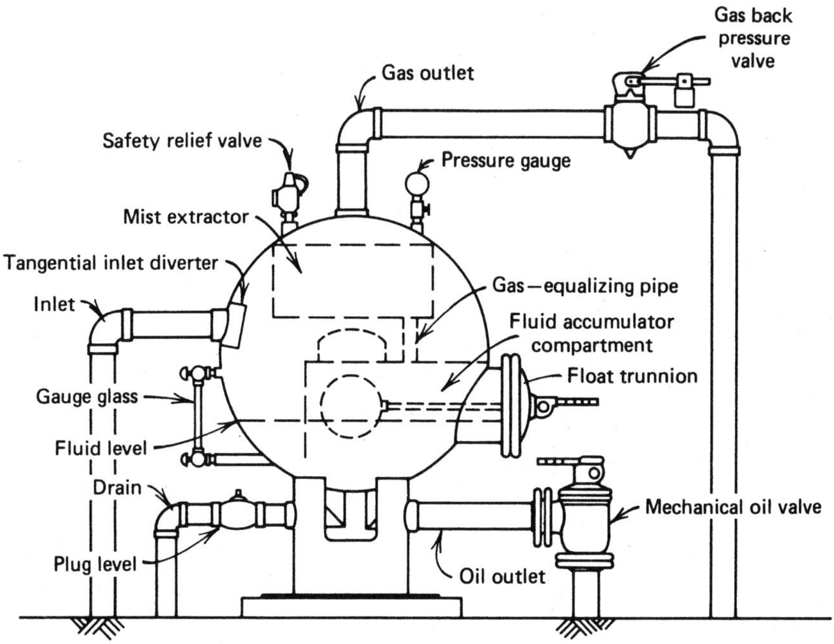

Figure 7–4 Spherical low-pressure separator (Sivalls 1977).

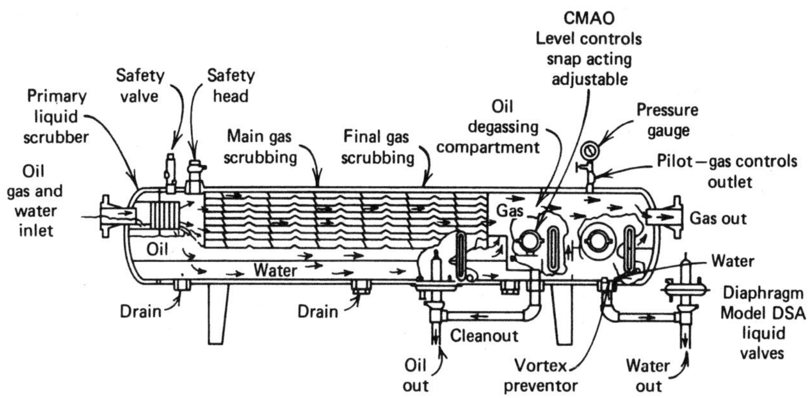

Figure 7–5 Conventional horizontal three-phase separator (Courtesy Petroleum Extension Services).

liquid covered in a separator. However, this is often untrue for gas condensate systems. There are optimum points in both cases beyond which further changes will not add to liquid recovery. Computer simulation (flash vaporization calculation) of phase behavior of the wellstream allows engineers to find the optimum pressure and temperature at which a separator should operate to give maximum liquid recovery. Sometimes it is not practical to operate at the optimum point. This is because storage system vapor losses may become too great under these optimum conditions.

At the wellhead separation facilities, operators tend to determine the optimum conditions for separators to maximize revenue. As the liquid hydrocarbon product is generally worth more than the gas, high liquid recovery is often desirable, provided that it can be handled in the available storage system. The operator can control operating pressure to some extent by use of backpressure valves. However, pipeline requirements for Btu content of the gas should also be considered as a factor affecting separator operation.

It is usually unfeasible to try to lower the operating temperature of a separator without adding expensive mechanical refrigeration equipment. However, an indirect heater can be used to heat the gas prior to pressure reduction of pipeline pressure in a choke. This is mostly applied to high-pressure wells. By carefully operating this indirect heater, the operator can prevent overheating the gas stream ahead of the choke. This adversely affects the temperature of the downstream separator.

7.2.4 Separator Design

Natural gas engineers normally do not perform detailed designing of separators but carry out selection of separators suitable for their operations from manufacturers' product catalogs. This section addresses how to determine separator specifications based on wellstream conditions. The specifications are used for separator selections.

7.2.4.1 Gas Capacity

The following empirical equations proposed by Souders-Brown are widely used for calculating gas capacity of oil/gas separators:

$$v = K\sqrt{\frac{\rho_L - \rho_g}{\rho_g}} \tag{7.1}$$

and

$$q = Av \tag{7.2}$$

where

A = total cross-sectional area of separator, ft^2

v = superficial gas velocity based on total cross-sectional area A, ft/s

q = gas flow rate at operating conditions, ft^3/s

ρ_L = density of liquid at operating conditions, lb$_m$/ft^3

ρ_g = density of gas at operating conditions, lb$_m$/ft^3

K = empirical factor

Table 7–1 presents K-values for various types of separators. Also listed in the table are K-values used for other designs such as mist eliminators and trayed towers in dehydration or gas sweetening units.

Table 7–1 *K*-Values Used for Designing Separators

Separator Type	K	Remarks
Vertical separators	0.06 to 0.35	
Horizontal separators	0.40 to 0.50	
Wire mesh mist eliminators	0.35	
Bubble cap trayed columns	0.16	24-in spacing
Volume tray columns	0.18	24-in spacing

Substituting Equation (7.1) into Equation (7.2) and applying real gas law gives

$$q_{st} = \frac{2.4D^2Kp}{z(T+460)}\sqrt{\frac{\rho_L - \rho_g}{\rho_g}} \tag{7.3}$$

where

q_{st} = gas capacity at standard conditions, MMscfd

D = internal diameter of vessel, ft

p = operation pressure, psia

T = operating temperature, °F

z = gas compressibility factor

It should be noted that Equation (7.3) is empirical. Height differences in vertical separators and length differences in horizontal separators are not considered. Field experience has indicated that additional gas capacity can be obtained by increasing height of vertical separators and length of horizontal separators. The separator charts (Sivalls 1977; Ikoku 1984) give more realistic values for the gas capacity of separators. In addition, for single-tube horizontal vessels, corrections must be made for the amount of liquid in the bottom of the separator. Although one-half full of liquid is more or less standard for most single-tube horizontal separators, lowering liquid level to increase the available gas space within the vessel can increase the gas capacity.

7.2.4.2 Liquid Capacity

Retention time of the liquid within the vessel determines liquid capacity of a separator. Adequate separation requires sufficient time to obtain an equilibrium condition between the liquid and gas phase at the temperature and pressure of separation. The liquid capacity of a separator relates to the retention time through the settling volume:

$$q_L = \frac{1440 V_L}{t} \tag{7.4}$$

where

q_L = liquid capacity, bbl/day

V_L = liquid settling volume, bbl

t = retention time, min

Table 7–2 presents *t*-values for various types of separators tested in fields. It is shown that temperature has a strong impact on three-phase separations at low pressures.

Table 7–2 Retention Time Required under Various Separation Conditions

<table>
<tr><th>Separation Condition</th><th>T (°F)</th><th>t (min.)</th></tr>
<tr><td>Oil/gas separation</td><td></td><td>1</td></tr>
<tr><td>High-pressure oil/gas/water separation</td><td></td><td>2 to 5</td></tr>
<tr><td rowspan="5">Low-pressure oil/gas/water separation</td><td>>100</td><td>5 to 10</td></tr>
<tr><td>90</td><td>10 to 15</td></tr>
<tr><td>80</td><td>15 to 20</td></tr>
<tr><td>70</td><td>20 to 25</td></tr>
<tr><td>60</td><td>25 to 30</td></tr>
</table>

Table 7–3 through Table 7–8 present liquid-settling volumes with the conventional placement of liquid-level controls for typical oil/gas separators.

Proper sizing of a separator requires the use of both Equation (7.3) for gas capacity and Equation (7.4) for liquid capacity. Experience shows that for high-pressure separators used for treating high gas/oil ratio wellstreams, the gas capacity is usually the controlling factor for separator selection. However, the reverse may be true for low-pressure separators used on wellstreams with low gas/oil ratios.

Table 7–3 Settling Volumes of Standard Vertical High-Pressure Separators (230 psi to 2,000 psi working pressure)

Size (D × H)	V_L (bbl)	
	Oil/Gas Separators	Oil/Gas/Water Separators
16" × 5'	0.27	0.44
16" × 7-1/2'	0.41	0.72
16" × 10'	0.51	0.94
20" × 5'	0.44	0.71

Table 7–3 Settling Volumes of Standard Vertical High-Pressure Separators (230 psi to 2,000 psi working pressure) (Continued)

Size (D × H)	V_L (bbl)	
	Oil/Gas Separators	Oil/Gas/Water Separators
20" × 7-1/2'	0.65	1.15
20" × 10'	0.82	1.48
24" × 5'	0.66	1.05
24" × 7-1/2'	0.97	1.68
24" × 10'	1.21	2.15
30" × 5'	1.13	1.76
30" × 7-1/2'	1.64	2.78
30" × 10'	2.02	3.54
36" × 7-1/2'	2.47	4.13
36" × 10'	3.02	5.24
36" × 15'	4.13	7.45
42" × 7-1/2'	3.53	5.80
42" × 10'	4.29	7.32
42" × 15'	5.80	10.36
48" × 7-1/2'	4.81	7.79
48" × 10'	5.80	9.78
48" × 15'	7.79	13.76
54" × 7-1/2'	6.33	10.12
54" × 10'	7.60	12.65
54" × 15'	10.12	17.70
60" × 7-1/2'	8.08	12.73
60" × 10'	9.63	15.83
60" × 15'	12.73	22.03
60" × 20'	15.31	27.20

Table 7–4 Settling Volumes of Standard Vertical Low-Pressure Separators (125 psi working pressure)

Size (D × H)	V_L (bbl)	
	Oil/Gas Separators	Oil/Gas/Water Separators
24" × 5'	0.65	1.10
24" × 7-1/2'	1.01	1.82
30" × 10'	2.06	3.75
36" × 5'	1.61	2.63
36" × 7-1/2'	2.43	4.26
36" × 10'	3.04	5.48
48" × 10'	5.67	10.06
48" × 15'	7.86	14.44
60" × 10'	9.23	16.08
60" × 15'	12.65	12.93
60" × 20'	15.51	18.64

Table 7–5 Settling Volumes of Standard Horizontal High-Pressure Separators (230 psi to 2,000 psi working pressure)

Size (D × L)	V_L (bbl)		
	1/2 Full	1/3 Full	1/4 Full
12-3/4" × 5'	0.38	0.22	0.15
12-3/4" × 7-1/2'	0.55	0.32	0.21
12-3/4" × 10'	0.72	0.42	0.28
16" × 5'	0.61	0.35	0.24
16" × 7-1/2'	0.88	0.50	0.34
16" × 10'	1.14	0.66	0.44
20" × 5'	0.98	0.55	0.38
20" × 7-1/2'	1.39	0.79	0.54
20" × 10'	1.80	1.03	0.70

Table 7–5 Settling Volumes of Standard Horizontal High-Pressure Separators (230 psi to 2,000 psi working pressure) (Continued)

Size (D × L)	V_L (bbl)		
	1/2 Full	1/3 Full	1/4 Full
24" × 5'	1.45	0.83	0.55
24" × 7-1/2'	2.04	1.18	0.78
24" × 10'	2.63	1.52	1.01
24" × 15'	3.81	2.21	1.47
30" × 5'	2.43	1.39	0.91
30" × 7-1/2'	3.40	1.96	1.29
30" × 10'	4.37	2.52	1.67
30" × 15'	6.30	3.65	2.42
36" × 7-1/2'	4.99	2.87	1.90
36" × 10'	6.38	3.68	2.45
36" × 15'	9.17	5.30	3.54
36" × 20'	11.96	6.92	4.63
42" × 7-1/2'	6.93	3.98	2.61
42" × 10'	8.83	5.09	3.35
42" × 15'	12.62	7.30	4.83
42" × 20'	16.41	9.51	6.32
48" × 7-1/2'	9.28	5.32	3.51
48" × 10'	11.77	6.77	4.49
48" × 15'	16.74	9.67	6.43
48" × 20'	21.71	12.57	8.38
54" × 7-1/2'	12.02	6.87	4.49
54" × 10'	15.17	8.71	5.73
54" × 15'	12.49	12.40	8.20
54" × 20'	27.81	16.08	10.68
60" × 7-1/2'	15.05	8.60	5.66
60" × 10'	18.93	10.86	7.17
60" × 15'	26.68	15.38	10.21
60" × 20'	34.44	19.90	13.24

Table 7-6 Settling Volumes of Standard Horizontal Low-Pressure Separators (125 psi working pressure)

Size (D × L)	V_L (bbl)		
	1/2 Full	1/3 Full	1/4 Full
24" × 5'	1.55	0.89	0.59
24" × 7-1/2'	2.22	1.28	0.86
24" × 10'	2.89	1.67	1.12
30" × 5'	2.48	1.43	0.94
30" × 7-1/2'	3.54	2.04	1.36
30" × 10'	4.59	2.66	1.77
36" × 10'	6.71	3.88	2.59
36" × 15'	9.76	5.66	3.79
48" × 10'	12.24	7.07	4.71
48" × 15'	17.72	10.26	6.85
60" × 10'	19.50	11.24	7.47
60" × 15'	28.06	16.23	10.82
60" × 20'	36.63	21.21	14.16

Table 7-7 Settling Volumes of Standard Spherical High-Pressure Separators (230 psi to 3,000 psi working pressure)

Size OD	V_L (bbl)
24"	0.15
30"	0.30
36"	0.54
42"	0.88
48"	1.33
60"	2.20

Table 7–8 Settling Volumes of Standard Spherical Low-Pressure Separators (1.25 psi)

Size OD	V_L (bbl)
41"	0.77
46"	1.02
64"	1.60

Example Problem 7.1

Calculate the minimum required size of a standard oil/gas separator for the following conditions. Consider both vertical and horizontal separators.

Gas flow rate: 5.0 MMscfd

Gas-specific gravity: 0.7

Condensate flow rate: 20 bbl/MMscf

Condensate gravity: 60 °API

Operating pressure: 800 psig

Operating temperature: 80 °F

Solution

The total required liquid flow capacity is (5)(20) = 100 bbl/day. Assuming a 20" × 7-1/2' vertical separator, Table 7–1 suggests an average K-value of 0.205. The spreadsheet program Hall-Yarborogh-z.xls gives z = 0.8427 and ρ_g = 3.38 lbm/ft^3 at 800 psig and 80 °F. Liquid density is calculated as:

$$\rho_L = 62.4 \frac{141.5}{131.5 + 60} = 46.11 \text{ lb}_\text{m}/\text{ft}^3$$

Equation (7.3) gives:

$$q_{st} = \frac{(2.4)(20/12)^2(0.205)(800)}{(0.8427)(80+460)} \sqrt{\frac{46.11 - 3.38}{3.38}} = 8.70 \text{ MMscfd}$$

Sivalls' chart gives 5.4 MMscfd.

From Table 7–3, a 20-in × 7-1/2-ft separator will handle the following liquid capacity:

$$q_L = \frac{1440(0.65)}{1.0} = 936 \text{ bbl/day}$$

which is much higher than the liquid load of 100 bbl/day.

Consider a 16-in × 5-ft horizontal separator and Equation (7.3) gives:

$$q_{st} = \frac{(2.4)(16/12)^2(0.45)(800)}{(0.8427)(80+460)}\sqrt{\frac{46.11-3.38}{3.38}} = 12.22 \text{ MMscfd}$$

If the separator is half full of liquid, it can still treat 6.11 MMscfd of gas. Sivalls' chart indicates that a 16-in × 5-ft horizontal separator will handle 5.1 MMscfd.

From Table 7–5, a 16-in × 5-ft horizontal separator will handle

$$q_L = \frac{1440(0.61)}{1.0} = 878 \text{ bbl/day}$$

which again is much higher than the liquid load of 100 bbl/day.

This example illustrates a case of high gas/oil ratio wellstreams where the gas capacity is the controlling factor for separator selection. It suggests that a smaller horizontal separator would be required and would be more economical. The selected separator should have at least a 1,000 psig working pressure.

7.3 Stage Separation

Stage separation is a process in which hydrocarbon mixtures are separated into vapor and liquid phases by multiple equilibrium flashes at consecutively lower pressures. A two-stage separation requires one separator and a storage tank, and a three-stage separation requires two separators and a storage tank. The storage tank is always counted as the final stage of vapor/liquid separation. Stage separation reduces the pressure a little at a time, in steps or stages, resulting in a more stable stock-tank liquid. Usually a stable stock-tank liquid can be obtained by a stage separation of not more than four stages.

In high-pressure gas-condensate separation systems, a stepwise reduction of the pressure on the liquid condensate can significantly increase the recovery of stock-tank liquids. Prediction of the performance of the various separators in a multistage separation system can be carried out with compositional computer models using the initial wellstream composition and the operating temperatures and pressures of the various stages.

Although three to four stages of separation theoretically increase the liquid recovery over a two-stage separation, the incremental liquid recovery rarely pays out the cost of the additional separators. It has been generally recognized that two stages of separation plus the stock tank are practically optimum. The increase in liquid recovery for two-stage separation over single-stage separation usually varies from 2 to 12 percent, although 20 to 25 percent increases in liquid recoveries have been reported.

The first-stage separator operating pressure is generally determined by the flow line pressure and operating characteristics of the well. The pressure usually ranges from 600 to 1,200 psi. In situations where the flow line pressure is greater than 600 psi, it is practical to let the first-stage separator ride the line or operate at the flow line pressure. Pressures at low-stage separations can be determined based on equal pressure ratios between the stages (Campbell 1976):

$$R_p = \left(\frac{p_1}{p_s}\right)^{\frac{1}{N_{st}}} \tag{7.5}$$

where

R_p = pressure ratio

N_{st} = number of stages–1

p_1 = first-stage or high-pressure separator pressure, psia

p_s = stock-tank pressure, psia

Pressures at the intermediate stages can be then designed with the following formula:

$$p_i = \frac{p_{i-1}}{R_p} \tag{7.6}$$

where p_i = pressure at stage i, psia.

7.4 Flash Calculation

Based on the composition of wellstream fluid, the quality of products from each stage of separation can be predicted by flash calculations, assuming phase equilibriums are reached in the separators. This requires the knowledge of equilibrium ratio defined as:

$$k_i = \frac{y_i}{x_i} \tag{7.7}$$

where

k_i = liquid/vapor equilibrium ratio of compound i

y_i = mole fraction of compound i in the vapor phase

x_i = mole fraction of compound i in the liquid phase

Accurate determination of k_i values requires computer simulators solving the Equation of State (EoS) for hydrocarbon systems. Ahmed (1989) presented a detailed procedure for solving the EoS. For pressures lower than 1,000 psia, a set of equations presented by Standing (1979) provides an easy and accurate means of determining k_i values. According to Standing, k_i can be calculated by:

$$k_i = \frac{1}{p} 10^{a+cF_i} \tag{7.8}$$

where

$$a = 1.2 + 4.5 \times 10^{-4} p + 1.5 \times 10^{-9} p^2 \quad \textbf{(7.9)}$$

$$c = 0.89 - 1.7 \times 10^{-4} p - 3.5 \times 10^{-8} p^2 \quad \textbf{(7.10)}$$

$$F_i = b_i \left(\frac{1}{T_{bi}} - \frac{1}{T} \right) \quad \textbf{(7.11)}$$

$$b_i = \frac{\log\left(\frac{p_{ci}}{14.7}\right)}{\frac{1}{T_{bi}} - \frac{1}{T_{ci}}} \quad \textbf{(7.12)}$$

where

p_c = critical pressure, psia

T_b = boiling point, °R

T_c = critical temperature, °R

Consider 1 mole of fed-in fluid and the following equation holds true on the basis of mass balance:

$$n_L + n_V = 1 \quad \textbf{(7.13)}$$

where

n_L = number of mole of fluid in the liquid phase

n_V = number of mole of fluid in the vapor phase

For compound i,

$$z_i = x_i n_L + y_i n_V \quad \textbf{(7.14)}$$

where z_i is the mole fraction of compound *i* in the fed-in fluid.

Combining Equation (7.7) and Equation (7.14) gives

$$z_i = x_i n_L + k_i x_i n_V \tag{7.15}$$

which yields

$$x_i = \frac{z_i}{n_L + k_i n_V} \tag{7.16}$$

Mass balance applied to Equation (7.16) requires

$$\sum_{i=1}^{N_c} x_i = \sum_{i=1}^{N_c} \frac{z_i}{n_L + k_i n_V} = 1 \tag{7.17}$$

where N_c is the number of compounds in the fluid.

Combining Equation (7.7) and Equation (7.14) also gives

$$z_i = \frac{y_i}{k_i} n_L + y_i n_V \tag{7.18}$$

which yields

$$y_i = \frac{z_i k_i}{n_L + k_i n_V} \tag{7.19}$$

Mass balance applied to Equation (7.19) requires

$$\sum_{i=1}^{N_c} y_i = \sum_{i=1}^{N_c} \frac{z_i k_i}{n_L + k_i n_V} = 1 \tag{7.20}$$

Subtracting Equation (7.20) from Equation (7.17) gives

$$\sum_{i=1}^{N_c} \frac{z_i}{n_L + k_i n_V} - \sum_{i=1}^{N_c} \frac{z_i k_i}{n_L + k_i n_V} = 0 \tag{7.21}$$

which can be rearranged to obtain

$$\sum_{i=1}^{N_c} \frac{z_i(1-k_i)}{n_L + k_i n_V} = 0 \tag{7.22}$$

Combining Equation (7.22) and Equation (7.13) results in

$$\sum_{i=1}^{N_c} \frac{z_i(1-k_i)}{n_V(k_i - 1) + 1} = 0 \tag{7.23}$$

If the values of k_i are known, Equation (7.23) can be used to solve for the number of mole of fluid in the vapor phase. Then, x_i and y_i can be calculated with Equation (7.16) and Equation (7.19), respectively. The apparent molecular weights of liquid phase ($MW_a{}^L$) and vapor phase ($MW_a{}^V$) can be calculated by

$$MW_a{}^L = \sum_{i=1}^{N_c} x_i MW_i \tag{7.24}$$

$$MW_a{}^V = \sum_{i=1}^{N_c} y_i MW_i \tag{7.25}$$

where MW_i is the molecular weight of compound i. With the apparent molecular weight of the vapor phase known, the specific gravity of the vapor phase can be determined, and the density of the vapor phase in lbm/ft^3 can be calculated by Equation (2.43), that is,

$$\rho_V = \frac{MW_a{}^V p}{zRT} \tag{7.26}$$

The liquid phase density in lb_m/ft^3 can be estimated by Standing's method (Standing 1981), that is,

$$\rho_L = \frac{62.4\gamma_o + 0.0136 R_s \gamma_g}{0.972 + 0.000147\left[R_s\sqrt{\frac{\gamma_g}{\gamma_o}} + 1.25(T-460)\right]^{1.175}} \quad \textbf{(7.27)}$$

where

γ_0 = specific gravity of stock-tank oil, water =1

γ_g = specific gravity of solution gas, air =1

R_s = gas solubility of the oil, scf/STB

Then the volumes of vapor and liquid phases can be calculated by:

$$V_{Vsc} = \frac{z n_V R T_{sc}}{p_{sc}} \quad \textbf{(7.28)}$$

$$V_L = \frac{n_L MW_a^L}{\rho_L} \quad \textbf{(7.29)}$$

where

V_{Vsc} = volume of vapor phase under standard condition, scf

R = gas constant, 10.73 ft^3-psia/lb mol-R

T_{sc} = standard temperature, 520 °R

p_{sc} = standard pressure, 14.7 psia

V_L = volume of liquid phase, ft^3

Finally, the gas oil ratio (GOR) can be calculated by dividing Equation (7.28) by Equation (7.29). Specific gravity and American Petroleum Institute (API) gravity of oil can be calculated based on liquid density from Equation (7.27).

Example Problem 7.2

Perform flash calculation under the following separator conditions:

Pressure: 600 psia

Temperature: 200 °F

Specific gravity of stock-tank oil: 0.90 water = 1

Specific gravity of solution gas: 0.70 air = 1

Gas solubility R_S: 500 scf/STB

Composition

Compound	Mole Fraction
C_1	0.6599
C_2	0.0869
C_3	0.0591
i-C_4	0.0239
n-C_4	0.0278
i-C_5	0.0157
n-C_5	0.0112
C_6	0.0181
C_{7+}	0.0601
N_2	0.0194
CO_2	0.0121
H_2S	0.0058

Solution

The flash calculation can be carried out using the spreadsheet program LP-Flash.xls. The results are shown in Table 7–9 and Table 7–10.

Table 7–9 Flash Calculation with Standing's Method for k_i-values

Flash Calculation for n_v = 0.8791			
Compound	z_i	k_i	$z_i(k_i-1)/[n_v(k_i-1)+1]$
C_1	0.6599	6.5255	0.6225
C_2	0.0869	1.8938	0.0435
C_3	0.0591	0.8552	–0.0098
i-C_4	0.0239	0.4495	–0.0255
n-C_4	0.0278	0.3656	–0.0399
i-C_5	0.0157	0.1986	–0.0426
n-C_5	0.0112	0.1703	–0.0343
C_6	0.0181	0.0904	–0.0822
C_{7+}	0.0601	0.0089	–0.4626
N_2	0.0194	30.4563	0.0212
CO_2	0.0121	3.4070	0.0093
H_2S	0.0058	1.0446	0.0002

Table 7–10 Flash Calculation with Standing's Method for k_i-values

Flash Calculation for n_L = 0.1209				
Compound	x_i	y_i	x_iMW_i	y_iMW_i
C_1	0.1127	0.7352	1.8071	11.7920
C_2	0.0487	0.0922	1.4633	2.7712
C_3	0.0677	0.0579	2.9865	2.5540
i-C_4	0.0463	0.0208	2.6918	1.2099
n-C_4	0.0629	0.0230	3.6530	1.3356
i-C_5	0.0531	0.0106	3.8330	0.7614
n-C_5	0.0414	0.0070	2.9863	0.5085
C_6	0.0903	0.0082	7.7857	0.7036
C_{7+}	0.4668	0.0042	53.3193	0.4766

Table 7–10 Flash Calculation with Standing's Method for k_i-values (Continued)

Flash Calculation for n_L = 0.1209				
Compound	x_i	y_i	x_iMW_i	y_iMW_i
N_2	0.0007	0.0220	0.0202	0.6156
CO_2	0.0039	0.0132	0.1709	0.5823
H_2S	0.0056	0.0058	0.1902	0.1987
Apparent molecular weight of liquid phase:			80.91	
Apparent molecular weight of vapor phase:			23.51	
Specific gravity of liquid phase:			0.76 water = 1	
Specific gravity of vapor phase:			0.81 air = 1	
Input vapor phase z-factor:			0.958	
Density of liquid phase:			47.19 lb_m/ft	
Density of vapor phase:			2.08 lb_m/ft^3	
Volume of liquid phase:			0.04 bbl	
Volume of vapor phase:			319.66 scf	
GOR:			8,659 scf/bbl	
API gravity of liquid phase:			56	

7.5 Low-Temperature Separation

Field experience and flash calculations prove that lowering the operating temperature of a separator increases the liquid recovery. The low-temperature separation process separates water and hydrocarbon liquids from the inlet wellstream and recovers more liquids from the gas than can be recovered with normal-temperature separators. It is also an efficient means of handling high-pressure gas and condensate at the wellhead. A low-temperature separation unit consists of a high-pressure separator, pressure-reducing chokes, and various pieces of heat exchange equipment. When the pressure is reduced by use of a choke, the fluid temperature decreases due to the Joule-Thomson or throttling effect. This is an irreversible adiabatic process in which the heat content of the gas remains the same across the choke but the pressure and temperature of the gas stream are reduced.

Generally at least 2,500 psi to 3,000 psi pressure drop is required from wellhead flowing pressure to pipeline pressure for a low-temperature separation unit to pay out in increased liquid recovery. The lower the operating temperature of the separator, the lighter the liquid recovery will be. The lowest operating temperature recommended for low-temperature units is usually around –20 °F. This is constrained by carbon steel embittlement, and high-alloy steels for lower temperatures are usually not economical for field installations. Low-temperature separation units are normally operated from 0 to 20 °F. The actual temperature drop per unit pressure drop is affected by several factors including composition of gas stream, gas and liquid flow rates, bath temperature, and ambient temperature. Temperature reduction in the process can be estimated using the equations presented in Chapter 5. Gas expansion pressures for hydrate formation can be found from the chart prepared by Katz (1945) (see Chapter 12). Liquid and vapor phase densities can be predicted by flash calculation.

Following the special requirement for construction of low-temperature separation units, the pressure-reducing choke is usually mounted directly on the inlet of the high-pressure separator. Hydrates form in the downstream of the choke due to the low gas temperature and fall to the bottom settling section of the separator. They are heated and melted by liquid heating coils located in the bottom of the separator.

7.6 References

Ahmed, T. *Hydrocarbon Phase Behavior*. Houston: Gulf Publishing Company, 1989.

Campbell, J. M. *Gas Conditioning and Processing*, Norman, Oklahoma: Campbell Petroleum Services, 1976.

Ikoku, C. U.: *Natural Gas Production Engineering*. New York: John Wiley & Sons, 1984.

Katz, D. L. "Prediction of Conditions for Hydrate Formation in Natural Gas." *Trans. AIME* **160** (1945): 140.

Sivalls, C. R. "Fundamentals of Oil and Gas Separation." Proceedings of the Gas Conditioning Conference, University of Oklahoma, Norman, Oklahoma, 1977.

Standing, M. B. "A Set of Equations for Computing Equilibrium Ratios of a Crude Oil/Natural Gas System at Pressures Below 1,000 psia." *Journal of Petroleum Technology,* Trans. AIME **31** (Sept. 1979): 1193.

Standing, M. B.: *Volume and Phase Behavior of Oil Field Hydrocarbon Systems*. 9th ed. Dallas: Society of Petroleum Engineers, 1981.

7.7 Problems

7-1 Calculate the minimum required size of a standard oil/gas separator for the following conditions. Consider vertical, horizontal, and spherical separators.

Gas flow rate: 4.0 MMscfd

Gas-specific gravity: 0.7

Condensate-gas ratio (CGR): 15 bbl/MMscf

Condensate gravity: 65 °API

Operating pressure: 600 psig

Operating temperature: 70 °F

7-2 A three-stage separation is proposed to treat a wellstream at a flow line pressure of 1,000 psia. Calculate pressures at each stage of separation.

7-3 Perform flash calculations under the following separator conditions. Plot GOR and API gravity against separator pressure.

Pressure: 100, 200, 300, 400, 500 psia

Temperature: 150 °F

Specific gravity of stock-tank oil: 0.85 water = 1

Specific gravity of solution gas: 0.65 air = 1

Gas solubility R_S: 600 scf/STB

Composition

Compound	Mole Fraction
C_1	0.6099
C_2	0.0869
C_3	0.0691
i-C_4	0.0339
n-C_4	0.0378
i-C_5	0.0257
n-C_5	0.0212
C_6	0.0181
C_{7+}	0.0601
N_2	0.0194
CO_2	0.0121
H_2S	0.0058

Chapter 8

Dehydration

8.1 Introduction

Natural gas to be transported by pipeline must meet certain specifications. In addition to specifications regarding delivery pressure, rate, and possibly temperature, other specifications include maximum water content (water dew point), maximum condensable hydrocarbon content (hydrocarbon dew point), allowable concentrations of contaminants such as H_2S, CO_2, mercaptans, minimum heating value, and cleanliness (allowable solids content). This chapter focuses on principles of field processing for removing water, H_2S, and CO_2, and selection of required equipment.

8.2 Dehydration of Natural Gas

The term dehydration means removal of water vapor. All natural gas downstream from the separators still contains water vapor to some degree. Water vapor is probably the most common undesirable impurity found in untreated natural gas. The main reason for removing water vapor from natural gas is that water vapor becomes liquid water under low-temperature and/or high-pressure conditions. Specifically, water content can affect long-distance transmission of natural gas due to the following facts:

- Liquid water and natural gas can form hydrates that may plug the pipeline and other equipment.
- Natural gas containing CO_2 and/or H_2S is corrosive when liquid water is present.

- Liquid water in a natural gas pipeline potentially causes slugging flow conditions resulting in lower flow efficiency of the pipeline.
- Water content decreases the heating value of natural gas being transported.

8.2.1 Water Content of Natural Gas Streams

Solubility of water in natural gas increases with temperature and decreases with pressure. Salt's presence in the liquid water reduces the water content of the gas. Water content of untreated natural gases is normally in the magnitude of a few hundred pounds of water per million standard cubic foot of gas (lb_m/MMscf); while gas pipelines normally require water content to be in the range of 6–8 lb_m/MMscf and even lower for pipelines in deep water.

The water content of natural gas is indirectly indicated by the dew point, defined as the temperature at which the natural gas is saturated with water vapor at a given pressure. At the dew point, natural gas is in equilibrium with liquid water; any decrease in temperature or increase in pressure will cause the water vapor to begin condensing. The difference between the dew point temperature of a water-saturated gas stream and the same stream after it has been dehydrated is called dew-point depression.

It is essential to accurately estimate the saturated water vapor content of natural gas in the design and operation of dehydration equipment. Several methods are available for this purpose including the correlations of McCarthy, Boyd, and Reid (1950) and McKetta and Wehe (1958). Dalton's law of partial pressures is valid for estimating water vapor content of gas at near-atmospheric pressures. Readings from the chart by McKetta and Wehe (1958) are presented in Table 8–1. Figure 8–1 shows a water content chart plotted in a different form from the chart by McKetta and Wehe (1958).

Example Problem 8.1

Estimate water content of a natural gas at a pressure of 3,000 psia and temperature of 150 °F.

Table 8–1 Water Content Readings from the Chart by McKetta and Wehe (1958) (lbm H_2O/MMcf @60 °F, 14.7 psia)

p (psia)	t (°F)															
	–60	–40	–20	0	20	40	60	80	100	120	140	160	180	200	240	280
15	3	10	27	70	170	380	750	1,550	3,000	5,500	9,500	17,000	28,000	46,000	90,000	200,000
25	2	6	16	45	100	220	480	900	1,750	3,100	5,800	9,500	15,000	27,000	50,000	110,000
50	1	3	9	24	58	120	250	450	850	1,500	2,700	4,700	7,200	13,000	23,000	50,000
100		2	5	13	30	63	130	240	480	750	1,400	2,200	3,800	6,600	12,000	24,000
200		1	3	7	16	35	70	130	250	400	700	1,200	1,900	3,000	6,000	12,000
300			2	5	11	24	47	90	170	290	480	800	1,300	2,000	4,000	8,500
400			2	4	9	20	37	70	135	220	370	600	1,000	1,500	3,000	6,200
500			1	3	7	16	30	60	105	180	300	500	840	1,200	2,600	5,000
600			1	3	7	14	26	50	90	160	250	450	700	1,000	2,100	4,200
800				2	5	11	20	40	75	130	200	350	550	800	1,700	3,300
1,000				2	5	10	18	34	60	105	180	300	470	690	1,400	2,800
1,500				2	3	7	14	25	46	80	130	220	340	500	1,000	2,000
2,000				1	3	6	12	21	38	67	110	180	260	400	800	1,600
3,000								18	30	52	85	130	200	300	600	1,150
4,000								16	26	45	75	110	180	255	500	950
5,000								15	24	40	69	100	160	230	450	800
6,000								14	22	37	61	95	150	200	400	750
8,000								13	21	34	55	85	130	180	350	650
10,000								12	20	32	50	79	125	170	340	600

Solution

The chart in Figure 8–1 gives water contents of:

$$C_{w140\ F} = 84\ \text{lb}_m/\text{MMcf}$$

$$C_{w160\ F} = 130\ \text{lb}_m/\text{MMcf}$$

Linear interpolation yields:

$$C_{w150\ F} = 107\ \text{lb}_m/\text{MMcf}$$

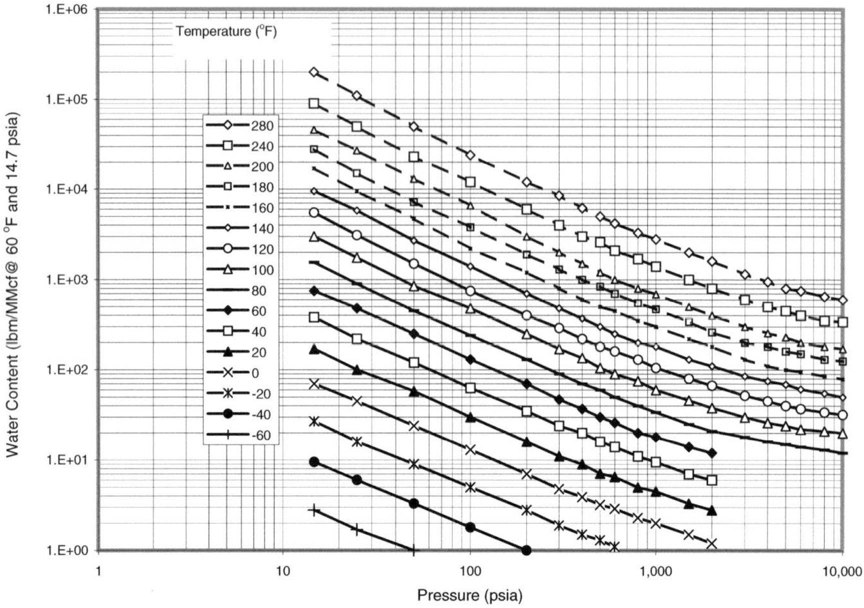

Figure 8–1 Water content of natural gases (Duplicated with data in the chart of McKetta and Wehe 1958).

8.2.2 Dehydration Systems

Dehydration systems used in the natural gas industry fall into four categories in principle: (a) direct cooling, (b) compression followed by cooling, (c) absorption, and (d) adsorption. Dehydration in the first two methods does not result in sufficiently low water contents to permit injection into a pipeline. Further dehydration by absorption or adsorption is often required.

Absorption is a process in which water vapor is removed from natural gas by bubbling the gas counter-currently through certain liquids that have a special attraction or affinity for water. Water vapor in the gas bubbles is entrained in the liquid and carried away by the liquid. Adsorption is a process in which gas flows through a bed of granular solids that have an affinity for water. The water is retained on the surface of the particles of the solid material. The vessel that allows either the absorption or adsorption process to take place is called the contactor or sorber. The liquid or solid that has affinity for water and is used in the contactor in connection with either of the processes is called the desiccant. Two major types of dehydration equipment in use today are the liquid desiccant dehydrator and the solid desiccant dehydrator. Each type of dehydrator has its advantages and disadvantages. These two types of dehydrators dehydrate practically all the natural gas moved through transmission lines.

8.2.2.1 Dehydration by Cooling

The ability of natural gas to contain water vapor decreases as the temperature is lowered at constant pressure. During the cooling process, the excess water in the vapor state becomes liquid and is removed from the system. Natural gas containing less water vapor at low temperature is output from the cooling unit. The gas dehydrated by cooling is still at its water dew point unless the temperature is raised again or the pressure is decreased. Cooling for the purpose of gas dehydration is sometimes economical if the gas temperature is unusually high. It is often a good practice that cooling is used in conjunction with other dehydration processes.

Gas compressors can be used partially as dehydrators. Because the saturation water content of gases decreases at higher pressure, some water is condensed and removed from gas at compressor stations by the compressor discharge coolers. Modern lean oil absorption gas plants use mechanical refrigeration to chill the inlet gas stream. Ethylene glycol is usually injected into the gas chilling section of the plant, which simultaneously dehydrates the gas and recovers liquid hydrocarbons, in a manner similar to the low-temperature separators.

8.2.2.2 Dehydration by Adsorption

Adsorption is defined as the ability of a substance to hold gases or liquids on its surface. In adsorption dehydration, the water vapor from the gas is concentrated and held at the surface of the solid desiccant by forces caused by residual valiancy. Solid desiccants have very large surface areas per unit weight to take advantage of these surface forces. The most common solid adsorbents used today are silica, alumina, and certain silicates known as molecular sieves. Dehydration plants can remove practically all water from natural gas using solid desiccants. Because of their great drying ability, solid desiccants are employed where higher efficiencies are required.

Figure 8–2 depicts a typical solid desiccant dehydration plant. The incoming wet gas should be cleaned by a filter separator to remove solid and liquid contaminants in the gas. The filtered gas flows downward during dehydration through one adsorber containing a desiccant bed. The down-flow arrangement reduces disturbance of the bed caused by the high gas velocity during the adsorption. While one adsorber is dehydrating, the other adsorber is being regenerated by a hot stream of inlet gas from the regeneration gas heater. A direct-fired heater, hot oil, steam, or an indirect heater can supply the necessary regeneration heat. The regeneration gas usually flows upward through the bed to ensure thorough regeneration of the bottom of the bed, which is the last area contacted by the gas being dehydrated. The hot regenerated bed is cooled by shutting off or bypassing the heater. The cooling gas then flows downward through the bed so that any water adsorbed from the cooling gas will be at the top of the bed and will not be desorbed into the gas during the dehydration step. The still hot regeneration gas and the cooling gas flow through the regeneration gas cooler to condense the desorbed water. Power-operated valves activated by a timing device switch the adsorbers between the dehydration, regeneration, and cooling steps.

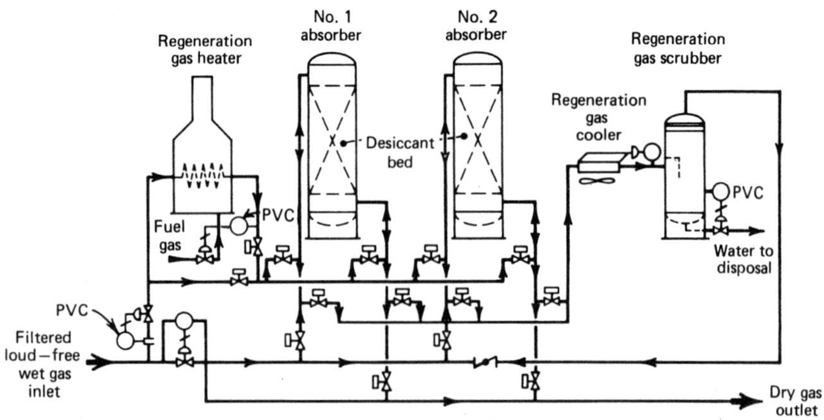

Figure 8–2 Flow diagram of a typical solid desiccant dehydration plant (Guenther 1979).

Under normal operating conditions, the usable life of a desiccant ranges from one to four years. Solid desiccants become less effective in normal use due to loss of effective surface area as they age. Abnormally fast degradation occurs through blockage of the small pores and capillary openings from lubricating oils, amines, glycols, corrosion inhibitors, and other contaminants, which cannot be removed during the regeneration cycle. Hydrogen sulfide can also damage the desiccant and reduce its capacity.

The advantages of solid-desiccant dehydration include:

- lower dew point, essentially dry gas (water content less than 1.0 lb/MMcf) can be produced
- higher contact temperatures can be tolerated with some adsorbents
- higher tolerance to sudden load changes, especially on startup
- quick startup after a shutdown
- high adaptability for recovery of certain liquid hydrocarbons in addition to dehydration functions

Operating problems with the solid-desiccant dehydration include:

- space adsorbents degenerate with use and require replacement
- dehydrating tower must be regenerated and cooled for operation before another tower approaches exhaustion. The maximum allowable time on dehydration gradually shortens because desiccant loses capacity with use

Although this type of dehydrator has high adaptability to sudden load changes, sudden pressure surges should be avoided because they may upset the desiccant bed and channel the gas stream resulting in poor dehydration. If a plant is operated above its rated capacity, high-pressure loss may cause some attrition to occur. Attrition causes fines, which may in turn cause excessive pressure loss and result in loss of capacity.

Replacing the desiccant should be scheduled and completed ahead of the operating season. To maintain continuous operation, this may require discarding the desiccant before its normal operating life is reached. To cut operating costs, the inlet part of the tower can be recharged and the remainder of the desiccant retained because it may still possess some useful life. Additional service life of the desiccant may be obtained if the direction of gas flow is reversed at a time when the tower would normally be recharged.

8.2.2.3 Dehydration by Absorption

Water vapor is removed from the gas by intimate contact with a hygroscopic liquid desiccant in absorption dehydration. The contact is usually achieved in packed or trayed towers. Glycols have been widely used as effective liquid desiccants. Dehydration by absorption with glycol is usually economically more attractive than dehydration by solid desiccant when both processes are capable of meeting the required dew point.

Glycols used for dehydrating natural gas are ethylene glycol (EG), diethylene glycol (DEG), triethylene glycol (TEG), and tetraethylene glycol (T_4EG). Normally a single type of pure glycol is used in a dehydrator, but sometimes a glycol blend is economically attractive. TEG has gained nearly universal acceptance as the most cost effective of the glycols due to its superior dew point depression, operating cost, and operation reliability. Triethylene glycol has been successfully used to dehydrate sweet and sour natural gases over wide ranges of operating conditions. Dew point depression of 40 °F to 140 °F can be achieved at a gas pressure ranging from 25 psig to 2500 psig and gas temperature between 40 °F and 160 °F. The dew point depression obtained depends on the equilibrium dew point temperature for a given TEG concentration and contact temperature. Increased glycol viscosity may cause problems at lower contact temperature. Thus, heating of the natural gas may be desirable. Very hot gas streams are often cooled prior to dehydration to prevent vaporization of TEG.

The feeding-in gas must be cleaned to remove all liquid water and hydrocarbons, wax, sand, drilling muds, and other impurities. These substances can cause severe foaming, flooding, higher glycol losses, poor efficiency, and increased maintenance in the dehydration tower or absorber. These impurities can be removed using an efficient scrubber, separator, or even a filter separator for very contaminated gases. Methanol, injected at the wellhead as a hydrate inhibitor, can cause several problems for glycol dehydration plants. It increases the heat requirements of the glycol regeneration system. Slugs of liquid methanol can cause flooding in the absorber. Methanol vapor vented to the atmosphere with the water vapor from the regeneration system is hazardous and should be recovered or vented at nonhazardous concentrations.

Figure 8–3 illustrates the process and flow through a typical glycol dehydrator. The dehydration process can be described as follows:

1 The feeding-in gas stream first enters the unit through an inlet gas scrubber to remove liquid accumulations. A two-phase inlet scrubber is normally required. If there is any liquid water in the gas stream, a three-phase inlet scrubber can be used to discharge the distillate and water from the vessel separately. A mist eliminator is normally in the scrubber to remove any entrained liquid particles from the gas stream leaving the top of the scrubber.

2 The wet gas is then introduced to the bottom of the glycol-gas contactor and allowed to flow upward through the trays, while glycol flows down through the column. The gas contacts the glycol on each tray and the glycol absorbs the water vapor from the gas steam.

3 The gas then flows down through a vertical glycol cooler, usually fabricated in the form of a concentric pipe heat exchanger, where the outlet dry gas aids in cooling the hot regenerated glycol before it enters the contactor. The dry gas then leaves the unit from the bottom of the glycol cooler.

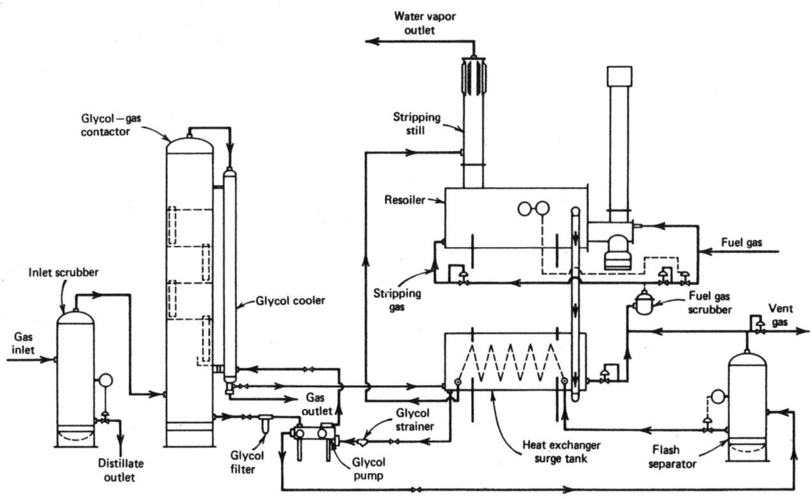

Figure 8–3 Flow diagram of a typical glycol dehydrator (Sivalls 1977).

4 The dry glycol enters the top of the glycol-gas contactor from the glycol cooler and is injected onto the top tray. The glycol flows across each tray and down through a downcomer pipe onto the next tray. The bottom tray downcomer is fitted with a seal pot to hold a liquid seal on the trays.

5 The wet glycol, which has now absorbed the water vapor from the gas stream, leaves the bottom of the glycol-gas contactor column, passes through a high-pressure glycol filter, which removes any foreign solid particles that may have been picked up from the gas stream, and enters the power side of the glycol pump.

6 In the glycol pump the wet high-pressure glycol from the contactor column pumps the dry regenerated glycol into the column. The wet glycol stream flows from the glycol pump to the inlet of the flash separator. The low-pressure flash separator allows for the release of the entrained solution gas, which must be used with the wet glycol to pump the dry glycol into the contactor.

7 The gas separated in the flash separator leaves the top of the flash separator vessel and can be used to supplement the fuel gas required for the reboiler. Any excess vent gas is discharged through a backpressure valve. The flash separator is equipped with a liquid level control and diaphragm motor valve that discharges the wet glycol stream through a heat exchange coil in the surge tank to preheat the wet glycol stream. If the wet glycol stream absorbs any liquid hydrocarbons in the contactor, it may be desirable to use a three-phase flash separator to separate the glycol from the liquid hydrocarbons before the stream enters the reboiler. Any liquid hydrocarbons present in the reboiler can cause undue glycol losses from the stripping still. The wet glycol stream leaves the heat exchange coil in the surge tank and enters the stripping still mounted on top of the reboiler at the feed point in the still. The stripping still is packed with a ceramic intalox saddle-type packing, and the glycol flows downward through the column and enters the reboiler. The wet glycol passing downward through the still is contacted by hot rising glycol and water vapors passing upward through the column. The water vapors released in the reboiler and stripped from the glycol in the stripping still pass upward through the still column through an atmospheric reflux condenser that provides a partial reflux for the

column. The water vapor then leaves the top of the stripping still column and is released to the atmosphere.

8 The glycol flows through the reboiler in essentially a horizontal path from the stripping still column to the opposite end. In the reboiler, the glycol is heated to approximately 350 °F to 400 °F to remove enough water vapor to reconcentrate it to 99.5% or higher. In field dehydration units, the reboiler is generally equipped with a direct-fired firebox, using a portion of the natural gas stream for fuel. In plant-type units, the reboiler may be fitted with a hot oil-heated coil or steam coil. A temperature control in the reboiler operates a fuel gas motor valve to maintain the proper temperature in the glycol. The reboiler is also generally equipped with a high-temperature safety overriding temperature controller to shut down the fuel gas system in case the primary temperature control should malfunction.

9 In order to provide extra-dry glycol (> 99%) it is usually necessary to add some stripping gas to the reboiler. A valve and small pressure regulator are generally provided to take a small amount of gas from the fuel gas system and inject it into the bottom of the reboiler through a spreader system. This stripping gas will "roll" the glycol in the reboiler to allow any pockets of water vapor to escape that might otherwise remain in the glycol due to its normal high viscosity. This gas will sweep the water vapor out of the reboiler and stripping still. By lowering the partial pressure of the water vapor in the reboiler and still column, the glycol can be reconcentrated to a higher percentage.

10 The reconcentrated glycol leaves the reboiler through an overflow pipe and passes into the shell side of the heat exchanger/surge tank. In the surge tank the hot reconcentrated glycol is cooled by exchanging heat with the wet glycol stream passing through the coil. The surge tank also acts as a liquid accumulator for feed for the glycol pump. The reconcentrated glycol flows from the surge tank through a strainer and into the glycol pump. From the pump it passes into the shell side of the glycol cooler mounted on the glycol-gas contactor. It then flows upward through the glycol cooler where it is further cooled and enters the column on the top tray.

Glycol dehydrators have several advantages including:

- low initial-equipment cost
- low-pressure drop across absorption towers
- makeup requirements may be added readily
- recharging of towers presents no problems
- the plant may be used satisfactorily in the presence of materials that would cause fouling of some solid adsorbents

Glycol dehydrators also present several operating problems including:

- Suspended matter, such as dirt, scale, and iron oxide, may contaminate glycol solutions.
- Overheating of solution may produce both low and high boiling decomposition products.
- The resultant sludge may collect on heating surfaces, causing some loss in efficiency, or, in severe cases, complete flow stoppage.
- When both oxygen and hydrogen sulfide are present, corrosion may become a problem because of the formation of acid material in the glycol solution.
- Liquids such as water, light hydrocarbons, or lubrication oils, in inlet gas may require installation of an efficient separator ahead of the absorber. Highly mineralized water entering the system with inlet gas may, over long periods, crystallize and fill the reboiler with solid salts.
- Foaming of solution may occur with a resultant carry-over of liquid. The addition of a small quantity of antifoam compound usually remedies this problem.
- Some leakage around the packing glands of pumps may be permitted because excessive tightening of packing may result in the scouring of rods. This leakage is collected and periodically returned to the system.

- Highly concentrated glycol solutions tend to become viscous at low temperatures and, therefore, are hard to pump. Glycol lines may solidify completely at low temperatures when the plant is not operating. In cold weather, continuous circulation of part of the solution through the heater may be advisable. This practice can also prevent freezing in water coolers.
- To start a plant, all absorber trays must be filled with glycol before good contact of gas and liquid can be expected. This may also become a problem at low-circulation rates because weep holes on trays may drain solution as rapidly as it is introduced.
- Sudden surges should be avoided in starting and shutting down a plant. Otherwise, large carry-over losses of solution may occur.

8.2.3 Glycol Dehydrator Design

Dehydrators with TEG in trays or packed-column contactors can be sized from standard models by using the following information:

- gas flow rate
- specific gravity of gas
- operating pressure
- Maximum working pressure of contact
- gas inlet temperature
- outlet gas water content required

One of the following two design criteria can be employed:

- Glycol to water ratio (GWR). A value of 2 to 6 gal TEG/lb_m H_2O removed is adequate for most glycol dehydration requirements. Very often 2.5 to 4 gal TEG/lb_m H_2O is used for field dehydrators.
- Lean TEG concentration from reconcentrator. Most glycol reconcentrators can output 99.0 to 99.9% lean TEG. A value of 99.5% lean TEG is utilized in most designs.

8.2.3.1 Inlet Scrubber

It is essential to have a good inlet scrubber for efficient operation of a glycol dehydrator unit. Two-phase inlet scrubbers are generally constructed with 7 1/2-ft shell heights. The required minimum diameter of a vertical inlet scrubber can be determined based on the operating pressure and required gas capacity using Figure 8–4 or Table 8–2, which was developed from the chart presented by Sivalls (1977).

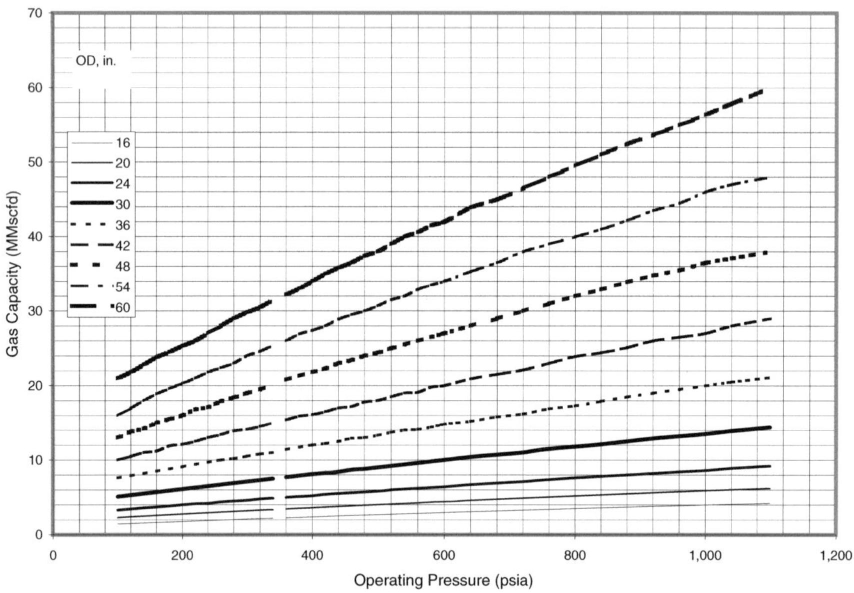

Figure 8–4 Gas capacity of vertical inlet scrubbers based on 0.7 specific gravity at 100 °F (Sivalls 1977).

8.2.3.2 Glycol-Gas Contactor

Glycol contactors are generally constructed with a standard height of 7 1/2 ft. The minimum required diameter of the contactor can be determined based on the gas capacity of the contactor for standard gas of 0.7 specific gravity at standard temperature 100 °F. If the gas is not the standard gas

Table 8–2 Gas Capacity in MMscfd of Vertical Inlet Scrubbers Based on 0.7 Specific Gravity at 100 °F (Sivalls 1977)

	Pressure (psia)												
OD (in.)	**100**	**120**	**140**	**160**	**180**	**200**	**220**	**240**	**260**	**280**	**300**	**320**	**340**
16	1.45	1.5	1.6	1.62	1.7	1.78	1.82	1.9	1.98	2.02	2.1	2.15	2.2
20	2.3	2.4	2.5	2.6	2.7	2.75	2.85	2.95	3.02	3.1	3.2	3.3	3.35
24	3.3	3.45	3.6	3.7	3.85	4	4.15	4.2	4.39	4.5	4.6	4.78	4.9
30	5.1	5.3	5.5	5.7	5.9	6.1	6.3	6.5	6.7	6.9	7.1	7.3	7.5
36	7.6	7.9	8.2	8.5	8.8	9.1	9.4	9.7	10	10.2	10.5	10.8	11
42	10	10.5	11	11.2	11.9	12.1	12.5	13	13.4	13.9	14.1	14.5	15
48	13	13.8	14.2	15	15.5	16	16.8	17.1	17.9	18.4	19	19.5	20
54	16	17	18	18.9	19.6	20.3	21	21.8	22.4	23	24	24.6	25.4
60	21	21.8	22.8	23.8	24.5	25.3	26	27	28	29	29.8	30.5	31.5
in.	**360**	**380**	**400**	**420**	**440**	**460**	**480**	**500**	**520**	**540**	**560**	**580**	**600**
16	2.25	2.3	2.4	2.45	2.5	2.58	2.6	2.7	2.72	2.8	2.85	2.9	2.95
20	3.45	3.53	3.6	3.7	3.8	3.85	3.95	4	4.1	4.2	4.25	4.38	4.4
24	5	5.1	5.2	5.4	5.5	5.6	5.7	5.8	6	6.1	6.2	6.3	6.4
30	7.7	7.9	8.1	8.25	8.4	8.7	8.8	9	9.2	9.4	9.6	9.8	10
36	11.4	11.7	12	12.2	12.5	12.9	13	13.3	13.7	14	14.1	14.4	14.8
42	15.4	15.9	16.1	16.5	17	17.2	17.8	18	18.5	19	19.1	19.8	20
48	20.8	21.3	21.8	22.3	23	23.4	24	24.4	25	25.5	26	26.5	27
54	26	26.9	27.4	28	28.9	29.5	30	30.7	31.5	32	32.9	33.5	34
60	32.2	33	34	35	35.8	36.5	37.4	38	39	40	40.6	41.5	42
in.	**640**	**680**	**720**	**760**	**800**	**840**	**880**	**920**	**960**	**1000**	**1040**	**1100**	
16	3.09	3.2	3.3	3.4	3.5	3.6	3.7	3.8	3.9	4	4.08	4.22	
20	4.6	4.75	4.9	5	5.2	5.35	5.5	5.6	5.79	5.9	6.02	6.21	
24	6.7	6.9	7.1	7.4	7.6	7.8	8	8.2	8.42	8.6	8.9	9.21	
30	10.4	10.7	11	11.5	11.8	12.1	12.5	12.9	13.2	13.5	13.9	14.4	
36	15.2	15.8	16.2	16.9	17.3	17.9	18.5	19	19.5	20	20.5	21.1	
42	20.9	21.5	22.1	23	23.9	24.4	25.1	26	26.5	27	28	29	
48	28	29	30	31	32	32.9	33.9	34.8	35.5	36.5	37.1	38	
54	35.3	36.5	38	39	40	41	42.2	43.5	44.5	46	47	48	
60	43.9	45	46.5	48	49.5	51	52.5	53.7	55	56.4	57.9	60	

and/or the operating temperature is different from the standard temperature, a correction should first be made using the following relation:

$$q_s = \frac{q}{C_t C_g} \tag{8.1}$$

where

q = gas capacity of contactor at operating conditions, MMscfd

q_s = gas capacity of contactor for standard gas (0.7 specific gravity) at standard temperature (100 °F), MMscfd

C_t = correction factor for operating temperature

C_g = correction factor for gas-specific gravity

The temperature and gas-specific gravity correction factors for trayed glycol contactors are given in Table 8–3 and Table 8–4, respectively. The temperature and specific gravity factors for packed glycol contactors are contained in Table 8–5 and Table 8–6, respectively. Once the gas capacity of the contactor for standard gas at standard temperature is calculated, the required minimum diameter of a trayed glycol contactor can be calculated using Table 8–7 or Figure 8–5.

Table 8–3 Temperature Correction Factors for Trayed Glycol Contactors (Sivalls 1977)

Operating Temperature (°F)	Correction Factor (C_t)
40	1.07
50	1.06
60	1.05
70	1.04
80	1.02
90	1.01
100	1.00
110	0.99
120	0.98

Table 8–4 Specific Gravity Correction Factors for Trayed Glycol Contactors (Sivalls 1977)

Gas-Specific Gravity (air = 1)	Correction Factor (C_g)
0.55	1.14
0.60	1.08
0.65	1.04
0.70	1.00
0.75	0.97
0.80	0.93
0.85	0.90
0.90	0.88

Table 8–5 Temperature Correction Factors for Packed Glycol Contactors (Sivalls 1977)

Operating Temperature (°F)	Correction Factor (C_t)
50	0.93
60	0.94
70	0.96
80	0.97
90	0.99
100	1.00
110	1.01
120	1.02

Table 8–6 Specific Gravity Correction Factors for Packed Glycol Contactors (Sivalls 1977)

Gas-Specific Gravity (air = 1)	Correction Factor (C_g)
0.55	1.13
0.60	1.08
0.65	1.04
0.70	1.00
0.75	0.97
0.80	0.94
0.85	0.91
0.90	0.88

Table 8–7 Gas Capacity in MMscfd for Trayed Glycol Contactors based on 0.7 Specific Gravity at 100 °F (Sivalls 1977)

	Pressure (psia)													
OD (in.)	**100**	**140**	**180**	**220**	**260**	**300**	**340**	**380**	**420**	**460**	**500**	**540**	**580**	**620**
12	1.16	1.27	1.37	1.45	1.55	1.65	1.74	1.83	1.92	2	2.08	2.15	2.22	2.28
15	1.61	1.8	1.98	2.14	2.3	2.43	2.6	2.7	2.85	3	3.1	3.2	3.33	3.45
18	2.38	2.65	2.85	3.05	3.25	3.45	3.6	3.8	3.95	4.15	4.3	4.44	4.59	4.7
20	3	3.3	3.55	3.8	4.03	4.25	4.5	4.75	4.95	5.17	5.34	5.56	5.7	5.9
24	4.75	5.1	5.5	5.83	6.2	6.58	6.9	7.2	7.6	7.9	8.2	8.5	8.8	9.1
30	7.5	8.2	8.8	9.4	10	10.6	11.2	11.8	12.3	12.9	13.3	13.8	14.2	14.8
36	11	12	12.9	13.8	14.7	15.4	16.3	17.2	18	18.8	19.4	20.1	20.9	21.5
42	14.9	16.2	17.4	18.5	19.8	21	22	23	24	25	26	27	28	29
48	19	20.8	22.6	24.2	26	27.5	29	30.5	32	33.5	35	36	37.5	38.9
54	24.5	26.5	28.8	31	33	35	37	39	40.5	42	44	46	47.8	49
60	31	34	36.5	39	41.5	44	46	48.5	51	53	55	57	59	61
66	36.8	40	43	46.5	50	52.5	56	58	61	64	66	69	72	74
72	44	48	52	56	59.8	63	66.5	70	73	76	79	82	85	88
OD (in.)	**660**	**700**	**740**	**780**	**820**	**860**	**900**	**940**	**980**	**1,020**	**1,060**	**1,100**	**1,140**	**1,200**
12	2.35	2.4	2.45	2.5	2.55	2.6	2.65	2.7	2.74	2.79	2.82	2.86	2.9	2.95
15	3.55	3.65	3.75	3.85	3.95	4.04	4.12	4.2	4.3	4.35	4.44	4.5	4.57	4.61
18	4.84	5	5.1	5.2	5.3	5.4	5.5	5.6	5.68	5.79	5.83	5.91	6	6.02
20	6	6.2	6.4	6.58	6.7	6.84	7	7.1	7.23	7.38	7.47	7.6	7.7	7.8
24	9.3	9.6	9.9	10.1	10.3	10.52	10.77	11	11.2	11.4	11.6	11.8	12	12.2
30	15.2	15.7	16	16.5	17	17.3	17.7	18	18.4	18.9	19.3	19.6	19.9	20.1
36	22.3	22.9	23.5	24	24.5	25.2	25.9	26.4	27	27.4	27.9	28.3	28.8	29.3
42	30	31	31.8	32.7	33.5	34.2	35	35.8	36.5	37.2	38	38.5	39.2	40.1
48	40	41	42.2	43.5	44.7	45.8	47	48	49	50	50.5	51.5	52.1	53
54	50.5	52.3	54	55.2	56.4	58	59.3	60.5	62	63	64	65	66	67.3
60	63	64.6	66.2	68.2	70	71.7	73	74.7	76	77.4	78.7	80.3	82	84
66	76	79	81	83	85	87	90	92	93.3	95	97	98.5	101	103.3
72	91	94	96	98	101	102.9	105.7	108	110	113	115	117	119	122

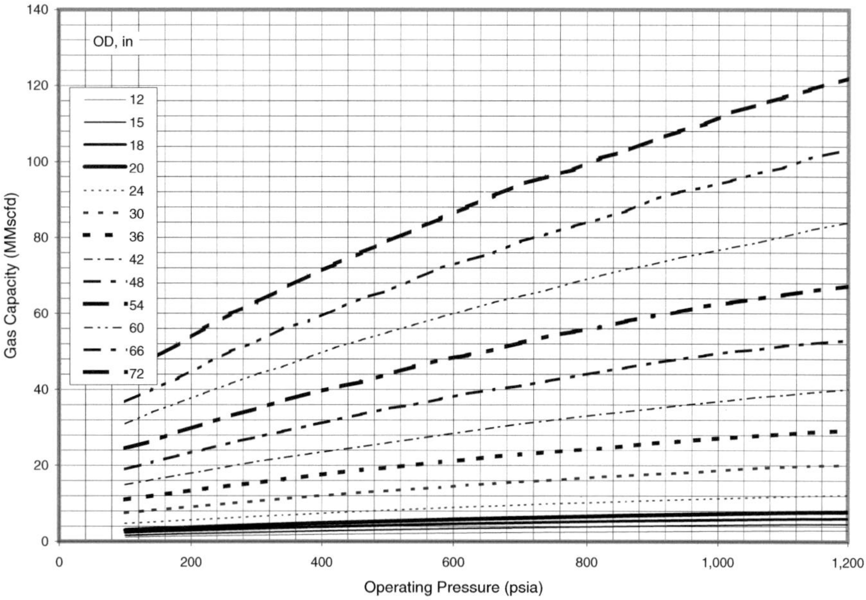

Figure 8–5 Gas capacity for trayed glycol contactors based on 0.7 specific gravity at 100 °F (Sivalls 1977).

The required minimum diameter of a packed glycol contactor can be determined based on the Table 8–8 or Figure 8–6.

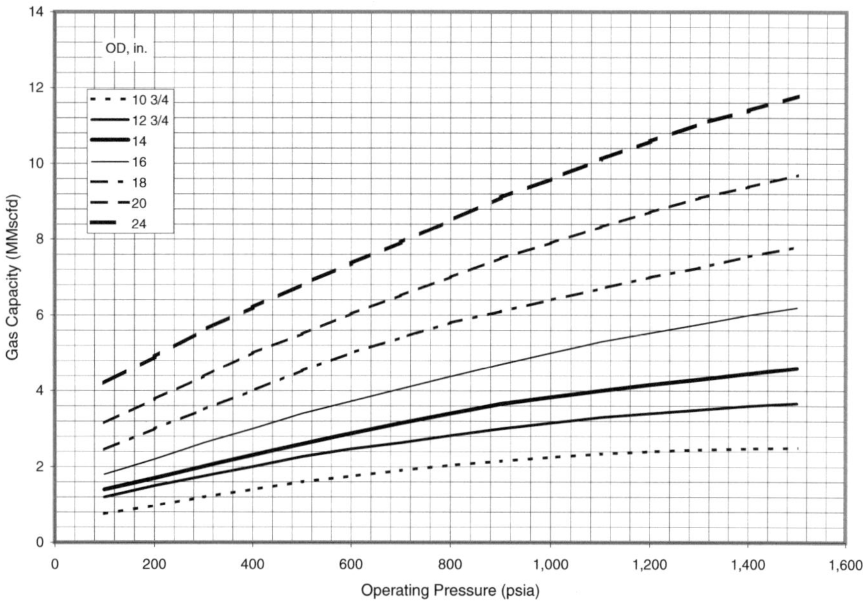

Figure 8–6 Gas capacity for packed glycol contactors based on 0.7 specific gravity at 100 °F (Sivalls 1977).

Table 8–8 Gas Capacity in MMscfd for Packed Glycol Contactors based on 0.7 Specific Gravity at 100 °F (Sivalls 1977)

Pressure (psia)	Contactor OD (in)						
	10 3/4	12 3/4	14	16	18	20	24
100	0.75	1.20	1.40	1.80	2.45	3.15	4.20
200	0.97	1.50	1.70	2.20	3.00	3.78	4.90
300	1.20	1.75	2.01	2.63	3.50	4.38	5.60
400	1.40	2.00	2.30	3.00	4.00	5.00	6.20
500	1.60	2.27	2.60	3.40	4.53	5.50	6.80
600	1.75	2.47	2.88	3.72	5.00	6.03	7.36
700	1.90	2.63	3.15	4.05	5.40	6.52	7.91
800	2.04	2.82	3.40	4.37	5.80	7.00	8.50
900	2.15	3.00	3.65	4.70	6.10	7.50	9.11
1,000	2.25	3.15	3.83	5.00	6.40	7.90	9.59
1,100	2.34	3.30	4.00	5.30	6.70	8.33	10.12
1,200	2.40	3.40	4.16	5.53	7.00	8.72	10.59
1,300	2.45	3.50	4.30	5.76	7.25	9.10	11.05
1,400	2.48	3.60	4.45	6.00	7.55	9.39	11.40
1,500	2.50	3.67	4.60	6.20	7.80	9.70	11.78

The required minimum height of packing of a packed contactor, or the minimum number of trays of a trayed contactor, can be determined based on Table 8–9 or Figure 8–7.

Table 8–9 The Required Minimum Height of Packing of a Packed Contactor, or the Minimum Number of Trays of a Trayed Contactor (Sivalls 1977)

Δt_d °F	Glycol to Water Ratio (gal TEG/lb$_m$ H_2O)												
	2	**2.5**	**3**	**3.5**	**4**	**4.5**	**5**	**5.5**	**6**	**6.5**	**7**	**7.5**	**8**
55	4.1	3.75	3.46	3.24	3.05	2.9	2.78	2.67	2.6	2.55	2.52	2.5	2.5
65	5.1	4.6	4.2	3.94	3.7	3.55	3.4	3.28	3.18	3.1	3.05	3	3
75	6	5.5	5.1	4.8	4.5	4.35	4.2	4.1	4	3.9	3.85	3.8	3.76
85	7.9	7	6.5	6.15	5.85	5.6	5.4	5.25	5.1	5	4.9	4.8	4.7
95	11.7	10.2	9.3	8.66	8.1	7.73	7.47	7.25	7	6.8	6.6	6.45	6.33

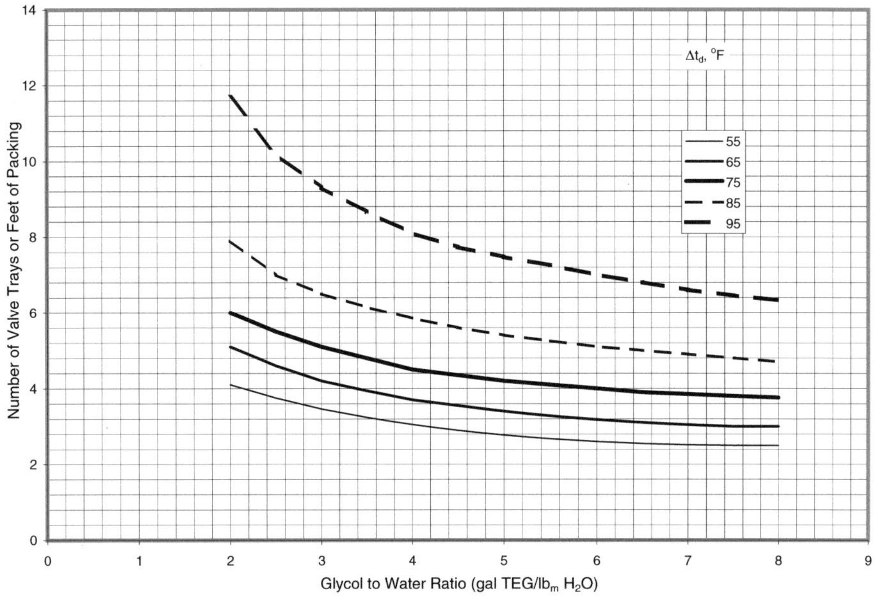

Figure 8–7 The required minimum height of packing of a packed contactor, or the minimum number of trays of a trayed contactor (Sivalls 1977).

It is also desired to know how much water will be removed from the glycol dehydration unit per hour. The water rate can be estimated by:

$$W_r = \frac{(C_{wi} - C_{wo})q}{24} \tag{8.2}$$

where

W_r = water to be removed, lb$_m$/hr

C_{wi} = water content of inlet gas, lb$_m$ H_2O/MMscf

C_{wo} = water content of outlet gas, lb$_m$ H_2O/MMscf

q = gas flow rate, MMscfd

Both C_{wi} and C_{wo} can be determined from Figure 8–1 based on the inlet and outlet dew point temperatures and the operating pressure.

Example Problem 8.2

Design a trayed-type glycol contactor for a field installation to meet the following requirements:

Gas flow rate: 12 MMscfd

Gas specific gravity: 0.75

Operating line pressure: 900 psig

Maximum working pressure of contactor: 1,440 psig

Gas inlet temperature: 90 °F

Outlet gas water content: 6 lb H_2O/MMscf

Design criteria: GWR = 3 gal TEG/lb_m H_2O with 99.5% TEG

Solution

Because the given gas is not a standard gas and the inlet temperature is not the standard temperature, corrections need to be made. Table 8–3 and Table 8–4 give $C_t = 1.01$ and $C_g = 0.97$. The gas capacity of contactor is calculated with Equation (8.1):

$$q_s = \frac{12}{(1.01)(0.97)} = 12.25 \text{ MMscfd.}$$

Table 8–7 gives contactor diameter:

$$D_C = 30 \text{ in}$$

Figure 8–1 gives water content of inlet gas:

$$C_{wi} = 50 \text{ lb}_\text{m}/\text{MMscf}$$

The required water content of outlet gas determines the dew point temperature of the outlet gas through Figure 8–1

$$t_{do} = 28 \text{ °F}$$

Therefore, the dew point depression is $\Delta t_d = 90 - 28 = 62$ °F.

Based on GWR = 3 gal TEG/lb_m H_2O and $\Delta t_d = 62$ °F, Figure 8–7 gives the number of trays rounded off to be four.

The rate of water to be removed is calculated by Equation (8.2) as:

$$W_r = \frac{(50-6)(12.25)}{24} = 22.46 \text{ lb}_\text{m}/\text{hour}$$

8.2.3.3 Glycol Reconcentrator

Sizing the various components of a glycol reconcentrator starts from calculating the required glycol circulation rate:

$$q_G = \frac{(GWR)C_{wi}q}{24} \tag{8.3}$$

where

q_G = glycol circulation rate, gal/hr

GWR = glycol to water ratio, gal TEG/lb_m H_2O

C_{wi} = water content of inlet gas, lb_m H_2O/MMscf

q = gas flow rate, MMscfd

8.2.3.4 Reboiler

The required heat load for the reboiler can be approximately estimated from the following equation:

$$H_t = 2{,}000 q_G \tag{8.4}$$

where H_t = total heat load on reboiler, Btu/h

Equation (8.4) is accurate enough for most high-pressure glycol dehydrator sizing. A more detailed procedure for determination of the required reboiler heat load can be found in literature (Ikoku 1984). The general overall size of the reboiler can be determined as follows:

$$A_{fb} = \frac{H_t}{7{,}000} \quad \textbf{(8.5)}$$

where A_{fb} is the total firebox surface area in ft^2.

8.2.3.5 Glycol Circulating Pump

The glycol circulating pump can be sized using the glycol circulation rate and the maximum operating pressure of the contactor. Commonly used glycol powered pumps utilize the rich glycol from the bottom of the contactor to power the pump and pump the lean glycol to the top of the contactor. The manufacturers of these pumps should be consulted to meet the specific needs of the glycol dehydrator.

8.2.3.6 Glycol Flash Separator

A glycol flash separator is usually installed downstream from the glycol pump to remove any entrained hydrocarbons from the rich glycol. A small 125 psi vertical two-phase separator is usually adequate for this purpose. The separator should be sized based on a liquid retention time in the vessel of at least five minutes.

$$V_s = \frac{q_G t_r}{60} \quad \textbf{(8.6)}$$

where

V_s = required settling volume in separator, gal

q_G = glycol circulation rate, gph

t_r = retention time, 5.0 min

Liquid hydrocarbon is not allowed to enter the glycol-gas contactor. If this is a problem a three-phase glycol flash separator should be used to keep these liquid hydrocarbons out of the reboiler and stripping still. Three-phase flash separators should be sized with a liquid retention time of 20 to 30 min. The hydrocarbon gas released from the flash separator can be piped to the reboiler to use as fuel gas and stripping gas. Based on the glycol circulation rate and the operating pressure of the contactor, the amount of gas available from the glycol pump can be determined.

8.2.3.7 Stripping Still

The size of the packed stripping still for the glycol reconcentrator can be determined based on the glycol-to-water circulation rate (gas TEG/lb_m H_2O) and the glycol circulation rate (gph). The required diameter for the stripping still is normally based on the required diameter at the base of the still using the vapor and liquid loading conditions at the base point. The vapor load consists of the water vapor and stripping gas flowing up through the still. The liquid load consists of the rich glycol stream and reflux flowing downward through the still column. One tray is normally sufficient for most stripping still requirements for TEG dehydration units. The amount of stripping gas required to reconcentrate the glycol is approximately 2 to 10 ft^3 per gal of glycol circulated.

8.3 Removal of Acid Gases

The H_2S and CO_2 in natural gas wellstreams are called acid gases because they form acids or acidic solutions in the presence of water. They have no heating value but cause problems to systems and the environment. H_2S is a toxic, poisonous gas and cannot be tolerated in gases that may be used for domestic fuels. H_2S in the presence of water is extremely corrosive and can cause premature failure of valves, pipeline, and pressure vessels. It can also cause catalyst poisoning in refinery vessels and requires expensive precautionary measures. Most pipeline specifications limit H_2S content to 0.25 g/100 ft^3 of gas (about 4 ppm).

Carbon dioxide is not as bad as H_2S and its removal is not always required. Removal of CO_2 may be required in gas going to cryogenic plants to prevent CO_2 solidification. Carbon dioxide is also corrosive in the presence of water. Most treating processes that remove H_2S will also remove CO_2. Therefore, the volume of CO_2 in the wellstream is added to the volume of H_2S to arrive at the total acid-gas volume to be removed.

The term sour gas refers to the gas containing H_2S in amounts above the acceptable industry limits. A sweet gas is a non-H_2S-bearing gas or gas that has been sweetened by treating. Some processes used for removing acid gases from natural gas are briefly described below.

8.3.1 Iron-Sponge Sweetening

The iron-sponge sweetening process is a batch process with the sponge being a hydrated iron oxide (Fe_2O_3) supported on wood shavings. The reaction between the sponge and H_2S is

$$2\ Fe_2O_3 + 6\ H_2S\ 2\ Fe_2S_3 + 6\ H_2O$$

The ferric oxide is present in a hydrated form. The reaction does not proceed without the water of hydration. The reaction requires the temperature be below approximately 120 °F or a supplemental water spray. Regeneration of the bed is sometimes accomplished by the addition of air continuously or by batch addition. The regeneration reaction is

$$2\ Fe_2O_3 + 3\ O_2\ 2\ Fe_2S_3 + 6\ S$$

The number of regeneration steps is limited due to the sulfur remaining in the bed. Eventually the beds have to be replaced.

8.3.2 Alkanolamine Sweetening

Alkanolamine encompasses the family of organic compounds of monoethanolamine (MEA), diethanolamine (DEA), and triethanolamine (TEA). These chemicals are used extensively for the removal of H_2S and

CO_2 from other gases and are particularly adapted for obtaining the low acid-gas residuals that are usually specified by pipelines. The alkanolamine process is not selective and must be designed for total acid-gas removal, even though CO_2 removal may not be required.

Typical reactions of acid gas with MEA are absorbing and regenerating. Absorbing reactions are:

$$MEA + H_2S \rightarrow MEA \text{ hydrosulfide} + \text{heat}$$

$$MEA + H_2O + CO_2 \rightarrow MEA \text{ carbonate} + \text{heat}$$

Regenerating reactions are:

$$MEA \text{ hydrosulfide} + \text{heat} \rightarrow MEA + H_2S$$

$$MEA \text{ carbonate} + \text{heat} \rightarrow MEA + H_2O + CO_2$$

MEA is preferred to either DEA or TEA solutions because it is a stronger base and is more reactive than either DEA or TEA. MEA has a lower molecular weight and thus requires less circulation to maintain a given amine to acid gas mole ratio. MEA also has greater stability and can be readily reclaimed from a contaminated solution by semicontinuous distillation.

8.3.3 Glycol/Amine Process

The glycol/amine process uses a solution composed of 10% to 30% weight MEA, 45% to 85% glycol, and 5% to 25% water for the simultaneous removal of water vapor, H_2S, and CO_2 from gas streams. The advantage of the process is that the combination dehydration and sweetening unit results in lower equipment cost than would be required with the standard MEA unit followed by a separate glycol/amine glycol dehydrator. The main disadvantages of the glycol/amine process include increased vaporization losses of MEA due to high regeneration temperatures, corrosion problems in the operating units, and limited applications for achieving low dew points.

8.3.4 Sulfinol Process

The sulfinol process uses a mixture of solvents allowing it to behave as both a chemical and physical solvent process. The solvent is composed of sulfolane, diisopropanolamine (DIPA), and water. The sulfolane acts as the physical solvent, while DIPA acts as the chemical solvent. The main advantages of sulfinol are low solvent circulation rates; smaller equipment and lower plant cost; low heat capacity of the solvent; low utility costs; low degradation rates; low corrosion rates; low foaming tendency; high effectiveness for removal of carbonyl sulfide, carbon disulfide, and mercaptans; low vaporization losses of the solvent; low heat-exchanger fouling tendency; and nonexpansion of the solvent when it freezes. Some of the disadvantages of sulfinol include absorption of heavy hydrocarbons and aromatics, and expense.

8.4 References

Guenther, J. D. "Natural gas Dehydration." Paper presented at the Seminar on Process Equipment and Systems on Treatment Platforms, Taastrup, Denmark, April 26, 1979.

Ikoku, C. U. *Natural Gas Production Engineering*. New York: John Wiley & Sons, 1984.

McCarth, E. L., W. L. Boyd, and L. S. Reid. "The Water Vapor Content of Essentially Nitrogen-Free Natural Gas Saturated at Various Conditions of Temperature and Pressure." *Trans. AIME* **189** (1950): 241–3.

McKetta, J. J. and W. L. Wehe. "Use This Chart for Water Content of Natural Gases." *Petroleum Refiner* **37** (1958): 153–4.

Sivalls, C. R. "Fundamentals of Oil and Gas Separation." Proceedings of the Gas Conditioning Conference, University of Oklahoma, Norman, Oklahoma, 1977.

8.5 Problems

8-1 Estimate water contents of a natural gas at a pressure of 2,000 psia and temperatures of 40, 80, 120, 160, 200, and 240 °F.

8-2 Design a glycol contactor for a field dehydration installation to meet the following requirements. Consider both trayed type and packed type contactors.

Gas flow rate: 10 MMscfd

Gas specific gravity: 0.65

Operating line pressure: 1,000 psig

Maximum working pressure of contactor: 1,440 psig

Gas inlet temperature: 90 °F

Outlet gas water content: 7 lb H_2O/MMscf

Design criteria: GWR = 3 gal TEG/lb_m H_2O with 99.5% TEG

Chapter 9

Compression and Cooling

9.1 Introduction

Portable compressors were first utilized in the late 1880s in the mining industry to drill in-mine pneumatic percussion boreholes (Singer, 1958a). Deep petroleum and natural wells were drilled utilizing portable air compressors in the 1920s (Singer, 1958b). With the advent of natural gas and its use as a fuel, the necessity arose of transporting natural gas from the gas well to the ultimate consumer. A compressor was unnecessary as long as the pressure at the gas well could force the gas through the pipeline to its destination. Compressors became essential because gas transmission pipelines extended great distances from the gas field.

When natural gas does not have sufficient potential energy to flow, a compressor station is needed. Five types of compressor stations are generally utilized in the natural gas production industry:

- Field gas-gathering stations to gather gas from wells in which pressure is insufficient to produce at a desired rate of flow into a transmission or distribution system. These stations generally handle suction pressures from below atmospheric pressure to 750 psig and volumes from a few thousand to many million cubic feet per day.
- Relay or main line stations to boost pressure in transmission lines. They compress generally large volumes of gas at a pressure range between 200 and 1,300 psig.
- Repressuring or recycling stations to provide gas pressures as high as 6,000 psig for processing or secondary oil recovery projects.
- Storage field stations to compress trunk line gas for injection into storage wells at pressures up to 4,000 psig.

- Distribution plant stations to pump gas from holder supply to medium- or high-pressure distribution lines at about 20 to 100 psig, or pump into bottle storage up to 2,500 psig.

9.2 Types of Compressors

The compressors used in today's natural gas production industry fall into two distinct types: reciprocating and rotary compressors. Reciprocating compressors are most commonly used in the natural gas industry. They are built for practically all pressures and volumetric capacities. As shown in Figure 9–1, reciprocating compressors have more moving parts and, therefore, lower mechanical efficiencies than rotary compressors. Each cylinder assembly of a reciprocation compressor consists of a piston, cylinder, cylinder heads, suction and discharge valves, and other parts necessary to convert rotary motion to reciprocation motion. A reciprocating compressor is designed for a certain range of compression ratios through the selection of proper piston displacement and clearance volume within the cylinder. This clearance volume can either be fixed or variable, depending on the extent of the operation range and the percent of load variation desired. A typical reciprocating compressor can deliver a volumetric gas flow rate up to 30,000 cubic feet per minute (cfm) at a discharge pressure up to 10,000 psig.

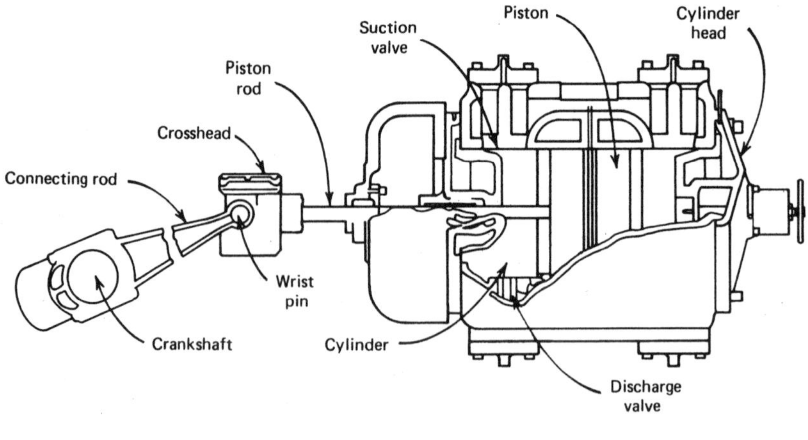

Figure 9–1 Elements of a typical reciprocating compressor (Courtesy of Petroleum Extension Services).

Rotary compressors are divided into two classes: the centrifugal compressor and the rotary blower. A centrifugal compressor (Figure 9–2 and Figure 9–3) consists of a housing with flow passages, a rotating shaft on which the impeller is mounted, bearings, and seals to prevent gas from escaping along the shaft. Centrifugal compressors have few moving parts because only the impeller and shaft rotate. Thus, its efficiency is high and lubrication oil consumption and maintenance costs are low. Cooling water is normally unnecessary because of lower compression ratio and lower friction loss. Compression rates of centrifugal compressors are lower because of the absence of positive displacement. Centrifugal compressors compress gas using centrifugal force. In this type of compressor, work is done on the gas by an impeller. Gas is then discharged at a high velocity into a diffuser where the velocity is reduced and its kinetic energy is converted to static pressure. Unlike reciprocating compressors, all this is done without confinement and physical squeezing. Centrifugal compressors with relatively unrestricted passages and continuous flow are inherently high-capacity, low-pressure ratio machines that adapt easily to series arrangements within a station. In this way, each compressor is required to develop only part of the station compression ratio. Typically, the volume is more than 100,000 cfm and discharge pressure is up to 100 psig.

A rotary blower is built of a casing in which one or more impellers rotate in opposite directions. Rotary blowers are primarily used in distribution systems where the pressure differential between suction and discharge is less than 15 psi. They are also used for refrigeration and closed regeneration of adsorption plants. The rotary blower has several advantages: large quantities of low-pressure gas can be handled at comparatively low horsepower, it has small initial cost and low maintenance cost, it is simple to install and easy to operate and attend, it requires minimum floor space for the quantity of gas removed, and it has almost pulsationless flow. As its disadvantages, it cannot withstand high pressures, it has noisy operation due to gear noise and clattering impellers, it improperly seals the clearance between the impellers and the casing, and it overheats if operated above safe pressures. Typically, rotary blowers deliver a volumetric gas flow rate of up to 17,000 cfm, and have a maximum intake pressure of 10 psig and a differential pressure of 10 psi.

When selecting a compressor, the pressure-volume characteristics and the type of driver must be considered. Small rotary compressors (vane or impeller type) are generally driven by electric motors. Large-volume positive compressors operate at lower speeds and are usually driven by

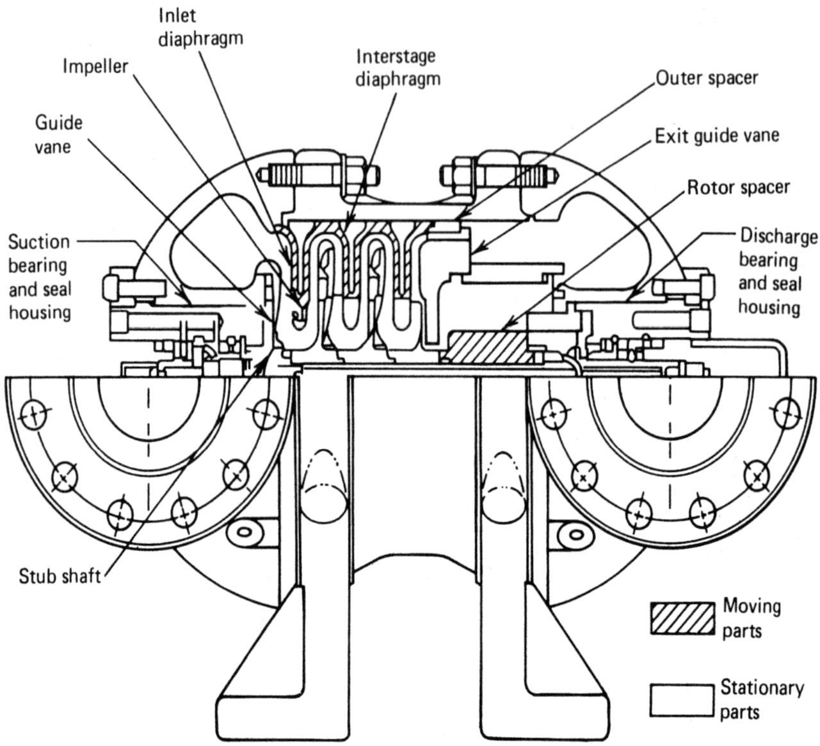

Figure 9–2 Cross section of a centrifugal compressor (Courtesy of Petroleum Extension Services).

steam or gas engines. They may be driven through reduction gearing by steam turbines or an electric motor. Reciprocation compressors driven by steam turbines or electric motors are most widely used in the natural gas industry as the conventional high-speed compression machine. Selection of compressors requires considerations of volumetric gas deliverability, pressure, compression ratio, and horsepower.

9.3 Selection of Reciprocating Compressors

Two basic approaches are used to calculate the horsepower theoretically required to compress natural gas. One is to use analytical expressions. In the case of adiabatic compression, the relationships are complicated and are usually based on the ideal-gas equation. When used for real gases

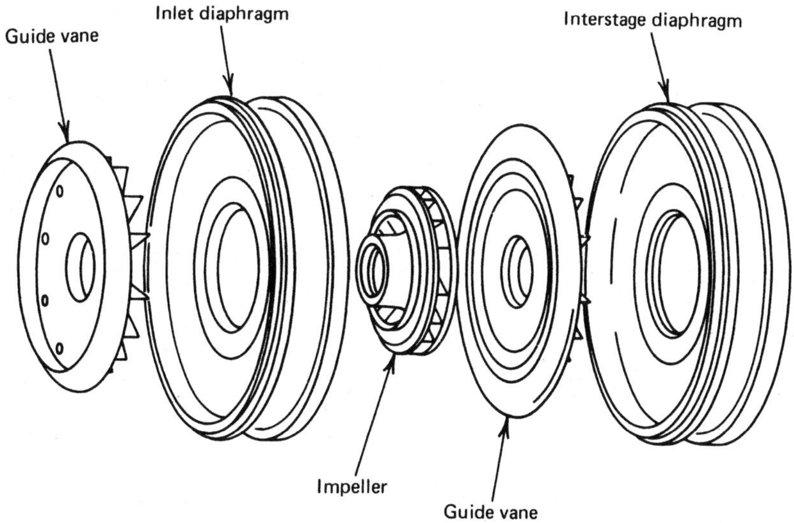

Figure 9–3 Internal parts of a centrifugal compressor (Courtesy of Petroleum Extension Services).

where deviation from ideal-gas law is appreciable, they are empirically modified to take into consideration the gas deviation factor. The second approach is the enthalpy-entropy or Mollier diagram for real gases. This diagram provides a simple, direct, and rigorous procedure for determining the horsepower theoretically necessary to compress the gas.

Even though in practice the cylinders in the reciprocating compressors may be water-cooled, it is customary to consider the compression process as fundamentally adiabatic; that is, to idealize the compression as one in which there is no cooling of the gas. Furthermore, the process is usually considered to be essentially a perfectly reversible adiabatic, that is, an isentropic process. Thus, in analyzing the performance of a typical reciprocating compressor, one may look upon the compression path following the general law:

$$pV^k = \text{a constant} \tag{9.1}$$

where k is an isentropic exponent given by the specific heat ratio

$$k = \frac{C_p}{C_v} \tag{9.2}$$

where C_p and C_v are specific heat at constant pressure and constant volume, respectively. For real natural gases in the gravity range $0.55 < \gamma_g < 1$, the following relationship can be used at approximately 150 °F:

$$k^{150^\circ F} \approx \frac{2.738 - \log \gamma_g}{2.328} \tag{9.3}$$

When a real gas is compressed in a single-stage compression, the compression is polytropic tending to approach adiabatic or constant-entropy conditions. Adiabatic-compression calculations give the maximum theoretical work or horsepower necessary to compress a gas between any two pressure limits, whereas isothermal-compression calculations give the minimum theoretical work or horsepower necessary to compress a gas. Adiabatic and isothermal work of compression thus give the upper and lower limits, respectively, of work or horsepower requirements to compress a gas. One purpose of intercoolers between multistage compressors is to reduce the horsepower necessary to compress the gas. The more intercoolers and stages, the closer the horsepower requirement approaches the isothermal value.

9.3.1 Volumetric Efficiency

The volumetric efficiency represents the efficiency of a compressor cylinder to compress gas. It may be defined as the ratio of the volume of gas actually delivered to the piston displacement, corrected to suction temperature and pressure. The principal reasons that the cylinder will not deliver the piston displacement capacity are wire-drawing, a throttling effect on the valves; heating of the gas during admission to the cylinder; leakage past valves and piston rings; and re-expansion of the gas trapped in the clearance-volume space from the previous stroke. Re-expansion has by far the greatest effect on volumetric efficiency.

The theoretical formula for volumetric efficiency is

$$E_v = 1 - (r^{1/k} - 1)C_l \tag{9.4}$$

where

E_v = volumetric efficiency, fraction

r = cylinder compression ratio

C_l = clearance, fraction

In practice, adjustments are made to the theoretical formula in computing compressor performance:

$$E_v = 0.97 - \left[\left(\frac{z_s}{z_d}\right) r^{1/k} - 1\right] C_l - e_v \tag{9.5}$$

where

z_s = gas deviation factor at suction of the cylinder

z_d = gas deviation factor at discharge of the cylinder

e_v = correction factor

In this equation, the constant 0.97 is a reduction of 1 to correct for minor inefficiencies such as incomplete filling of the cylinder during the intake stroke. The correction factor e_v is to correct for the conditions in a particular application that affect the volumetric efficiency and for which the theoretical formula is inadequate.

9.3.2 Stage Compression

The ratio of the discharge pressure to the inlet pressure is called pressure ratio. The volumetric efficiency becomes less, and mechanical stress limitation becomes more pronounced as pressure ratio increases. Natural gas is usually compressed in stages with the pressure ratio per stage being less than 6. In field practice, the pressure ratio seldom exceeds 4 when boosting gas from low pressure for processing or sale. When the total compression ratio is greater than this, more stages of compression are used to reach high pressures.

The total power requirement is a minimum when the pressure ratio in each stage is the same. This may be expressed in equation form as:

$$r = \left(\frac{p_d}{p_s} \right)^{1/N_s} \tag{9.6}$$

where

p_d = final discharge pressure, absolute

p_s = suction pressure, absolute

N_s = number of stages required

As large compression ratios result in gas being heated to undesirably high temperatures, it is common practice to cool the gas between stages and, if possible, after the final stage of compression (Figure 9–4).

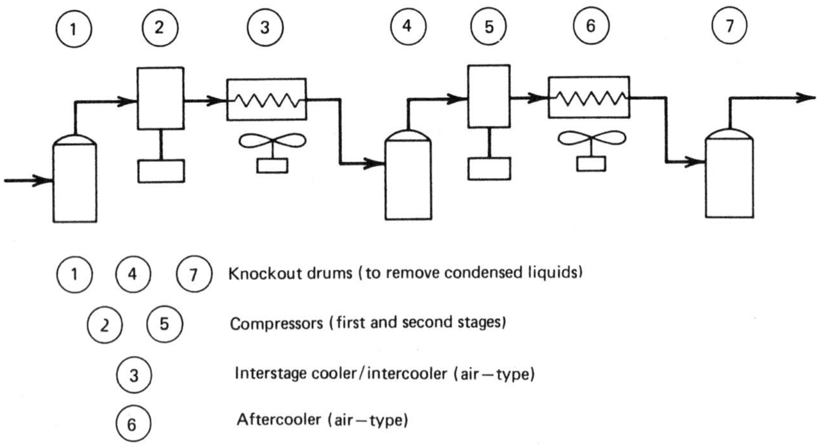

Figure 9–4 Two-stage compression arrangement with intercoolers and aftercoolers (Ikoku 1984).

9.3.3 Isentropic Horsepower

The computation is based on the assumption that the process is ideal isentropic or perfectly reversible adiabatic. The total ideal horsepower for a given compression is the sum of the ideal work computed for each stage of compression. The ideal isentropic work can be determined for each stage of compression in a number of ways. One simple and rapid way to solve a compression problem is by using the Mollier diagram (Figure 9–5 to Figure 9–10). This is done by tracing the increase in enthalpy from the cylinder suction pressure and temperature to its discharge pressure along the path of constant entropy (Figure 9–11). This involves some care in handling and converting the various units such as cubic feet per minute, pounds of vapor, British thermal units, and horsepower, but it is a simple and straightforward method.

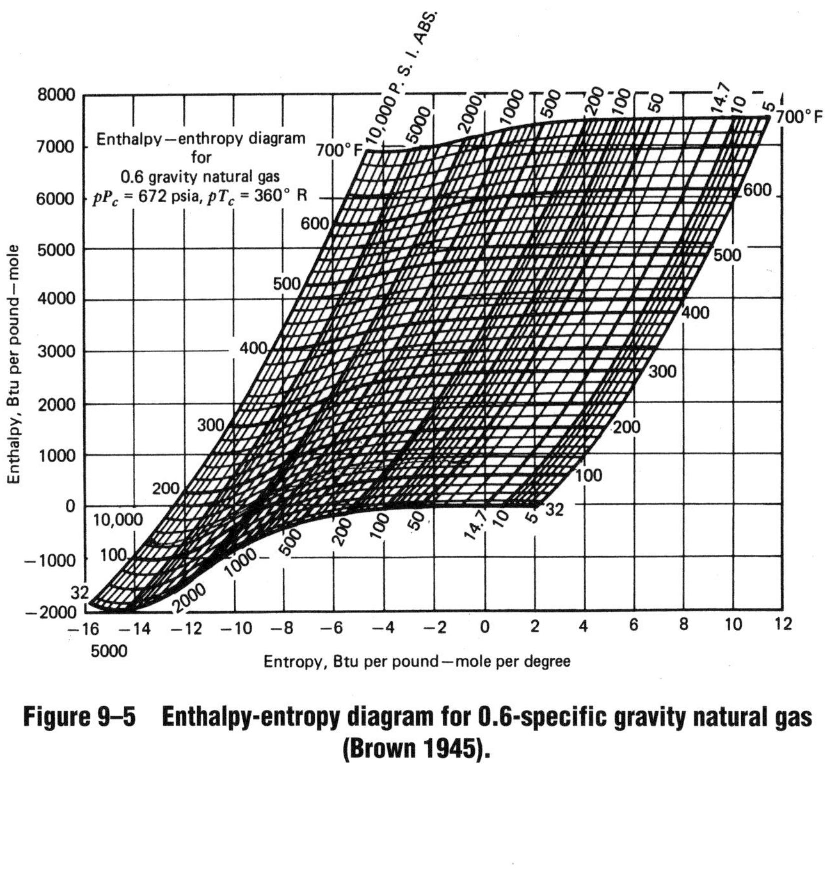

Figure 9–5 Enthalpy-entropy diagram for 0.6-specific gravity natural gas (Brown 1945).

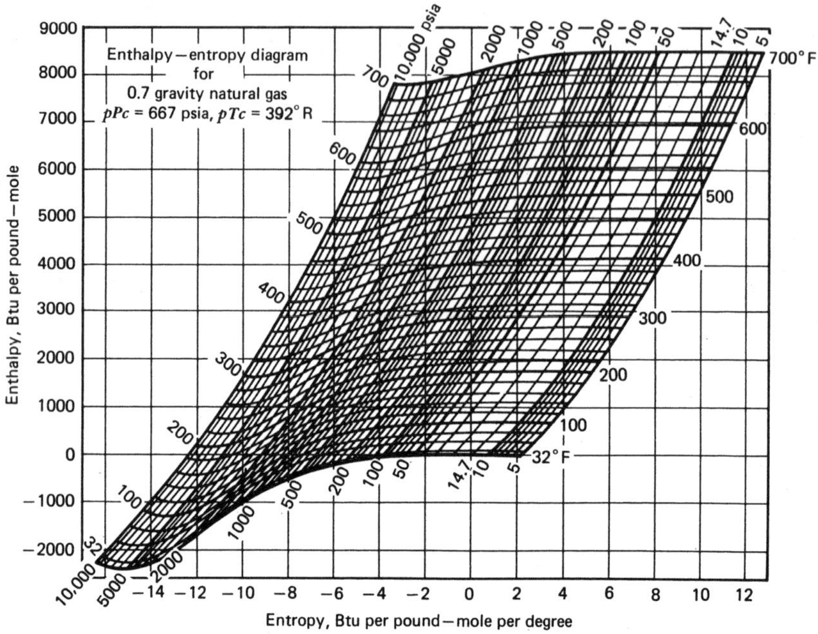

Figure 9–6 Enthalpy-entropy diagram for 0.70-specific gravity natural gas (Brown 1945).

Another approach commonly used is to calculate the horsepower for each stage from the isentropic work formula:

$$r = \left(\frac{p_d}{p_s} \right)^{1/N_s} \tag{9.7}$$

where

w = theoretical shaft work required to compress the gas, ft-lb$_f$/lb$_m$

T_1 = suction temperature of the gas, °R

γ_g = gas-specific gravity, air = 1

p_1 = suction pressure of the gas, psia

p_2 = pressure of the gas at discharge point, psia

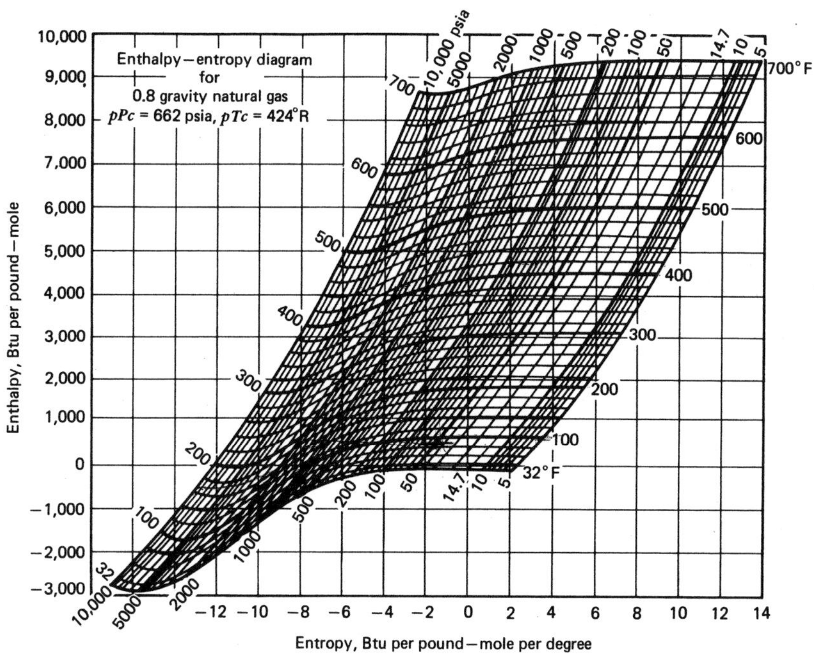

Figure 9–7 Enthalpy-entropy diagram for 0.80-specific gravity natural gas (Brown 1945).

When the deviation from ideal-gas behavior is appreciable, Equation (9.7) is empirically modified. One such modification is:

$$w = \frac{k}{k-1} \frac{53.241 T_1}{\gamma_g} \left[\left(\frac{p_2}{p_1} \right)^{Z_1(k-1)/k} - 1 \right] \tag{9.8}$$

or, in terms of power,

$$Hp_{MM} = \frac{k}{k-1} \frac{3.027 p_b}{T_b} T_1 \left[\left(\frac{p_2}{p_1} \right)^{Z_1(k-1)/k} - 1 \right] \tag{9.9}$$

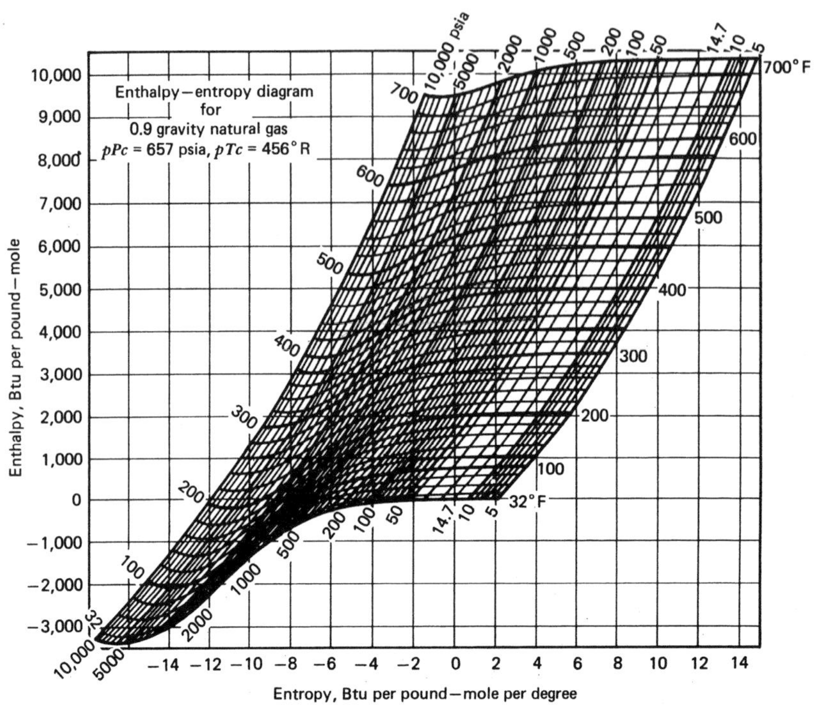

Figure 9–8 Enthalpy-entropy diagram for 0.90-specific gravity natural gas (Brown 1945).

where

Hp_{MM} = required theoretical compression power, hp/MMcfd

p_b = base pressure for specifying 1 MMcfd of gas, psia

T_b = base temperature for specifying 1 MMcfd of gas, °R

z_1 = compressibility factor at suction conditions

The theoretical adiabatic horsepower obtained by the proceeding equations can be converted to brake horsepower (Hp_b) required at the end of prime mover of the compressor using an overall efficiency factor, E_o. The brake horsepower is the horsepower input into the compressor. The efficiency factor E_o consists of two components: compression efficiency (compressor-valve losses) and the mechanical efficiency of the compressor. The overall efficiency of a compressor depends on a number of

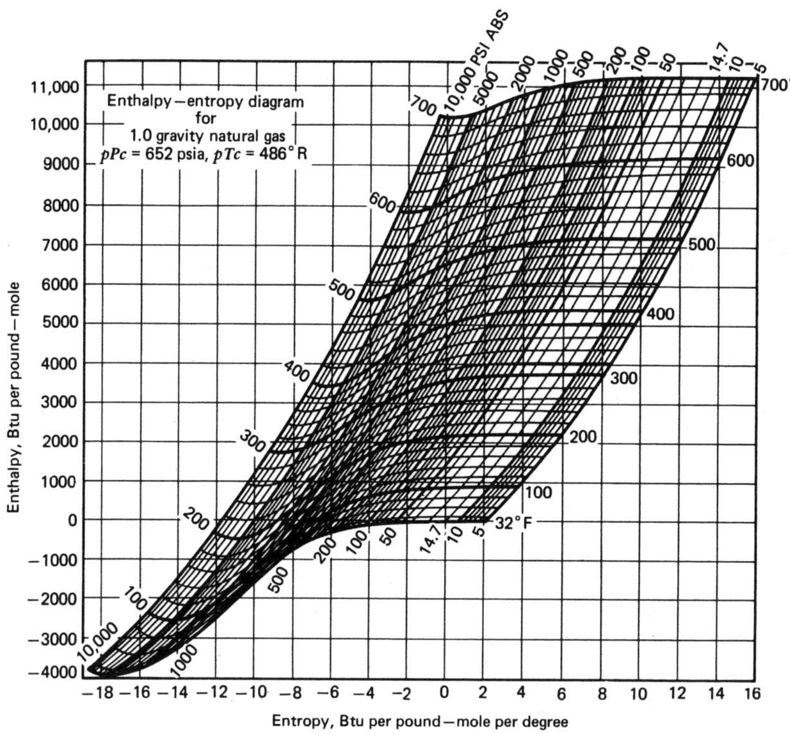

Figure 9–9 Enthalpy-entropy diagram for 1.0-specific gravity natural gas (Brown 1945).

factors, including design details of the compressor, suction pressure, speed of the compressor, compression ratio, loading, and general mechanical condition of the unit. In most modern compressors, the compression efficiency ranges from 83 to 93%. The mechanical efficiency of most modern compressors ranges from 88 to 95%. Thus, most modern compressors have an overall efficiency ranging from 75 to 85%, based on the ideal isentropic compression process as a standard. The actual efficiency curves can be obtained from the manufacturer. Applying these factors to the theoretical horsepower gives

$$\mathrm{Hp_b} = \frac{\mathrm{qHp_{MM}}}{E_o} \tag{9.10}$$

where q is the gas flow rate in MMscfd.

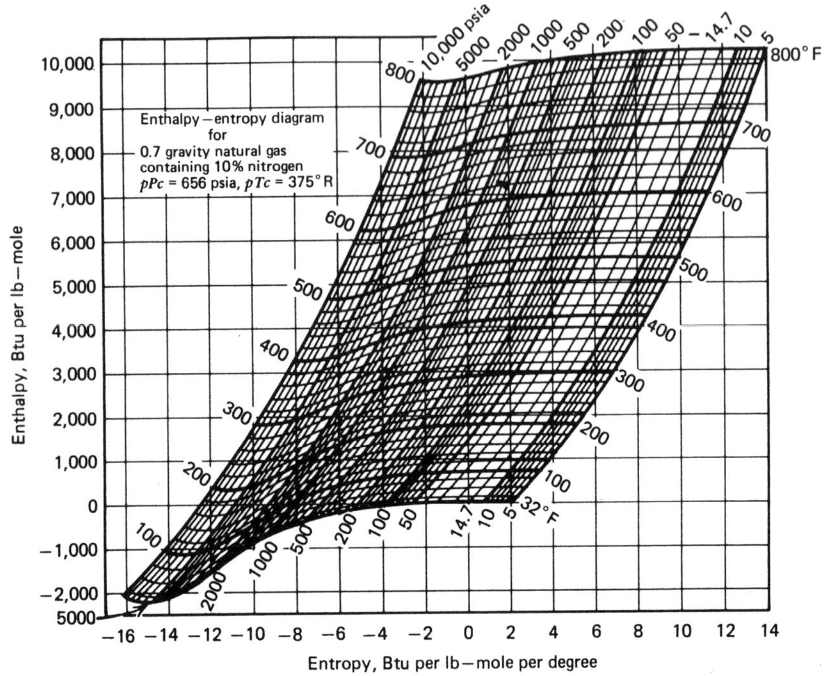

Figure 9–10 Enthalpy-entropy diagram for 0.70-specific gravity natural gas containing 10% nitrogen (Brown 1945).

The discharge temperature for real gases can be calculated by:

$$T_2 = T_1 \left(\frac{p_2}{p_1} \right)^{z_1 (k-1)/k} \tag{9.11}$$

Calculation of the heat removed by intercoolers and aftercoolers can be accomplished using constant pressure specific heat data:

$$\Delta H = n_G \overline{C}_p \Delta T \tag{9.12}$$

where

n_G = number of lb-mole of gas

$\overline{C}_p$ = specific heat under constant pressure evaluated at cooler operating pressure and the average temperature, Btu/lb-mol-°F

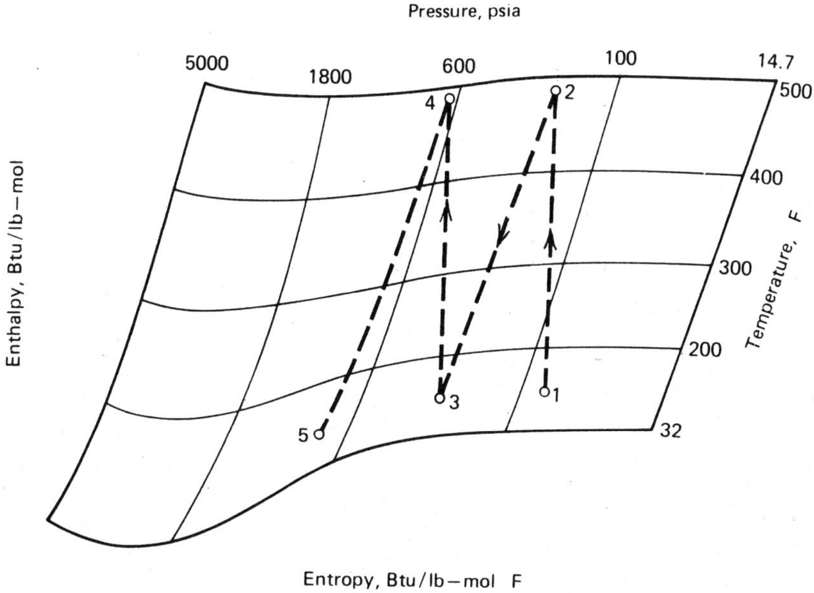

Figure 9–11 Adiabatic work of compressions from the enthalpy-entropy diagram (Katz and Lee 1990).

Example Problem 9.1

Assuming the overall efficiency is 0.80, calculate the theoretical and brake horsepower required to compress 1 MMcfd of a 0.6-specific gravity natural gas from 100 psia and 80 °F to 1,600 psia. If intercoolers cool the gas to 80 °F, what is the heat load on the intercoolers and what is the final gas temperature?

Solution

The overall compression ratio is:

$$r = \frac{1600}{100} = 16$$

Because this is greater than six, more than one-stage compression is required. Using two stages of compression gives:

$$r = \left(\frac{1600}{100}\right)^{1/2} = 4$$

The gas is compressed from 100 psia to 400 psia in the first stage, and from 400 psia to 1,600 psia in the second stage. Based on gas-specific gravity, the following gas property data can be obtained using the methods in Chapter 2:

$$T_c = 358\ °R$$

$$p_c = 671 \text{ psia}$$

$$T_r = 1.51$$

$$p_{r,1} = 0.149 \text{ at } 100 \text{ psia}$$

$$p_{r,2} = 0.595 \text{ at } 400 \text{ psia}$$

$$z_1 = 0.985 \text{ at } 80\ °F \text{ and } 100 \text{ psia}$$

$$z_2 = 0.940 \text{ at } 80\ °F \text{ and } 400 \text{ psia}$$

First stage:

$$Hp_{MM} = \frac{1.28}{0.28}\left(3.027 \times \frac{14.7}{520}\right) 540\left[(4)^{0.985(0.28/1.28)} - 1\right] = 73.3 \text{ hp/MMcfd}$$

Second stage:

$$Hp_{MM} = \frac{1.28}{0.28}\left(3.027 \times \frac{14.7}{520}\right) 540\left[(4)^{0.940(0.28/1.28)} - 1\right] = 69.5 \text{ hp/MMcfd}$$

Total theoretical compression work = 73.3 + 69.5 = 142.8 hp/MMcfd.

Required brake horsepower is

$$\text{Hp}_\text{b} = \frac{(1)(142.8)}{(0.8)} = 178.5 \text{ hp.}$$

Number of moles of gas is

$$n_G = \frac{1{,}000{,}000}{378.6} = 2.640 \times 10^3 \text{ lb-mole/day}$$

Gas temperature after the first stage of compression is

$$T_2 = (540)(4)^{0.985(0.28/1.28)} = 728\ °\text{R} = 268\ °\text{F}$$

The average cooler temperature is $\dfrac{268+80}{2} = 174\ °\text{F}$

$$\bar{C}_p \text{ at } 174\ °\text{F and } 400 \text{ psia} = 10.15\ \frac{\text{Btu}}{\text{lb-mol }°\text{F}}$$

$$\text{Cooler load} = 2.639\times10^3(10.15)(268-80)\ = 5.036\times10^6\ \text{Btu/day}$$

Final gas temperature:

$$T_d = (540)(4)^{0.940(0.28/1.28)} = 718\ °\text{R} = 258\ °\text{F}$$

It can be shown that the results obtained using the analytical expressions compare very well to those obtained from the Mollier diagram.

9.4 Selection of Centrifugal Compressors

Although the adiabatic compression process can be assumed in centrifugal compression, polytropic compression process is commonly considered as the basis for comparing centrifugal compressor performance. The process is expressed as:

$$pV^n = \text{a constant} \tag{9.13}$$

where n denotes the polytropic exponent. The isentropic exponent k applies to the ideal frictionless adiabatic process, while the polytropic exponent n applies to the actual process with heat transfer and friction. The n is related to k through polytropic efficiency E_p:

$$\frac{n-1}{n} = \frac{k-1}{k} \times \frac{1}{E_p} \tag{9.14}$$

The polytropic efficiency of centrifugal compressors is nearly proportional to the logarithm of gas flow rate in the range of efficiency between 0.7 and 0.75. The polytropic efficiency chart presented by Rollins (1973) can be represented by the following correlation:

$$E_p = 0.61 + 0.03\log(q_1) \tag{9.15}$$

where q_1 = gas capacity at the inlet condition, cfm.

There is a lower limit of gas flow rate below which severe gas surge occurs in the compressor. This limit is called surge limit. The upper limit of gas flow rate is called stone-wall limit which is controlled by compressor horsepower.

The procedure of preliminary calculations for selection of centrifugal compressors is summarized as follows:

1 Calculate compression ratio based on the inlet and discharge pressures:

$$r = \frac{p_2}{p_1} \tag{9.16}$$

2 Based on the required gas flow rate under standard conditions (q), estimate the gas capacity at inlet condition (q_1) by ideal gas law:

$$q_1 = \frac{p_b}{p_1}\frac{T_1}{T_b}q \tag{9.17}$$

3 Find a value for the polytropic efficiency E_p from the manufacturer's manual based on q_1.

4 Calculate polytropic ratio $(n-1)/n$ using Equation (9.14):

$$R_p = \frac{n-1}{n} = \frac{k-1}{k} \times \frac{1}{E_p} \tag{9.18}$$

5 Calculate discharge temperature by:

$$T_2 = T_1\, r^{R_p} \tag{9.19}$$

6 Estimate gas compressibility factor values at inlet and discharge conditions.

7 Calculate gas capacity at the inlet condition (q_1) by real gas law:

$$q_1 = \frac{z_1 p_b}{z_2 p_1} \frac{T_1}{T_b} q \tag{9.20}$$

8 Repeat steps 2 through 7 until the value of q_1 converges within an acceptable deviation.

9 Calculate gas horsepower by:

$$Hp_g = \frac{q_1 p_1}{229E_p} \left(\frac{z_1 + z_2}{2z_1} \right) \left(\frac{r^{R_p} - 1}{R_p} \right) \tag{9.21}$$

Some manufacturers present compressor specifications using polytropic head in lb_f-ft/lb_m defined as:

$$H_g = RT_1 \left(\frac{z_1 + z_2}{2} \right) \left(\frac{r^{R_p} - 1}{R_p} \right) \tag{9.22}$$

where R is the gas constant given by $1{,}544/MW_a$ in psia-ft^3/lb$_m$-°R. The polytropic head relates to the gas horsepower by

$$Hp_g = \frac{\dot{m} H_g}{33{,}000 E_p} \tag{9.23}$$

where $\dot{m}$ is mass flow rate in lb$_m$/min.

10 Calculate gas horsepower by:

$$Hp_b = Hp_g + \Delta Hp_m \tag{9.24}$$

where ΔHp_m is mechanical power losses, which is usually taken as 20 horsepower for bearing, and 30 horsepower for seals.

Example Problem 9.2

Size a centrifugal compressor for the following given data:

Gas-specific gravity: 0.68

Gas-specific heat ratio: 1.24

Gas flow rate: 144 MMscfd at 14.7 psia and 60 °F

Inlet pressure: 250 psia

Inlet temperature: 100 °F

Discharge pressure: 600 psia

Polytropic efficiency: $E_p = 0.61 + 0.03 \log(q_1)$

Solution

Calculate compression ratio based on the inlet and discharge pressures:

$$r = \frac{600}{250} = 2.4$$

Calculate gas flow rate in scfm:

$$q = \frac{144,000,000}{(24)(60)} = 100,000 \text{ scfm}$$

Based on the required gas flow rate under standard condition (q), estimate the gas capacity at inlet condition (q_1) by ideal gas law:

$$q_1 = \frac{(14.7)}{(250)} \frac{(560)}{(520)} (100,000) = 6,332 \text{ cfm}$$

Find a value for the polytropic efficiency based on q_1:

$$E_p = 0.61 + 0.03 \log(6,332) = 0.724$$

Calculate polytropic ratio $(n-1)/n$:

$$R_p = \frac{n-1}{n} = \frac{1.24-1}{1.24} \times \frac{1}{0.724} = 0.2673$$

Calculate discharge temperature by:

$$T_2 = (560)\ (2.4)^{0.2673} = 707.7\ °R = 247.7\ °F$$

Estimate gas compressibility factor values at inlet and discharge conditions:

$z_1 = 0.97$ at 250 psia and 100 °F

$z_2 = 0.77$ at 600 psia and 247.7 °F

Calculate gas capacity at the inlet condition (q_1) by real gas law:

$$q_1 = \frac{(0.97)(14.7)}{(0.77)(250)} \frac{(560)}{(520)} (100{,}000) = 7{,}977 \text{ cfm}$$

Use the new value of q_1 to calculate E_p:

$$E_p = 0.61 + 0.03 \log(7{,}977) = 0.727$$

Calculate the new polytropic ratio $(n-1)/n$:

$$R_p = \frac{n-1}{n} = \frac{1.24-1}{1.24} \times \frac{1}{0.727} = 0.2662$$

Calculate the new discharge temperature:

$$T_2 = (560)\ (2.4)^{0.2662} = 707\ °R = 247\ °F$$

Estimate the new gas compressibility factor value:

$$z_2 = 0.77 \text{ at } 600 \text{ psia and } 247\ °F$$

Because z_2 did not change, q_1 remains the same value of 7,977 cfm.

Calculate gas horsepower:

$$Hp_g = \frac{(7{,}977)(250)}{(229)(0.727)}\left(\frac{0.97+0.77}{2(0.97)}\right)\left(\frac{2.4^{0.2662}-1}{0.2662}\right) = 10{,}592 \text{ hp}$$

Calculate gas apparent molecular weight:

$$MW_a = (0.68)(29) = 19.72$$

Calculated gas constant:

$$R = \frac{1{,}544}{19.72} = 78.3 \text{ psia-ft}^3/\text{lb}_\text{m}\text{-°R}$$

Calculate polytropic head:

$$H_g = (78.3)(560)\left(\frac{0.97+0.77}{2}\right)\left(\frac{2.4^{0.2662}-1}{0.2662}\right) = 37{,}610 \text{ lb}_\text{f}\text{-ft/lb}_\text{m}$$

Calculate gas horsepower:

$$Hp_b = 10{,}592 + 50 = 10{,}642 \text{ hp}$$

9.5 Selection of Rotary Blowers

A rotary positive blower employs two symmetrical impellers rotating in a fixed relationship with each other and in opposite directions within an elongated cylinder. As each lobe of an impeller passes the blower inlet, it traps a quantity of air equal to exactly one-fourth the displacement of the blower. The entrapment occurs four times during each revolution, moving from the gas inlet to the outlet. Timing gears position the impellers accurately in relationship to each other, maintaining minute clearances, which allow the rotary positive blower to operate at high volumetric efficiency without internal seal or lubrication. Because of these minute clearances, a certain amount of gas escapes past the operating clearances back to the suction side of the blower. This leakage, defined as slip, is a constant for any given blower at a given pressure. It is expressed in revolutions per minute by dividing the leakage volume per minute by the displacement per revolution. Rotary blowers are available in capacities ranging from 5 to 30,000 cfm and pressures of up to 12 psig in single stage. In some sizes, two-stage machines are available for pressures of up to 20 psig.

Total operating speed of a blower, within size range, is determined by the following:

$$N_t = \frac{q_d}{V_{dis}} + N_{sl} \tag{9.25}$$

where

N_t = total operating speed, rpm

q_d = desired gas capacity, cfm

V_{dis} = displacement, ft^3/revolution

N_{sl} = slip speed, rpm

The total horsepower consumed is expressed as:

$$\text{hp} = 0.005\, V_{dis}\, N_t\, \Delta p \tag{9.26}$$

where Δp is the pressure differential in psi. The temperature rise can be calculated from Equation (9.11).

Most requirements can be met with a single machine of the required capacity and are suitable to produce the required pressure. The positive displacement blower can be adapted to variable-capacity requirements if provided with a variable-speed transmission or driver. Capacity control can also be provided by installing multiple units of identical or different capacities.

9.6 References

Brown, G. G. "A Series of Enthalpy-Entropy Charts for Natural Gases." *Trans. AIME* **60** (1945): 65.

Katz, D. L. and R. L. Lee. *Natural Gas Engineering—Production and Storage*. New York: McGraw-Hill Publishing Company, 1990.

Ikoku, C. U. *Natural Gas Production Engineering*. New York: John Wiley & Sons, 1984.

Rollins, J. P. *Compressed Air and Gas Handbook*. New York: Compressed Air and Gas Institute, 1973.

Singer, C. *History of Technology* **4** London: Oxford Press, 1958 (a).

Singer, C. *History of Technology* **6** London: Oxford Press, 1958 (b).

9.7 Problems

9-1 For a reciprocating compressor, calculate the theoretical and brake horsepower required to compress 50 MMcfd of a 0.7-specific gravity natural gas from 200 psia and 70 °F to 2,500 psia. If intercoolers cool the gas to 90 °F, what is the heat load on the intercoolers and what is the final gas temperature? Assuming the overall efficiency is 0.75.

9-2 For a reciprocating compressor, calculate the theoretical and brake horsepower required to compress 30 MMcfd of a 0.65-specific gravity natural gas from 100 psia and 70 °F to 2,000 psia. If intercoolers and endcoolers cool the gas to 90 °F, what is the heat load on the coolers? Assuming the overall efficiency is 0.80.

9-3 For a centrifugal compressor, use the following data to calculate required input horsepower and polytropic head:

Gas-specific gravity: 0.70

Gas-specific heat ratio: 1.30

Gas flow rate: 50 MMscfd at 14.7 psia and 60 °F

Inlet pressure: 200 psia

Inlet temperature: 70 °F

Discharge pressure: 500 psia

Polytropic efficiency: $E_p = 0.61 + 0.03\log(q_1)$

9-4 For the data given in Problem 9-3, calculate required brake horsepower if a reciprocating compressor is used.

Chapter 10

Volumetric Measurement

10.1 Introduction

Natural gas is transported in pipelines with continuous flow from the gas reservoir to its ultimate user. Accurate measurement of the total quantity of gas that has passed through a given section of pipe over a period of time is of paramount importance to both gas sellers and purchasers. The commonly used method of measuring natural gas is by volume. Because natural gas is compressible (volume depends on pressure and temperature), to measure gas in meaningful terms by the volume method, first specifying the base, or standard, pressure and temperature is of fundamental importance. In other words, the pressure and temperature of the reference or base cubic foot must be established. Most operators account for gas in units of 1,000 cubic feet at predefined standard conditions (pressure and temperature), commonly referred to as one Mscf. However, the standard condition is defined differently from area to area. Since 1967, the American Petroleum Institute (API) and the American Gas Association (AGA) have been using 14.73 psia and 60 °F as their standard conditions.

This chapter first focuses on measurement of gas volumetric flow rate with orifice meters and then presents an introduction to other measuring methods.

10.2 Measurement with Orifice Meters

Orifice meters are the most common equipment used in the natural gas industry for measurement of natural gas flow rate. As illustrated in Figure 10–1, an orifice meter consists of a thin flat plate with an accurately machined circular hole that is centered in a pair of flanges or other

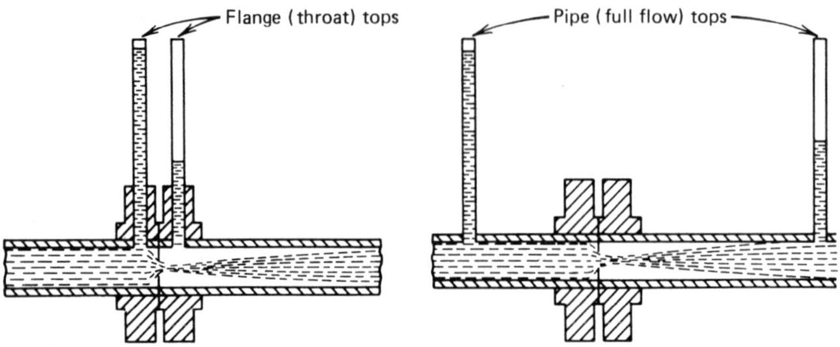

Figure 10–1 Arrangements of two types of orifice meters: flange taps and pipe taps (Ikoku 1984).

plate-holding device in a straight section of smooth pipe. Pressure tap connections are provided on the upstream and downstream sides of the plate so that the pressure drop or differential pressure may be measured.

The typical orifice meter consists primarily of a thin stainless steel plate about 3/16-in thick, with a hole in the center that is placed in the flow line. Placing an orifice in a pipe in which there is a gas flow causes a pressure difference across the orifice. This pressure difference and the absolute pressure in the line at a specified "tap" location are recorded continuously and are later translated into rate of flow.

The advantages of the orifice meter are accuracy, ruggedness, simplicity, ease of installation and maintenance, range capacity, low cost, acceptance for gas measurement by the joint AGA-ASME committee, and availability of standard tables of meter factors.

An orifice meter is composed of two major elements: 1) the primary element for producing differential pressure, and 2) the secondary element for measuring the pressures. As shown in Figure 10–2, the primary element consists of a meter tube, orifice plate-holding device, orifice plate, pressure taps, and straightening vanes which is a device that may be inserted in the upstream section of the meter tube to reduce swirling in the gas stream.

The secondary element is a gauge (or gauges) connected with tubing to the upstream and downstream pressure taps of the primary element. One part indicates or records the difference between the pressures on each side of the orifice plate and the other part indicates or records one of these

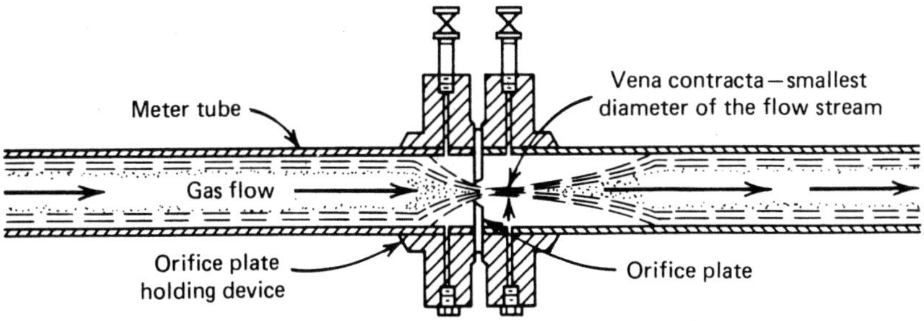

Figure 10–2 Primary element of an orifice meter (Ikoku 1984).

pressures. Recording differential and static pressure gauges, using circulate charts with printed scales, are extensively used and they provide a permanent record. Integrating differential gauges are also made, in both the indicating and recording type that register the flow in uncorrected cubic feet.

10.2.1 Orifice Equation

The basis for the orifice-meter equation is the first law of thermal dynamics. Derivation of the equation can be found in a number of publications such as that by Ikoku (1984). For the calculation of the quantity of gas, AGA (1956) recommends the formula:

$$q_h = C'\sqrt{h_w p_f} \tag{10.1}$$

where

q_h = quantity rate of flow at base conditions, cfh

C' = orifice flow constant

h_w = differential pressure in inches of water at 60 °F

p_f = absolute static pressure, psia

The orifice flow constant C' is expressed in the following equation:

$$C' = (F_b)(F_r)(Y)(F_{pb})(F_{tb})(F_{tf})(F_g)(F_{pv})(F_m)(F_l)(F_a) \tag{10.2}$$

where

F_b = basic orifice factor, cfh

F_r = Reynolds number factor

Y = expansion factor

F_{pb} = pressure base factor

F_{tb} = temperature base factor

F_{tf} = flowing temperature factor

F_g = specific gravity factor

F_{pv} = supercompressibility factor

F_m = manometer factor for mercury meter

F_l = gauge location factor

F_a = orifice thermal expansion factor

The basic orifice factor, F_b, is dependent on the location of the taps, the internal diameter of the run, and the size of the orifice. Tables for the basic orifice factor are presented in Appendix C of this book.

The Reynolds number factor, F_r, is dependent on the pipe diameter and the viscosity, density, and velocity of the gas. It is expressed as:

$$F_r = 1 + \frac{b}{\sqrt{h_w p_f}} \tag{10.3}$$

where the values of b are given in Appendix C of this book.

The expansion factor, Y, depends on the expansion of gas through the orifice. The density of the stream changes because of the pressure drop and the adiabatic temperature change. The expansion factor Y corrects for the variation in density. It is a function of the differential pressure, the absolute pressure, the diameter of the pipe, the diameter of the orifice, and the type of taps. Tables for Y values are presented in Appendix C of this book.

The pressure base factor, F_{pb}, is a direct application of Boyle's law in the correction for the difference in base from 14.73 psia. The pressure base is set by contract:

$$F_{pb} = \frac{14.73}{p_b} \tag{10.4}$$

The temperature base factor, F_{tb}, would be used in a direct application of Charles's law to correct for the base temperature change from 60 °F. Gas measured at one base temperature will have a different calculated volume if it is sold to a customer on a different base. That is, if the gas is measured at a base temperature of 60 °F and sold at a base temperature of 70 °F, the company must correct the volume to the contract temperature or, in this case, lose money. It is clear that the absolute temperature of the base (60 °F) divided by the absolute temperature of the contract will give a factor that should be applied to correct the meter reading to the terms of the contract temperature.

$$F_{tb} = \frac{t_b + 460}{520} \tag{10.5}$$

The flowing temperature factor, F_{tf}, corrects the effects of temperature variation. The flowing temperature has two effects on the volume. A higher temperature means a lighter gas so that flow will increase. Also, a higher temperature causes the gas to expand, which reduces the flow. The combined effect is to cause the quantity of flow of a gas to vary inversely as the square root of the absolute flow temperature. The F_{tf} is usually applied to the average temperature during the time gas is passing. The temperature may be taken by recording charts or by periodic indicating thermometer readings.

$$F_{tf} = \sqrt{\frac{520}{t + 460}} \tag{10.6}$$

where t = fluid temperature, °F

The specific gravity factor, F_g, is used to correct for changes in the specific gravity and should be based on the actual flowing specific gravity of the gas as determined by test. The specific gravity may be determined continuously by a recording gravitometer or by gravity balance on a daily, weekly, or monthly schedule, or as often as necessary to meet conditions of the contract. The basic orifice factor is determined by air with a specific gravity of 1. With a given force applied on a gas, a larger quantity of lightweight gas can be pushed through an orifice than a heavier gas. To make the basic orifice factor usable for any gas, the proper correction for the specific gravity of the gas being measured must be applied. This factor varies inversely as the square root of specific gravity.

$$F_g = \sqrt{\frac{1}{\gamma_g}} \tag{10.7}$$

The supercompressibility factor, F_{pv}, corrects for the fact that gas does not follow the ideal gas laws. It varies with temperature, pressure, and specific gravity. The development of the general hydraulic flow equation involves the actual density of the fluid at the point of measurement. In the measurement of gas, this depends on the flowing pressure and temperature compared to base pressure and temperature. It is necessary to apply the law for an ideal gas. All gases deviate from this ideal gas law to a greater or lesser extent. The actual density of a gas under high pressure is usually greater than the theoretical density obtained by calculation of the ideal gas law. This deviation has been termed supercompressibility. A factor to account for this supercompressibility is necessary in the measurement of some gases. This factor is particularly appreciable at high line pressures.

$$F_{pv} = \sqrt{\frac{1}{z}} \tag{10.8}$$

The manometer factor, F_m, is used with mercury differential gauges and compensates for the column of compressed gas opposite the mercury leg. Usually, this is not considered for pressures below 500 psia, nor is it required for mercury-less differential gauges. The weight of the gas column over the mercury reservoir of orifice meter gauges, introduces an error in determining the differential pressure across the orifice, unless some adjustment is made. This error is consistently in one direction and becomes increasingly important with increasing pressure. The correction varies with ambient temperature, static pressure, and specific gravity. Because the correction is very small, usually some average conditions are selected and a factor is agreed on.

$$F_m = \sqrt{\frac{62.3663 - \dfrac{p_{atm} + \dfrac{h_w}{27.707}}{192.4}}{62.3663}} \tag{10.9}$$

The gauge location factor, F_l, is used where orifice meters are installed at locations other than 45° latitude and sea-level elevation. It may affect the total flow of gas as recorded by the orifice meter.

$$F_l = \sqrt{\frac{g}{32.17405}} \tag{10.10}$$

where

$$g = 3.2808 \times 10^{-2}\left(9.7801855 \times 10^{2} - 2.8247 \times 10^{-3} L + 2.029 \times 10^{-3} L^{2} - 1.5058 \times 10^{-5} L^{3} - 9.4 \times 10^{-5} H\right) \tag{10.11}$$

where

L = latitude, deg.

H = elevation above sea level, ft

The orifice thermal expansion factor, F_a, is introduced to correct for the error resulting from expansion or contraction of the orifice operating at temperatures appreciably different from the temperature at which the orifice was bored.

$$F_a = 1 + 1.8 \times 10^{-5}(t - t_m) \tag{10.12}$$

where t_m = temperature during orifice boring, °F

10.2.2 Recording Charts

Although digital recording has been utilized in the industry for natural gas metering, round charts are still used extensively on all kinds of recording instruments associated with gas measurement. Circular charts for recording differential and static pressure gauges are usually 12 inches in diameter.

Two principal types of meter charts are widely used: 1) the uniform scale direct reading chart for the differential pressure in inches of water and the static pressure in psi, and 2) the chart that reads the square root. Clocks turn the charts at the desired speed, one turn each time period.

10.2.2.1 Direct-Reading Charts

In this type of scale, the lines are spaced an equal distance apart. The scale value of each line, in terms of the full range of the instrument with which it is used, should be 1, 2, or 5 units, or some multiple of these. In many cases, the differential pressure and static pressure are recorded on a chart with a common spacing. For example, if a 50-unit chart is used on a gauge having a differential pressure range of 100 inches of water and a static pressure range of 500 psig, then each circular line on the chart represents 2 inches of water pressure and 10 psig.

10.2.2.2 Square Root Charts

A square root chart with actual recordings is shown in Figure 10–3. The recorded differential of a typical flow pattern normally shows a weaving line on the chart. The smoother line on the chart represents the static pressure. This scale shows the square root of the percentage of the full-scale range of the gauge, or as represented by the full scale of the chart. A reading at full scale or full range of the gauge will be 10, the square root of 100. Using the 100-in, 500-psi gauge, a chart reading of 5 would represent a differential pressure of 25 inches of water or a static pressure of 125 psia.

For square root charts, a chart factor may be defined as

$$\text{Chart factor} = \sqrt{\frac{R_h}{100}} \tag{10.13}$$

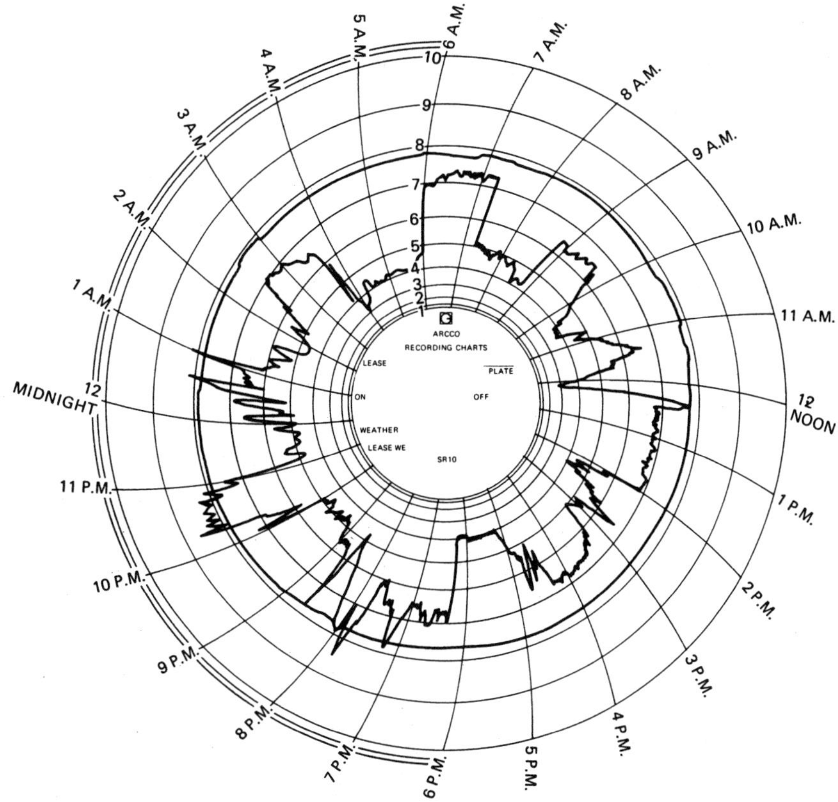

Figure 10–3 Square root chart with actual recordings (Ikoku 1984).

and

$$\text{Chart factor} = \sqrt{\frac{R_p}{100}} \tag{10.14}$$

where

R_h = differential pressure range, in

R_p = static pressure range, psi

Then

$$\text{Actual parameter value} = (\text{chart reading} \times \text{chart factor})^2$$

thus,

$$h_w = \left(\frac{\text{chart reading}}{10}\right)^2 R_h \tag{10.15}$$

and

$$p_f = \left(\frac{\text{chart reading}}{10}\right)^2 R_p \tag{10.16}$$

Example Problem 10.1

A 50-in × 100 lb gauge has a differential pressure range of $R_h = 50$ inches and static pressure range of $R_p = 100$ psi. If a square root chart shows a reading of 7.2 for differential pressure and 9.4 for static pressure, calculate differential pressure and static pressure.

Solution

Actual differential pressure:

$$h_w = \left(\frac{7.2}{10}\right)^2 (50) = 25.92 \text{ in water}$$

Actual static pressure:

$$p_f = \left(\frac{9.4}{10}\right)^2 (100) = 88.36 \text{ psia}$$

Because the static pressure to be used in all gas computation is the absolute pressure, when a square root chart is used the static pen is set so that, theoretically, it should read zero only if subjected to a pressure of absolute zero.

10.2.3 Computation of Volumes

After the differential pressure, static pressure, and temperature data at the field location have been recorded on charts, the latter must be picked up and taken to some location for processing. For standard gauges, depending on the chart rotation, this requires trips to the field location every day, every other day, every third day, or every week. With the advent of automatic changers, this is no longer necessary. Charts for several days may be loaded at one time. At the completion of recording, the chart automatically changes, and several fully recorded charts may be picked up at one time. This saves much chart-changing time and allows more accurate chart recording because faster rotating charts are economically feasible. At the central chart processing locations, the charts are integrated or scanned to obtain chart units per period of operation (usually 24 hours). These chart units must then be converted to volume by use of the proper basic orifice factor and all the related factors. The most proficient manner of doing this is by programming the rather complex calculations on a computer.

Example Problem 10.2

Calculate the hourly gas flow rate for the conditions given as follows:

Base conditions: Gas field in Texas, $p_b = 14.65$ psia, $t_b = 60$ °F

Meter pipe: 4-in schedule 40 (4.026-in ID), flange taps, static pressure measured upstream taps

Orifice plate: Stainless steel, 1.5-in measured at 20 °C

Recorder: 100-in water column differential, 1,000 psia static spring

Readings:

Elevation: 500 ft

Atmospheric pressure: 14.4 psia

Flowing temperature: 100 °F

Gas-specific gravity: 0.6

Differential pressure: 65-in water column

Static pressure: 641 psig

Solution

Based on pipe ID D = 4.026 inches and orifice diameter d = 1.5 inches, Appendix C, Table C–1 gives:

$$F_b = 460.80$$

Based on pipe ID D = 4.026 inches and orifice diameter d = 1.5 inches, Appendix C, Table C–2 gives b = 0.0336. Thus,

$$F_r = 1 + \frac{0.0336}{\sqrt{(65)(655.4)}} = 1.0002$$

Based on pipe ID D = 4.026 inches, orifice diameter d = 1.5 inches,

$$\beta = d/D = (1.5)/(4.026) = 0.373$$

and

$$h_w/p_f = (65)/(655.4) = 0.098$$

Appendix C, Table C–3 with interpolation gives Y = 0.9988.

The pressure base factor:

$$F_{pb} = \frac{14.73}{14.65} = 1.0055$$

The temperature base factor:

$$F_{tb} = \frac{60 + 460}{520} = 1.0000$$

The flowing temperature factor:

$$F_{tf} = \sqrt{\frac{520}{100 + 460}} = 0.9636$$

The specific gravity factor:

$$F_g = \sqrt{\frac{1}{0.60}} = 1.2910$$

Spreadsheet Hall-Yarborogh-z.xls gives z = 0.917 at 655.4 psia and 100 °F for a 0.6 specific gravity gas. The supercompressibility factor is calculated as:

$$F_{pv} = \sqrt{\frac{1}{0.917}} = 1.0443$$

The manometer factor:

$$F_m = \sqrt{\frac{62.3663 - \dfrac{14.4 + \dfrac{65}{27.707}}{192.4}}{62.3663}} = 0.9993$$

The gravitational acceleration at the given location is calculated with Equation (10.11) to be $g = 32.1418$ ft/s2. The gauge location factor is:

$$F_l = \sqrt{\frac{32.1418}{32.17405}} = 0.9995$$

The orifice thermal expansion factor is:

$$F_a = 1 + 1.8 \times 10^{-5}(100 - 68) = 1.0006$$

Therefore,

$$C' = (460.80)(1.0002)(0.9988)(1.0055)(1.0000)(0.9636)$$
$$(1.2910)(1.0443)(0.9993)(0.9995)(1.0006)$$

$$C' = 600.66$$

$$q_h = 600.66\sqrt{(65)(655.4)} = 125{,}100 \text{ scfh}$$

10.2.4 Selection of Orifice Meter

It is necessary to gather the following information about the characteristics and conditions of the flow to be metered: maximum peak hourly rate, minimum hourly rate, metering gauge pressure required and available, and permissible pressure variations. The quantity of gas flowing through an orifice at constant pressure varies as the square root of the differential pressure. Accordingly, for half of a given rate of flow, the differential pressure will be one-fourth of that for the given rate. Because of mechanical and installation limitations, it has been considered impractical to construct a differential gauge that will continuously record pressures with acceptable accuracy below about one-sixteenth of its maximum range. Therefore, the working range of one orifice plate and one differential gauge is from maximum capacity to about one-sixteenth of maximum. The maximum and minimum capacity can be changed by changing the orifice size.

10.3 Other Methods of Measurement

In addition to the orifice meters, natural gas can be measured using a variety of other measurement techniques including displacement meters, turbine meters, venturi meters, flow nozzles, critical low provers, elbow meters, and rotameters. Factors affecting the selection of the measurement method include desired accuracy, expected useful life, range of flow, pressure, temperature, initial cost, costs of operation, and acceptability by others involved.

10.3.1 Displacement Metering

Displacement meters fall into two categories: (a) reciprocating displacement and (b) rotary displacement. Displacement metering relies on a piston moving in a cylinder. A quantity of gas is taken into the cylinder through the inlet port to occupy the space displaced by the piston in a stroke. On the return stroke, the gas is discharged out of the cylinder through the outlet. The volume of space the discharged gas occupied while in the cylinder is equal to the piston displacement. Where the volume of the piston displacement is known, it is a simple matter to connect a counter to the piston rod that will tally the piston displacement for each compression stroke. Because the volume of gas discharged is equal to the total piston displacement, the counter will indicate a measured volume of gas. The pressure and temperature of the gas in the cylinder will be supplied to the cylinder through the inlet port. If a thermometer and pressure gauge are added to the cylinder, these conditions may be observed. From this information, Boyle's and Charles's law formulas can be applied to the volume of gas discharged as indicated on the counter in order to convert this volume to the equivalent quantity of gas at base temperature and pressure conditions. The equation to use is:

$$q_b = \frac{pT_b r}{p_b T z} \tag{10.17}$$

where

q_b = quantity of gas at base conditions, ft^3

p = pressure of gas, psia

p_b = pressure base, psia

T = temperature of gas, °R

T_b = temperature base, °R

z = gas deviation factor at p and T

r = counter registration, ft^3

The initial reading of the index is subtracted form the final reading to obtain the registration during any period. The displacement meter formula can be rewritten for this procedure:

$$q_b = (r_2 - r_1)\frac{pT_b}{p_b T z} \tag{10.18}$$

where

r_2 = final index reading, ft^3

r_1 = initial index reading, ft^3

The most common type of displacement meter has diaphragms separating the measuring compartments. These usually have four measuring compartments and two diaphragms. The movement of a diaphragm from one side to the other allows one compartment to fill while the second is discharging.

Rotary displacement meters rely on an entirely different mechanical principle than that of the reciprocating displacement meters. It uses two metal impellers of the same size. These impellers rotate on individual shafts and are designed and spaced to rotate tangentially to each other. They are enclosed in a cylindrical case. Gas flowing through the meter rotates the impellers and, because the close-off volume between an impeller and the case is fixed, a definite volume of gas will pass through the meter with each revolution of the impellers. By connecting an index to the shaft of an impeller, the volume of gas may be registered by this index.

10.3.2 Turbine Meter

A turbine meter uses the flowing gas as a driving force impacting to a laded rotor. With appropriate gearing, revolutions of the rotor can be converted to volume. Accuracy curves are usually developed for each turbine meter, and proving or calibration techniques as available. To get sustained accuracy and trouble-free operation, filters are almost a necessity ahead of turbine meters.

10.3.3 Elbow Meter

Elbow meters use centrifugal force in the curve of a pipe elbow to measure flow. For accuracy, calibrations with some other acceptable measurement as a standard are needed. Accuracy is not usually the objective when elbow meters are used. Relatively little pressure loss of differential pressure is created. Because of this, the meters are used primarily for control or other operations.

10.4 Natural Gas Liquid Measurement

The conventional gauging of tanks and various metering techniques can be used for field measurement of natural gas liquids. The orifice meter is sometimes used. Installation and operation requirements are about the same as for gas. The following formula may be used in calculations:

$$q_h = C'\sqrt{h_w} \tag{10.19}$$

where

q_h = rate of liquid flow, gph

C' = orifice constant ($F_b \times F_g \times F_r$)

h_w = differential pressure, inches of water

F_b = basic orifice factor

F_g = specific gravity factor

F_r = Reynolds number factor

Equation (10.10) can be written in terms of weight as:

$$\dot{W} = SND^2 F_a F_m F_c F_{pv} \sqrt{\gamma_f h_w} \tag{10.20}$$

where

$\dot{W}$ = rate of flow, lb_m/day

S = a value determined from the bore of the orifice and internal diameter of the metering tube

N = combined constant for weight-flow measurement (68,045 when $\dot{W}$ is in pounds/day)

D = ID of tube, inches

F_a = orifice thermal expansion factor

F_m = manometer factor (1 for bellows-type meter)

F_c = viscosity factor (usually assumed equal to one)

F_{pv} = supercompressibility factor

γ_f = specific gravity of liquid stream at flowing temperature and pressure as determined by gravitometer readings

h_w = differential pressure, inches of water

For simplicity, Equation (10.20) may be written as (Foxboro 1961):

$$\dot{W} = 68{,}045SD^2\sqrt{\gamma_f h_w} \tag{10.21}$$

Accuracy suffers when it is necessary to make measurements of two-phase (gas and liquid) for operation and allocation purposes. Certain precautions should be taken to arrive at acceptable measurements of a two-phase stream:

- Maintain pressure and temperature as high as possible at the meter.
- Use a free-water knockout ahead of the meter.
- A vertical meter run may be utilized to improve the differential-pressure relationship to the volume.
- Use test data from periodic full-scale separator tests to calibrate the meter.
- Connect manifold lead lines to bottom of bellows-type meter with self-draining pots installed above orifice fitting.

10.5 References

American Gas Association. "Orifice Metering of Natural Gas." Gas Measurement Committee Report No. 3. New York, 1956.

Foxboro, P. K. *Principles and Practices of Flow Meter Engineering*. New York: McGraw-Hill, 1961.

Ikoku, C. U.: *Natural Gas Production Engineering*. New York: John Wiley & Sons, Inc., 1984.

10.6 Problems

10-1 For the chart shown in Figure 10–3, assuming a differential pressure range of $R_h = 50$ inches and static pressure range of $R_p = 100$ psi, calculate the average differential pressure and static pressure between 1:00 pm. and 2:00 pm.

10-2 Calculate the hourly gas flow rate for the conditions given as follows:

Base conditions: Gas field in Oklahoma, $p_b = 14.65$ psia, $t_b = 60$ °F.

Meter pipe: 4-in schedule 40 (4.026-in ID), flange taps, static pressure measured upstream taps.

Orifice plate: Stainless steel, 1.500 inches measured at 20 °C.

Recorder: 100-in water column differential, 1,000 psia static spring.

Readings:

Elevation: 450 ft.

Atmospheric pressure: 14.5 psia

Flowing temperature: 95 °F

Gas-specific gravity: 0.65

Differential pressure: 75-in water column

Static pressure: 750 psia

10-3 Calculate the hourly gas flow rate for the readings obtained in Problem 10-1. Assume the conditions given as follows:

Base conditions: Gas field in Texas, $p_b = 14.65$psia, $t_b = 60$ °F

Meter pipe: 4-in schedule 40 (4.026-in ID), flange taps, static pressure measured upstream taps

Orifice plate: stainless steel, 1.5 inches measured at 20 °C

Recorder: 50-in water column differential, 1,000 psia static spring

Readings:

Elevation: 650 ft

Atmospheric pressure: 14.6 psia

Flowing temperature: 95 °F

Gas-specific gravity: 0.70

10-4 For the chart shown in Figure 10–3, assuming a differential pressure range of $R_h = 100$ inches and static pressure range of $R_p = 1{,}000$ psi, calculate the differential pressures and static pressures over the 24 hours.

10-5 Calculate the daily gas flow rate for the readings obtained in Problem 10-4. Assume the conditions given as follows:

Base conditions: Gas field in Texas, $p_b = 14.65$ psia, $t_b = 60$ °F

Meter pipe: 4-in schedule 40 (4.026-in ID), flange taps, static pressure measured upstream taps

Orifice plate: Stainless steel, 1.5 inches measured at 20 °C

Recorder: 100-in water column differential, 1,000 psia static spring

Readings:

Elevation: 350 ft

Atmospheric pressure: 14.7 psia

Flowing temperature: 85 °F

Gas-specific gravity: 0.75

Chapter 11

Transportation

11.1 Introduction

Unlike other products that are packaged and transported by vehicles, commercial natural gas is continuously transported through pipelines. The transmission of natural gas to the consumer may be divided into three distinct pipeline units: the gathering system, the main trunk line, and the distribution lines. This chapter focuses on design and operation of natural gas pipelines in onshore and offshore fields.

11.2 Pipeline Design

Many factors must be considered in the design of long-distance gas pipelines. These include the nature and volume of the gas to be transmitted, the length of the line, the type of terrain to be crossed, and maximum elevation of the route. After a gas compression station is located and sized, the gathering system is designed. This involves the location of the wells, the ability of right of way, the amount of gas to be handled, the distance to be transported, and the pressure difference between the field and the main transmission line. The gas wells are generally located in groups around a geological structure or within the defined limits of a pool or gas reservoir. In a new field, the gather system must be large enough to handle the production of additional leases. The gathering system is made up of branches that lead into trunk lines. The trunk line is small at the most distant well and, as more wells along the line are attached to it, the line must be larger to accommodate the greater volume of gas. In addition to the gathering system and major trunk pipelines, there is also a network of smaller-diameter feeder and transmission mains that may carry gas to consumption centers. In addition, complex systems of still smaller-diameter distribution piping run to individual homes, shops, and factories.

A successful design of a transmission system requires a team of well-trained and experienced engineering and legal staff. Complex engineering studies are needed to decide on diameter, yield strength, and pumping horsepower required to give the optimum results for any particular pipeline transmission system. Computer programs that enable high-pressure gas transmission networks to be dynamically simulated on a digital computer have been developed and are used by gas pipeline operating companies and service companies. Several designs are usually made so that the economical one can be selected. The maximum capacity of a pipeline is limited by its initial parameters of construction. In general, the tendency is to use higher transmission pressures and strong materials of construction. For economic operation, it is important to preserve full pipeline utilization.

11.2.1 Sizing Pipelines

The capacity of gas transmission of a pipeline is controlled mainly by its size. Complex equations have been developed for sizing natural gas pipelines in various flow conditions. The Weymouth equation, the Panhandle equation, and the Modified-Panhandle equation are used for relating the volume transmitted through a gas pipeline to the various factors involved, thus deciding the optimum pressure and pipe dimensions. From equations of this type, various combinations of pipe diameter and wall thickness for a desired rate of gas throughout can be calculated. An optimum balance is sought between pipe tonnage and pumping horsepower.

11.2.1.1 Definition of Friction Factor

In addition to the size of a pipeline, the pressure and capacity of gas transmission of a pipeline are primarily limited by the resistance to flow from the pipe wall. Flow of natural gas in pipelines always results in some mechanical energy being converted into heat. The so-called lost work, represents all energy losses resulting from irreversibilities of the flowing stream. In the case of single-phase gas flow in a pipe, these irreversibilities consist primarily of friction losses: internal losses due to viscosity effects and losses due to the roughness of the inner wall of the pipeline.

Under turbulent flow conditions that always exist in natural gas transmission pipelines, nature does not allow the energy losses of actual systems to be predicted theoretically; they must be determined by actual experiment and then correlated as some function of the flow variables. The lost work is usually calculated using a friction factor by dimensional analysis. It can be shown that the friction factor is a function of the Reynolds number and of the relative roughness of pipe.

The first law of thermal dynamics (conservation of energy) is the theoretical basis of most fluid flow equations. Under steady-state fluid and heat flow conditions, the conservation of energy gives the expression for the pressure gradient due to viscous shear or frictional losses as (Ikoku 1984; Katz and Lee 1990):

$$\left(\frac{dp}{dL}\right)_f \equiv \rho \frac{d(lw)}{dL} \tag{11.1}$$

where

p = pressure, lb_f/ft^2

ρ = fluid density, lb_m/ft^3

lw = mechanical energy (loss of work) converted to heat, ft-lb_f/lb_m

L = pipe length, ft

The equation that relates lost work per unit length of pipe and the flow variables is (Ikoku 1984; Katz and Lee, 1990):

$$\frac{d(lw)}{dL} = \frac{fu^2}{2g_c D} \tag{11.2}$$

where

u = flow velocity, ft/sec

g_c = gravitational conversion factor = 32.17 lb_m ft/lb_f sec^2

D = pipe diameter, ft

f = Moody friction factor

Substituting Equation (11.2) into Equation (11.1) gives:

$$\left(\frac{dp}{dL}\right)_f = \frac{f\rho u^2}{2g_c D} \tag{11.3}$$

which is a differential equation governing frictional pressure drop in a pipe. It must be noted that fluid density and velocity are functions of local pressure. A similar equation, using the Fanning friction factor, f', is

$$\left(\frac{dp}{dL}\right)_f = \frac{2f'\rho u^2}{g_c D} \tag{11.4}$$

11.2.1.2 Reynolds Number

The Reynolds number (N_{Re}) is defined as the ratio of fluid momentum force to viscous shear force. The Reynolds number can be expressed as a dimensionless group defined as

$$N_{Re} = \frac{Du\rho}{\mu} \tag{11.5}$$

where

D = pipe ID, ft

u = fluid velocity, ft/sec

ρ = fluid density, lb_m/ft^3

μ = fluid viscosity, lb_m/ft-sec

The Reynolds number can be used as a parameter to distinguish between laminar and turbulent fluid flow. The change from laminar to turbulent flow is usually assumed to occur at a Reynolds number of 2,100 for flow in a circular pipe. If U.S. field units of ft for diameter, ft/sec for velocity, lb_m/ft^3 for density and centipoises for viscosity are used, the Reynolds number equation becomes

$$N_{Re} = 1,488\frac{Du\rho}{\mu} \tag{11.6}$$

If a gas of specific gravity γ_g and viscosity μ (cp) is flowing in a pipe with an inner diameter D (in) at flow rate q (Mcfd) measured at base conditions of T_b (°R) and p_b (psia), the Reynolds number can be expressed as:

$$N_{Re} = \frac{711 p_b q \gamma_g}{T_b D \mu} \quad \textbf{(11.7)}$$

As T_b is 520 °R and p_b varies only from 14.4 psia to 15.025 psia in the United States, the value of $711p_b/T_b$ varies between 19.69 and 20.54. For all practical purposes, the Reynolds number for natural gas flow problems may be expressed as

$$N_{Re} = \frac{20 q \gamma_g}{\mu D} \quad \textbf{(11.8)}$$

where

q = gas flow rate at 60 °F and 14.73 psia, Mcfd

γ_g = gas-specific gravity (air = 1)

μ = gas viscosity at in-situ temperature and pressure, cp

D = pipe diameter, in

11.2.1.3 Relative Roughness

The frictional losses of fluid energy and pressure depend on the roughness of the inside wall of a pipe. Wall roughness is a function of pipe material, method of manufacture, and the environment to which it has been exposed. From a microscopic sense, wall roughness is not uniform, and thus the distance from the peaks to valleys on the wall surface will vary greatly. The absolute roughness, ε, of a pipe wall is defined as the mean protruding height of relatively uniformly distributed and sized, tightly packed sand grains that would give the same pressure gradient behavior as the actual pipe wall.

Analysis has suggested that the effect of roughness is not due to its absolute dimensions, but to its dimensions relative to the inside diameter of the pipe. Relative roughness, e_D, is defined as the ratio of the absolute roughness to the pipe internal diameter:

$$e_D = \frac{\varepsilon}{D} \tag{11.9}$$

where ε and D have the same unit.

The absolute roughness is not a directly measurable property for a pipe, which makes the selection of value of pipe wall roughness difficult. The way to evaluate the absolute roughness is to compare the pressure gradients obtained from the pipe of interest with a pipe that is sand-roughened. If measured pressure gradients are available, the friction factor and Reynolds number can be calculated and an effective e_D obtained from the Moody diagram. This value of e_D should then be used for future predictions until updated. If no information is available on roughness, a value of $\varepsilon = 0.0006$ inches is recommended for tubing and line pipes.

11.2.1.4 Equations for Friction Factor

Fluid flow ranges in nature between two extremes: laminar or streamline flow and turbulent flow. Within this range are four distinct regions: laminar, critical, transition, and turbulent flow. Figure 11–4 is a Moody friction factor chart covering the full range of flow conditions. It is a log-log graph of (log f) versus (log N_{Re}). Due to the characteristics of the complex nature of the curves, the equation for the friction factor in terms of the Reynolds number and relative roughness varies for each of the four regions.

11.2.1.4.1 Laminar Single-Phase Flow

The friction factor for laminar flow can be determined analytically. The Hagen-Poiseuille equation for laminar flow is

$$\left(\frac{dp}{dL}\right)_f = \frac{32\mu u}{g_c D^2} \tag{11.10}$$

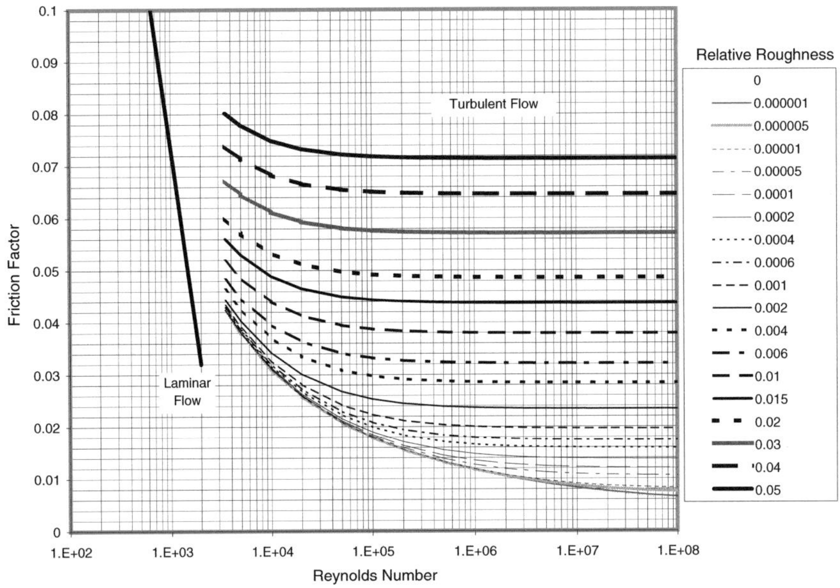

Figure 11–1 Moody friction factor chart (Moody 1944).

Equating the frictional pressure gradients given by Equation (11.3) and Equation (11.10) gives

$$\frac{f\rho u^2}{2g_c D} = \frac{32\mu u}{g_c D^2} \tag{11.11}$$

which yields

$$f = \frac{64\mu}{du\rho} = \frac{64}{N_{Re}} \tag{11.12}$$

11.2.1.4.2 Turbulent Single-Phase Flow

Studies of turbulent flow have shown that the velocity profile and pressure gradient are very sensitive to the characteristics of the pipe wall, that is, the roughness of the wall. Although a number of empirical correlations for friction factors are available, only the most accurate ones are presented.

For smooth wall pipes in the turbulent flow region, Drew, Koo, and McAdams (1930) presented the most commonly used correlation:

$$f = 0.0056 + \frac{0.5}{N_{Re}^{\ 0.32}} \tag{11.13}$$

which is valid over a wide range of Reynolds numbers,

$$3 \times 10^3 < N_{Re} < 3 \times 10^6$$

For rough wall pipes in the turbulent flow region, the effect of wall roughness on friction factor depends on the relative roughness and Reynolds number. Nikuradse (1933) friction factor correlation is still the best one available for fully developed turbulent flow in rough pipes:

$$\frac{1}{\sqrt{f}} = 1.74 - 2\log(2e_D) \tag{11.14}$$

This equation is valid for large values of the Reynolds number where the effect of relative roughness is dominant. Guo and Ghalambor (2002) showed that the Nikuradse friction factor correlation is also valid for gas flows with solid particles and liquid mist.

The correlation that is used as the basis for modern friction factor charts was proposed by Colebrook (1938):

$$\frac{1}{\sqrt{f}} = 1.74 - 2\log\left(2e_D + \frac{18.7}{N_{Re}\sqrt{f}}\right) \tag{11.15}$$

which is applicable to smooth pipes and to flow in transition and fully rough zones of turbulent flow. It degenerates to the Nikuradse correlation at large values of the Reynolds number. Equation (11.15) is not explicit in f. However, values of f can be obtained by an numerical procedure such as Newton-Raphson iteration. An explicit correlation for friction factor was presented by Jain (1976):

$$\frac{1}{\sqrt{f}} = 1.14 - 2\log\left(e_D + \frac{21.25}{N_{Re}^{\ 0.9}}\right) \tag{11.16}$$

This correlation is comparable to the Colebrook correlation. For relative roughness between 10^{-6} and 10^{-2} and the Reynolds number between 5×10^3 and 10^8, the errors were within ±1.0% when compared with the Colebrook correlation. Therefore, Equation (11.16) is recommended for all calculations requiring friction factor determination of turbulent flow.

11.2.1.5 Theoretical Pipeline Equations

Designing a long-distance pipeline for transportation of natural gas requires knowledge of flow formulas for calculating capacity and pressure requirements. There are several equations in the petroleum industry for calculating the flow of gases in pipelines. In the early development of the natural gas transmission industry, pressures were low and the equations used for design purposes were simple and adequate. However, as pressure increased to meet higher capacity demands, more complicated equations were developed to meet the new requirements. Probably the most common pipeline flow equation is the Weymouth equation, which is generally preferred for smaller-diameter lines ($D \leq 15$ in). The Panhandle equation and the Modified Panhandle equation are usually better for larger-sized transmission lines.

Based on the first law of thermal dynamics, the total pressure gradient is made up of three distinct components:

$$\frac{dp}{dL} = \frac{g}{g_c}\rho \sin\theta + \frac{f\rho u^2}{2g_c D} + \frac{\rho u du}{g_c dL} \tag{11.17}$$

where $\frac{g}{g_c}\rho \sin\theta$ is the component due to elevation or potential energy change; $\frac{f\rho u^2}{2g_c D}$ is the component due to frictional losses; and $\frac{\rho u du}{g_c dL}$ is the component due to acceleration or kinetic energy change.

The elevation component is pipe angle dependent. It is zero for horizontal flow. The friction loss component applies to any type of flow at any pipe angle and causes a pressure drop in the direction of flow. The acceleration component causes a pressure drop in the direction of velocity increase in any flow condition in which velocity changes occurs. It is zero for constant-area, incompressible flow.

Equation (11.17) applies for any fluid in steady-state, one-dimensional flow for which ρ, f, and u can be defined. It is in differential equation form and would have to be integrated to yield pressure drop as a function of flow rate, pipe diameter, and fluid properties.

Consider steady-state flow of dry gas in a constant-diameter, horizontal pipeline. The mechanical energy equation, Equation (11.3), becomes

$$\frac{dp}{dL} = \frac{f \rho u^2}{2 g_c D} = \frac{p(MW)_a}{zRT} \frac{f u^2}{2 g_c D} \tag{11.18}$$

which serves as a base for development of many pipeline equations. The difference in these equations originated from the methods used in handling the z-factor and friction factor. Integrating Equation (11.18) gives

$$\int dp = \frac{(MW)_a f u^2}{2 R g_c D} \int \frac{p}{zT} dL \tag{11.19}$$

If temperature is assumed to be constant at average value in a pipeline, $\bar{T}$, and gas deviation factor, $\bar{z}$, is evaluated at average temperature and average pressure, $\bar{p}$, Equation (11.19) can be evaluated over a distance L between upstream pressure, p_1, and downstream pressure, p_2:

$$p_1^2 - p_2^2 = \frac{25 \gamma_g q^2 \bar{T} \bar{z} f L}{D^5} \tag{11.20}$$

where

γ_g = gas gravity (air = 1)

q = gas flow rate, MMscfd (at 14.7 psia, 60 °F)

$\bar{T}$ = average temperature, °R

$\bar{z}$ = gas deviation factor at $\bar{T}$ and $\bar{p}$

$\bar{p} = (p_1 + p_2)/2$

L = pipe length, ft

D = pipe internal diameter, in.

f = Moody friction factor

Equation (11.20) may be written in terms of flow rate measured at arbitrary base conditions (T_b and p_b):

$$q = \frac{CT_b}{p_b}\sqrt{\frac{\left(p_1^2 - p_2^2\right)D^5}{\gamma_g \bar{T}\bar{z}fL}} \tag{11.21}$$

where C is a constant with a numerical value that depends on the units used in the pipeline equation. If L is in miles and q is in scfd, $C = 77.54$.

The use of Equation (11.21) involves an iterative procedure. The gas deviation factor depends on pressure and the friction factor depends on flow rate. This problem prompted several investigators to develop pipeline flow equations that are noniterative or explicit. This has involved substitutions for the friction factor f. The specific substitution used may be diameter-dependent only (Weymouth equation) or Reynolds number dependent only (Panhandle equations).

11.2.1.5.1 Weymouth Equation for Horizontal Flow

Equation (11.21) takes the following form when the unit of scfh for gas flow rate is used:

$$q_h = \frac{3.23T_b}{p_b}\sqrt{\frac{1}{f}}\sqrt{\frac{\left(p_1^2 - p_2^2\right)D^5}{\gamma_g \bar{T}\bar{z}L}} \tag{11.22}$$

where $\sqrt{\frac{1}{f}}$ is called transmission factor. The Moody friction factor may be a function of flow rate and pipe roughness. If flow conditions are in the fully turbulent region, Equation (11.16) degenerates to

$$f = \frac{1}{\left[1.14 - 2\log\left(e_D\right)\right]^2} \tag{11.23}$$

where f depends only on the relative roughness, e_D. When flow conditions are not completely turbulent, f depends on the Reynolds number also:

$$N_{Re} \approx \frac{0.48 q_h \gamma_g}{\mu D} \tag{11.24}$$

Therefore, use of Equation (11.22) and Equation (11.24) requires a trial-and-error procedure to calculate q_h. To eliminate the trial-and-error procedure, Weymouth proposed that f vary as a function of diameter in inches as follows:

$$f = \frac{0.032}{D^{1/3}} \tag{11.25}$$

With this simplification, Equation (11.22) reduces to

$$q_h = \frac{18.062 T_b}{p_b} \sqrt{\frac{\left(p_1^2 - p_2^2\right) D^{16/3}}{\gamma_g \bar{T} \bar{z} L}} \tag{11.26}$$

which is the form of the Weymouth equation commonly used in the natural gas industry.

The use of the Weymouth equation for an existing transmission line or for the design of a new transmission line involves a few assumptions including no mechanical work, steady flow, isothermal flow, constant compressibility factor, horizontal flow, and no kinetic energy change. These assumptions can affect accuracy of calculation results.

In the study of an existing pipeline, the pressure-measuring stations should be placed so that no mechanical energy is added to the system between stations. No mechanical work is done on the fluid between the points at which the pressures are measured. Thus, the condition of no mechanical work can be fulfilled.

Steady flow in pipeline operation seldom, if ever, exists in actual practice because pulsations, liquid in the pipeline, and variations in input or output gas volumes cause deviations from steady-state conditions. Deviations from steady-state flow are the major cause of difficulties experienced in pipeline flow studies.

The heat of compression is usually dissipated into the ground along a pipeline within a few miles downstream from the compressor station. Otherwise, the temperature of the gas is very near that of the containing pipe and, as pipelines usually are buried, the temperature of the flowing gas is not influenced appreciably by rapid changes in atmospheric temperature. Therefore, the gas flow can be considered isothermal at an average effective temperature without causing significant error in long-pipeline calculations.

The compressibility of the fluid can be considered constant and an average effective gas deviation factor may be used. When the two pressures p_1 and p_2 lie in a region where z is essentially linear with pressure, then it is accurate enough to evaluate $\overline{z}$ at the average pressure $\overline{p} = (p_1 + p_2)/2$. One can also use the arithmetic average of the z with $\overline{z} = (z_1 + z_2)/2$ where z_1 and z_2 are obtained at p_1 and p_2, respectively. On the other hand, should p_1 and p_2 lie in the range where z is not linear with pressure (double-hatched lines), the proper average would result from determining the area under the z-curve and dividing it by the difference in pressure:

$$\overline{z} = \frac{\int_{p_1}^{p_2} z dp}{(p_1 - p_2)} \tag{11.27}$$

where the numerator can be evaluated numerically. Also, $\overline{z}$ can be evaluated at an average pressure given by

$$\overline{p} = \frac{2}{3}\left(\frac{p_1^{\ 3} - p_2^{\ 3}}{p_1^{\ 2} - p_2^{\ 2}}\right) \tag{11.28}$$

Regarding the assumption of horizontal pipeline, in actual practice, transmission lines seldom, if ever, are horizontal, so that factors are needed in Equation (11.26) to compensate for changes in elevation. With the trend to higher operating pressures in transmission lines, the need for these factors is greater than is generally realized. This issue of correction for change in elevation is addressed in the next section.

If the pipeline is long enough, the changes in the kinetic-energy term can be neglected. The assumption is justified for work with commercial transmission lines.

Example Problem 11.1

For the following data given for a horizontal pipeline, predict gas flow rate in cubic ft/hr through the pipeline.

D = 12.09 in

L = 200 mi

e = 0.0006 in

T = 80 °F

γ_g = 0.70

T_b = 520 °R

p_b = 14.7 psia

p_1 = 600 psia

p_2 = 200 psia

Solution

The problem can be solved using Equation (11.22) with the trial-and-error method for friction factor and the Weymouth equation without the Reynolds number-dependent friction factor.

The average pressure is:

$$\bar{p} = (200 + 600)/2 = 400 \text{ psia}$$

With $\bar{p}$ = 400 psia, T = 540 °R, and γ_g = 0.70, Brill-Beggs-Z.xls gives:

$$\bar{z} = 0.9188$$

With $\bar{p}$ = 400 psia, T = 540 °R, and γ_g = 0.70, Carr-Kobayashi-Burrows Viscosity.xls gives:

$$\mu = 0.0099 \text{ cp}$$

Relative roughness:

$$e_D = 0.0006/12.09 = 0.00005$$

A. Trial-and-Error Calculation:

First trial:

$$q_h = 500{,}000 \text{ cfh}$$

$$N_{Re} = \frac{0.48(500{,}000)(0.7)}{(0.0099)(12.09)} = 1{,}403{,}733$$

$$\frac{1}{\sqrt{f}} = 1.14 - 2\log\left(0.00005 + \frac{21.25}{(1{,}403{,}733)^{0.9}}\right)$$

$$f = 0.01223$$

$$q_h = \frac{3.23(520)}{14.7}\sqrt{\frac{1}{0.01223}}\sqrt{\frac{\left(600^2 - 200^2\right)(12.09)^5}{(0.7)(540)(0.9188)(200)}}$$

$$q_h = 1{,}148{,}450 \text{ cfh}$$

Second trial:

$$q_h = 1{,}148{,}450 \text{ cfh}$$

$$N_{Re} = \frac{0.48(1{,}148{,}450)(0.7)}{(0.0099)(12.09)} = 3{,}224{,}234$$

$$\frac{1}{\sqrt{f}} = 1.14 - 2\log\left(0.00005 + \frac{21.25}{(3{,}224{,}234)^{0.9}}\right)$$

$$f = 0.01145$$

$$q_h = \frac{3.23(520)}{14.7}\sqrt{\frac{1}{0.01145}}\sqrt{\frac{\left(600^2 - 200^2\right)(12.09)^5}{(0.7)(540)(0.9188)(200)}}$$

$$q_h = 1{,}186{,}759 \text{ cfh}$$

Third trial:

$$q_h = 1{,}186{,}759 \text{ cfh}$$

$$N_{Re} = \frac{0.48(1{,}186{,}759)(0.7)}{(0.0099)(12.09)} = 3{,}331{,}786$$

$$\frac{1}{\sqrt{f}} = 1.14 - 2\log\left(0.00005 + \frac{21.25}{(3{,}331{,}786)^{0.9}}\right)$$

$$f = 0.01143$$

$$q_h = \frac{3.23(520)}{14.7}\sqrt{\frac{1}{0.01143}}\sqrt{\frac{\left(600^2 - 200^2\right)(12.09)^5}{(0.7)(540)(0.9188)(200)}}$$

$$q_h = 1{,}187{,}962 \text{ cfh}$$

which is close to the assumed 1,186,759 cfh

B. Using the Weymouth equation:

$$q_h = \frac{18.062(520)}{14.7}\sqrt{\frac{\left(600^2 - 200^2\right)(12.09)^{16/3}}{(0.7)(540)(0.9188)(200)}}$$

$$q_h = 1{,}076{,}035 \text{ cfh}$$

To speed up trial-and-error calculations, a spreadsheet program, PipeCapacity.xls, was developed. The solution given by the spreadsheet is shown in Table 11–1.

11.2.1.5.2 Weymouth Equation for Nonhorizontal Flow

Gas transmission lines are often nonhorizontal. Account should be taken of substantial pipeline elevation changes. Considering gas flow from

Table 11-1 Input Data and Results Given by PipeCapacity.xls[a]

Instructions: 1) Update input data; 2) Run Macro Solution and view results.	
Input Data	
Pipe ID:	12.09 in
Pipe roughness:	0.0006 in
Pipeline length:	200 mi
Average temperature:	80 °F
Base temperature:	60 °F
Base pressure:	14.7 psia
Inlet pressure:	600 psia
Outlet pressure:	200 psia
Gas properties:	
Gas-specific gravity:	0.7 air = 1
Mole fraction of N_2:	0
Mole fraction of CO_2:	0
Mole fraction of H_2S:	0
Calculated Parameter Values	
Pseudocritical pressure:	
Pseudocritical temperature:	389.14 °R
Uncorrected gas viscosity at 14.7 psia:	0.010079 cp
N_2 correction for gas viscosity at 14.7 psia:	0.000000 cp
CO_2 correction for gas viscosity at 14.7 psia:	0.000000 cp
H_2S correction for gas viscosity at 14.7 psia:	0.000000 cp
Corrected gas viscosity at 14.7 psia (μ1):	0.010079 cp
Gas viscosity:	0.009899 cp
Pseudoreduced temperature:	1.34
Average pressure (psia):	400 psia
Pseudoreduced pressure:	0.599
Average z-factor:	0.9188
Pipe relative roughness:	0.000050
A) Use Reynolds number dependent friction factor:	
Pipeline flow capacity:	1,188,000 cfh
Reynolds number:	3,335,270
Friction factor:	0.01143
Objective function:	0
B) Use the Weymouth equation:	
Pipeline flow capacity:	1,076,035 cfh

a. This program computes the capacity of gas pipelines (see Example Problem 11.1).

point 1 to point 2 in a nonhorizontal pipe, the first law of thermal dynamics gives:

$$\int_1^2 v dp + \frac{g}{g_c}\Delta z + \int_1^2 \frac{fu^2}{2g_c D} dL = 0 \quad \textbf{(11.29)}$$

Based on the pressure gradient due to the weight of gas column

$$\frac{dp}{dz} = \frac{\rho_g}{144} \quad \textbf{(11.30)}$$

and real gas law, $\rho_g = \frac{p(MW)_a}{zRT} = \frac{29\gamma_g p}{zRT}$, Weymouth (1912) developed the following equation:

$$q_h = \frac{3.23T_b}{p_b}\sqrt{\frac{\left(p_1^2 - e^s p_2^2\right)D^5}{f\gamma_g \bar{T}\bar{z}L}} \quad \textbf{(11.31)}$$

where e = 2.718 and

$$s = \frac{0.0375\gamma_g \Delta z}{\bar{T}\bar{z}} \quad \textbf{(11.32)}$$

and Δz is equal to outlet elevation minus inlet elevation (note that Δz is positive when outlet is higher than inlet). A general and more rigorous form of the Weymouth equation with compensation for elevation is

$$q_h = \frac{3.23T_b}{p_b}\sqrt{\frac{\left(p_1^2 - e^s p_2^2\right)D^5}{f\gamma_g \bar{T}\bar{z}L_e}} \quad \textbf{(11.33)}$$

where L_e is the effective length of the pipeline. For a uniform slope, L_e is defined as

$$L_e = \frac{(e^s - 1)L}{s} \tag{11.34}$$

For a nonuniform slope (where elevation change cannot be simplified to a single section of constant gradient), an approach in steps to any number of sections, n, will yield

$$L_e = \frac{(e^{s_1} - 1)}{s_1} L_1 + \frac{e^{s_1}(e^{s_2} - 1)}{s_2} L_2 + \frac{e^{s_1 + s_2}(e^{s_3} - 1)}{s_3} L_3 + \ldots\ldots + \sum_{i=1}^{n} \frac{e^{\sum_{j=1}^{i-1} s_j} (e^{s_i} - 1)}{s_i} L_i \tag{11.35}$$

where

$$s_i = \frac{0.0375\gamma_g \Delta z_i}{\bar{T}\bar{z}} \tag{11.36}$$

11.2.1.5.3 Panhandle A Equation—Horizontal Flow

The Panhandle A pipeline flow equation assumes the following Reynolds number dependent friction factor:

$$f = \frac{0.085}{N_{\mathrm{Re}}{}^{0.147}} \tag{11.37}$$

The resultant pipeline flow equation is thus

$$q = 435.87 \frac{D^{2.6182}}{\gamma_g^{0.4604}} \left(\frac{T_b}{p_b} \right)^{1.07881} \left[\frac{\left(p_1^2 - p_2^2\right)}{\bar{T}\bar{z}L} \right]^{0.5394} \tag{11.38}$$

where q is the gas flow rate in cfd measured at T_b and p_b, and other terms are the same as in the Weymouth equation.

11.2.1.5.4 Panhandle B Equation—Horizontal Flow (Modified Panhandle)

Panhandle B equation is the most widely used equation for long transmission and delivery lines. It assumes that f varies as:

$$f = \frac{0.015}{N_{\text{Re}}^{0.0392}} \tag{11.39}$$

and it takes the following resultant form:

$$q = 737D^{2.530}\left(\frac{T_b}{p_b}\right)^{1.02}\left[\frac{\left(p_1^2 - p_2^2\right)}{\bar{T}\bar{z}L\gamma_g^{0.961}}\right]^{0.510} \tag{11.40}$$

11.2.1.5.5 Clinedinst Equation—Horizontal Flow

The Clinedinst equation rigorously considers the deviation of natural gas from ideal gas through integration. It takes the following form:

$$q = 3973.0\frac{z_b T_b p_{pc}}{p_b}\sqrt{\frac{D^5}{\bar{T} fL\gamma_g}\left(\int_0^{p_{r1}} \frac{p_r}{z}\,dp_r - \int_0^{p_{r2}} \frac{p_r}{z}\,dp_r\right)} \tag{11.41}$$

where

q = volumetric flow rate, Mcfd

p_{pc} = pseudocritical pressure, psia

D = pipe internal diameter, in

L = pipe length, ft

p_r = pseudoreduced pressure

$\bar{T}$ = average flowing temperature, °R

γ_g = gas gravity, air = 1.0

z_b = gas deviation factor at T_b and p_b, normally accepted as 1.0

Based on Equation (2.5) for pseudocritical pressure (Wichert and Aziz 1972), the values of the integral function $\int_0^{p_r} \frac{p_r}{z} dp_r$ have been calculated for various gas-specific gravity values. The results are presented in Appendix B.

11.2.1.6 Practical Pipeline Equations

11.2.1.6.1 Pipeline Efficiency

All pipeline flow equations were developed for perfectly clean lines filled with gas. In actual pipelines, water, condensates, and sometimes crude oil accumulate in low spots in the line. There are often scales and even "junk" left in the line. The net result is that the flow rates calculated for the 100 percent efficient cases are often modified by multiplying them by an efficiency factor *E*. The efficiency factor expresses the actual flow capacity as a fraction of the theoretical flow rate. An efficiency factor ranging from 0.85 to 0.95 would represent a "clean" line. Table 11–2 presents typical values of efficiency factors.

Table 11–2 Typical Values of Efficiency Factors

Type of Line	Liquid Content (gal/MMcf)	Efficiency *E*
Dry-gas field	0.1	0.92
Casing-head gas	7.2	0.77
Gas and condensate	800	0.6

11.2.1.6.2 Transmission Factor

In addition to the pipeline efficiency E, the transmission factor $\sqrt{\frac{1}{f}}$ in Equation (11.22) is used for further tuning the theoretical pipeline flow equations. The transmission factor has long been the most difficult to evaluate. Thus, the literature contains many different empiricaltransmission factors that have been used to meet the needs of pipeline engineers. Table 11–3 presents some transmission factors that are the most significant and have either best stood the test of usage or have strong foundations in basic flow theories.

Table 11–3 Transmission Factors for Pipeline Flow Equations

Flow Equation	Transmission Factor
Smooth pipe (laminar flow)	$2\log\left(\sqrt{f}N_{\mathrm{Re}}\right)+0.3$
Weymouth	$1.10\times5.6D^{0.167}$
Panhandle A	$0.92\times3.44N_{\mathrm{Re}}^{0.073}$
Panhandle B	$0.90\times8.25N_{\mathrm{Re}}^{0.0196}$
Rough pipe (fully turbulent flow)	$2\log\left(\frac{3.7}{e_D}\right)$

11.2.1.6.3 Empirical Pipeline Equation

A general noniterative pipeline flow equation is written as

$$q = a_1 E \left(\frac{T_b}{p_b} \right)^{a_2} \left[\frac{\left(p_1^2 - p_2^2 \right)}{\bar{T} \, \bar{z} \, L} \right]^{a_3} \left(\frac{1}{\gamma_g} \right)^{a_4} D^{a_5} \qquad \textbf{(11.42)}$$

where the units are q in cfd measured at T_b and p_b, $\bar{T}$ in °R, p in psia, L in miles, and D in inches. The values of the constants are given in Table 11–4 for the different pipeline flow equations.

Table 11–4 Constants for Empirical Pipeline Flow Equations

Equation	a_1	a_2	a_3	a_4	a_5
Weymouth	433.5	1	0.5	0.5	2.667
Panhandle A	435.87	1.0788	0.5394	0.4604	2.618
Panhandle B	737	1.02	0.51	0.49	2.53

11.2.1.7 Series, Parallel, and Looped Pipelines

It is often desirable to increase the throughput of an existing pipeline by gathering gas from new gas wells. A similar type of problem may arise when an existing pipeline must be "pressure derated" because of age (corrosion, etc.) but this pipelinc is dcsircd to maintain the same throughput. A common economical solution to these problems is to place one or more lines in parallel, either partially or throughout the whole length, or to replace a portion of the line with a larger one. This requires calculations involving flow in series, parallel, and series-parallel (looped) lines. The philosophy involved in deriving the special relationships used in the solution of complex transmission systems is to express the various lengths and diameters of the pipe in the systems as equivalent lengths of common diameter or equivalent diameter of a common length, there equivalent means that both lines will have the same capacity with the same totally pressure drop. For simplicity, illustrative examples will be based on the Weymouth equation.

11.2.1.7.1 Pipelines in Series

Consider a three-segment pipeline in a series of total length L depicted in Figure 11–2(a). Applying the Weymouth equation to each of the three segments gives:

$$p_1^2 - p_2^2 = \frac{\gamma_g \bar{T}\bar{z}L_1}{D_1^{16/3}}\left(\frac{q_h p_b}{18.062T_b}\right)^2 \tag{11.43}$$

$$p_2^2 - p_3^2 = \frac{\gamma_g \bar{T}\bar{z}L_2}{D_2^{16/3}}\left(\frac{q_h p_b}{18.062T_b}\right)^2 \tag{11.44}$$

$$p_3^2 - p_4^2 = \frac{\gamma_g \bar{T}\bar{z}L_3}{D_3^{16/3}}\left(\frac{q_h p_b}{18.062T_b}\right)^2 \tag{11.45}$$

Adding equations (11.43), (11.44), and (11.45) gives:

$$p_1^2 - p_4^2 = \gamma_g \bar{T}\bar{z}\left(\frac{L_1}{D_1^{16/3}} + \frac{L_2}{D_2^{16/3}} + \frac{L_3}{D_3^{16/3}}\right)\left(\frac{q_h p_b}{18.062T_b}\right)^2 \tag{11.46}$$

or

$$q_h = \frac{18.062T_b}{p_b}\sqrt{\frac{p_1^2 - p_4^2}{\gamma_g \bar{T}\bar{z}\left(\frac{L_1}{D_1^{16/3}} + \frac{L_2}{D_2^{16/3}} + \frac{L_3}{D_3^{16/3}}\right)}} \tag{11.47}$$

Capacity of a single-diameter (D_1) pipeline is expressed as:

$$q_1 = \frac{18.062T_b}{p_b}\sqrt{\frac{p_1^2 - p_4^2}{\gamma_g \bar{T}\bar{z}\left(\frac{L}{D_1^{16/3}}\right)}} \tag{11.48}$$

Dividing Equation (11.47) by Equation (11.49) yields:

$$\frac{q_t}{q_1} = \sqrt{\frac{\left(\frac{L}{D_1^{16/3}}\right)}{\left(\frac{L_1}{D_1^{16/3}} + \frac{L_2}{D_2^{16/3}} + \frac{L_3}{D_3^{16/3}}\right)}} \tag{11.49}$$

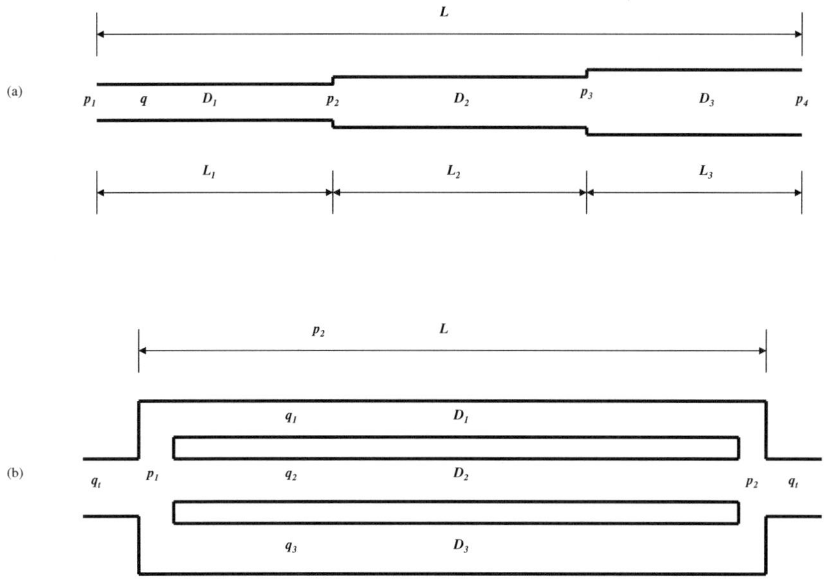

Figure 11–2 Sketch of series pipeline.

11.2.1.7.2 Pipelines in Parallel

Consider a three-segment pipeline in parallel as depicted in Figure 11–2(b). Applying the Weymouth equation to each of the three segments gives:

$$q_1 = 18.062 \frac{T_b}{p_b} \sqrt{\frac{(p_1^2 - p_2^2) D_1^{16/3}}{\gamma_g \bar{T} \bar{z} L}} \tag{11.50}$$

$$q_2 = 18.062 \frac{T_b}{p_b} \sqrt{\frac{(p_1^2 - p_2^2) D_2^{16/3}}{\gamma_g \bar{T} \bar{z} L}} \tag{11.51}$$

$$q_3 = 18.062 \frac{T_b}{p_b} \sqrt{\frac{(p_1^2 - p_2^2) D_3^{16/3}}{\gamma_g \bar{T} \bar{z} L}} \tag{11.52}$$

Adding equations (11.50), (11.51), and (11.52) gives:

$$q_t = q_1 + q_2 + q_3 = 18.062 \frac{T_b}{p_b} \sqrt{\frac{(p_1^2 - p_2^2)}{\gamma_g \bar{T} \bar{z} L}} \left(\sqrt{D_1^{16/3}} + \sqrt{D_2^{16/3}} + \sqrt{D_3^{16/3}} \right) \tag{11.53}$$

Dividing Equation (11.53) by Equation (11.50) yields:

$$\frac{q_t}{q_1} = \frac{\sqrt{D_1^{16/3}} + \sqrt{D_2^{16/3}} + \sqrt{D_3^{16/3}}}{\sqrt{D_1^{16/3}}} \tag{11.54}$$

11.2.1.7.3 Looped Pipelines

Consider a three-segment looped pipeline depicted in Figure 11–3. Applying Equation (11.53) to the first two (parallel) segments gives:

$$q_t = q_1 + q_2 = 18.062 \frac{T_b}{p_b} \sqrt{\frac{(p_1^2 - p_3^2)}{\gamma_g \bar{T} \bar{z} L_1}} \left(\sqrt{D_1^{16/3}} + \sqrt{D_2^{16/3}} \right) \tag{11.55}$$

or

$$p_1^2 - p_3^2 = \frac{\gamma_g \bar{T} \bar{z} L_1}{\left(\sqrt{D_1^{16/3}} + \sqrt{D_2^{16/3}} \right)^2} \left(\frac{q_t p_b}{18.062 T_b} \right)^2 \tag{11.56}$$

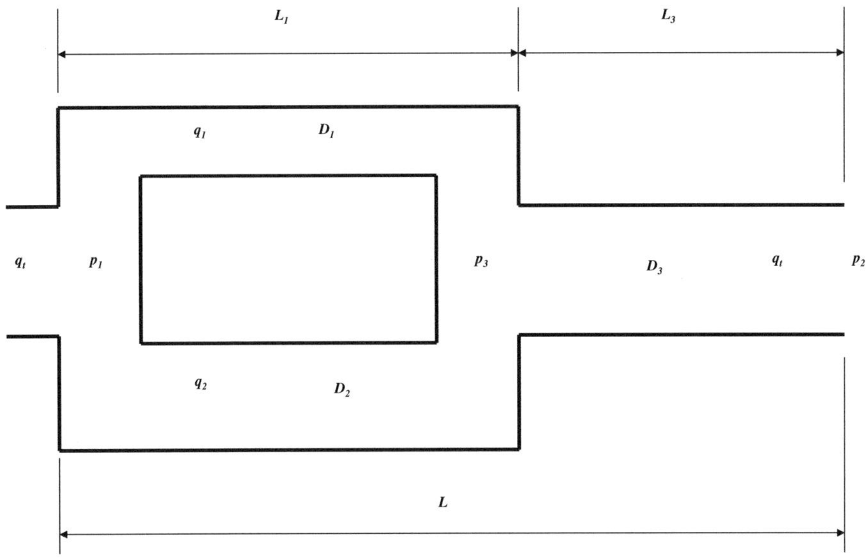

Figure 11–3 Sketch of looped pipeline.

Applying the Weymouth equation to the third segment (with diameter D_3) yields:

$$p_3^2 - p_2^2 = \frac{\gamma_g \bar{T}\bar{z}L_3}{D_3^{16/3}}\left(\frac{q_t p_b}{18.062T_b}\right)^2 \quad \textbf{(11.57)}$$

Adding equations (11.56) and (11.57) results in:

$$p_1^2 - p_2^2 = \gamma_g \bar{T}\bar{z}\left(\frac{q_t p_b}{18.062T_b}\right)^2 \left(\frac{L_1}{\left(\sqrt{D_1^{16/3}} + \sqrt{D_2^{16/3}}\right)^2} + \frac{L_3}{D_3^{16/3}}\right) \quad \textbf{(11.58)}$$

or

$$q_t = \frac{18.062T_b}{p_b}\sqrt{\frac{\left(p_1^2 - p_2^2\right)}{\gamma_g \bar{T}\bar{z}\left(\frac{L_1}{\left(\sqrt{D_1^{16/3}} + \sqrt{D_2^{16/3}}\right)^2} + \frac{L_3}{D_3^{16/3}}\right)}} \quad \textbf{(11.59)}$$

Capacity of a single-diameter (D_3) pipeline is expressed as:

$$q_3 = \frac{18.062T_b}{p_b}\sqrt{\frac{p_1^2 - p_2^2}{\gamma_g \bar{T}\bar{z}\left(\frac{L}{D_3^{16/3}}\right)}} \tag{11.60}$$

Dividing Equation (11.59) by Equation (11.60) yields:

$$\frac{q_t}{q_3} = \sqrt{\frac{\left(\frac{L}{D_3^{16/3}}\right)}{\left(\frac{L_1}{\left(\sqrt{D_1^{16/3}} + \sqrt{D_2^{16/3}}\right)^2} + \frac{L_3}{D_3^{16/3}}\right)}} \tag{11.61}$$

Let Y be the fraction of looped pipeline and X be the increase in gas capacity, Equation (11.61) can be rearranged as:

$$Y = \frac{1 - \frac{1}{(1+X)^2}}{1 - \frac{1}{\left(1 + R_D^{2.31}\right)^2}} \tag{11.62}$$

where R_D is ratio of the looping pipe diameter to the original pipe diameter. Equation (11.62) can be rearranged to solve for X explicitly:

$$X = \frac{1}{\sqrt{1 - Y\left(1 - \frac{1}{\left(1 + R_D^{2.31}\right)^2}\right)}} - 1 \tag{11.63}$$

The effects of looped line on the increase of gas flow rate for various pipe diameter ratios are shown in Figure 11–4.

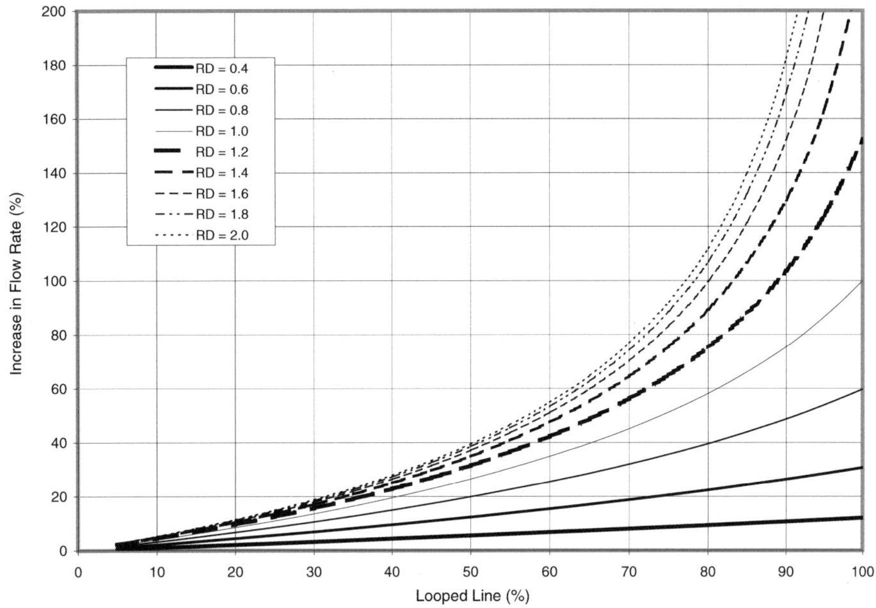

Figure 11–4 Effects of looped line and pipe diameter ratio on the increase of gas flow rate.

Example Problem 11.2

Consider a 4-in pipeline that is 10 miles long. Assuming that the compression and delivery pressures will maintain unchanged, calculate gas capacity increases by using the following measures of improvement: (a) Replace three miles of the 4-in pipeline by a 6-in pipeline segment; (b) Place a 6-in parallel pipeline to share gas transmission; and (c) Loop three miles of the 4-in pipeline with a 6-in pipeline segment.

Solution

(a) This problem can be solved with Equation (11.49).

$L = 10$ mi

$L_1 = 7$ mi

$L_2 = 3$ mi

$D_1 = 4$ in

$D_2 = 6$ in

$$\frac{q_t}{q_1} = \sqrt{\frac{\left(\frac{10}{4^{16/3}}\right)}{\left(\frac{7}{4^{16/3}} + \frac{3}{6^{16/3}}\right)}} = 1.1668, \text{ or } 16.68\% \text{ increase in flow}$$

capacity

(b) This problem can be solved with Equation (11.54).

$D_1 = 4$ in

$D_2 = 6$ in

$$\frac{q_t}{q_1} = \frac{\sqrt{4^{16/3}} + \sqrt{6^{16/3}}}{\sqrt{4^{16/3}}} = 3.9483, \text{ or } 294.83\% \text{ increase in flow}$$

capacity

(c) This problem can be solved with Equation (11.61).

$L = 10$ mi

$L_1 = 7$ mi

$L_2 = 3$ mi

$D_1 = 4$ in

$D_2 = 6$ in

$$\frac{q_t}{q_3} = \sqrt{\frac{\left(\frac{10}{4^{16/3}}\right)}{\left(\frac{L_1}{\left(\sqrt{4^{16/3}} + \sqrt{6^{16/3}}\right)^2} + \frac{L_3}{4^{16/3}}\right)}} = 1.1791, \text{ or } 17.91\%$$

increase in flow capacity

Similar problems can also be solved using the spreadsheet program LoopedLines.xls. Table 11–5 shows the solution of Example Problem 11.2 given by the spreadsheet.

Table 11–5 Input Data and Solution Given by LoopedLines.xls[(a)]

Input Data		
Original pipe ID:	4	in
Total pipeline length:	10	mi
Series pipe ID:	4 and 6	in
Segment lengths:	7 and 3	mi
Parallel pipe ID:	4 and 6	in
Looped pipe ID:	4, 6, and 4	in
Segment lengths:	3 and 7	mi
Solution		
Capacity improvement by series pipelines:		
$\frac{q_t}{q_1} = \sqrt{\frac{\left(\frac{L}{D_1^{16/3}}\right)}{\left(\frac{L_1}{D_1^{16/3}} + \frac{L_2}{D_2^{16/3}} + \frac{L_3}{D_3^{16/3}}\right)}}$	=	1.1668
Capacity improvement by parallel pipelines:		
$\frac{q_t}{q_1} = \frac{\sqrt{D_1^{16/3}} + \sqrt{D_2^{16/3}} + \sqrt{D_3^{16/3}}}{\sqrt{D_1^{16/3}}}$	=	3.9483
Capacity improvement by looped pipelines:		
$\frac{q_t}{q_3} = \sqrt{\frac{\left(\frac{L}{D_3^{16/3}}\right)}{\left(\frac{L_1}{\left(\sqrt{D_1^{16/3}} + \sqrt{D_2^{16/3}}\right)^2} + \frac{L_3}{D_3^{16/3}}\right)}}$	=	1.1791

a. This spreadsheet computes capacities of series, parallel, and looped pipelines (see Example Problem 11.2).

11.2.2 Pipeline Wall Thickness

Wall thickness design for steel pipelines is governed by U.S. Codes ASME/ANSI B32.8. Other codes such as Z187 (Canada), DnV (Norway), and IP6 (United Kingdom), have essentially the same requirements, but should be checked by the readers.

Except for large diameter pipes (over 30 in), material grade is usually taken as X-60 or X-65 (414 or 448 MPa) for high-pressure pipelines. Higher grades can be selected in special cases. Lower grades such as X-42, X-52, or X-56 can be selected for low-pressure, large diameter pipelines to reduce material cost, or in cases that require high ductility for improved impact resistance. Pipe types are: seamless, submerged arc welded (SAW or DSAW), electric resistance welded (ERW), and spiral welded.

Except in specific cases, only seamless or SAW pipe are to be used, with seamless being the preference for diameters of 12 inches or less. If ERW pipe is used, special inspection provisions such as full body ultrasonic testing are required. Spiral weld pipe is very unusual for gas pipelines and should be used only for low-pressure water or outfall lines.

11.2.2.1 Design Procedure

Determination of pipeline wall thickness is based on the design-internal pressure or the external hydrostatic pressure. Maximum longitudinal stresses and combined stresses are sometimes limited by applicable codes and shall be checked for installation and operation. However, these criteria are not normally used for wall thickness determination. Increasing the wall thickness can sometimes ensure hydrodynamic stability of subsea pipelines in lieu of other stabilization methods (e.g., weight coating). This is not normally economic except in deepwater where the presence of concrete may interfere with preferred installation method. Bai (2001) presents a Design Through Analysis (DTA) method for pipeline sizing. Guo et al. (2005) published a comprehensive description of pipeline design, installation, and operations for offshore and deepwater development. In this chapter we recommend the following procedure for designing pipeline wall thickness:

1 Calculate the minimum wall thickness required for the design-internal pressure.

2 Calculate the minimum wall thickness required to withstand external pressure.

3 Add wall thickness allowance for corrosion, if applicable, to the maximum of the preceeding.

4 Select next highest nominal wall thickness. Note: In certain cases, it may be desirable to order a nonstandard wall. This can be done for large orders.

5 Check selected wall thickness for hydrotest condition.

6 Check for handling practice, that is, pipeline handling is difficult for *D/t* larger than 50; welding of wall thickness less than 0.3 in (7.6 mm) requires special provisions.

11.2.2.2 Design Codes

11.2.2.2.1 Pipeline Design for Internal Pressure

Two pipeline codes typically used for design are ASME B31.8 (1989) and DnV (1981). ASME B31.8 is for all gas lines and two-phase flowlines in North America. DnV is for oil, gas, and two-phase flow pipelines in the North Sea. All these codes can be used in other areas when no other code is available.

The nominal pipeline wall thickness (t_{NOM}) can be calculated as follows:

$$t_{NOM} = \frac{P_d D}{2E_w \eta \sigma_y F_t} + t_a \tag{11.64}$$

where P_d is the design internal pressure defined as the difference between the internal pressure (P_i) and external pressure (P_e); D is nominal outside diameter; t_a is thickness allowance for corrosion; and, σ_y is the specified minimum yield strength.

Most codes allow credit for external pressure. This credit should be used whenever possible, although care should be exercised for oil export lines to account for head of fluid and for lines that traverse from deep to shallow water.

DnV 1981 defines P_i as the Maximum Allowable Operating Pressure (MAOP) under normal conditions, indicating that surge pressure up to 110% MAOP is acceptable. In some cases, P_i is defined as Wellhead Shut-In Pressure (WSIP) or specified by the operators.

In Equation (11.64), the weld efficiency factor (E_w) is 1.0 for seamless, ERW, and DSAW pipes. The temperature derating factor (F_t) is equal to 1.0 for temperatures under 250 °F. The usage factor (η) is defined in Table 11–6 for gas lines.

The under-thickness due to manufacturing tolerance is taken into account in the design factor. There is no need to add any allowance for fabrication to the wall thickness calculated with Equation (11.64).

Table 11–6 Design and Hydrostatic Pressure Definitions and Usage Factors for Gas Lines

Gas	ASME B31.8 1989 Edition 1990 Addendum	DnV 1981
Normal Operations		
P_d (a)	$P_i - P_e$ [A842.221]	$P_i - P_e$ [4.2.2.2]
η for pipeline	0.72 [A842.221]	0.72 [4.2.2.1]
η for riser sections (b)	0.5 [A842.221]	0.5 [4.2.2.1]
P_h	1.25 P_i (c) [A847.2]	1.25 P_d [8.8.4.3]

a. Credit can be taken for external pressure for gathering lines or flowlines when the MAOP (P_i) is applied at the wellhead or at the seabed. For export lines, when P_i is applied on a platform deck, the head of fluid shall be added to P_i for the pipeline section on the seabed (particularly for two-phase flow).

b. Including prefabricated or retrofit sections and pipeline section in a J-tube.

c. ASME B31.8 imposes $P_h = 1.4\ P_i$ for offshore risers but allows onshore testing of prefabricated portions.

11.2.2.2.2 Pipeline Design for External Pressure

Different practices can be found in the industry using different external pressure criteria. As a rule of thumb or unless qualified thereafter, it is recommended to use propagation criteria for pipeline diameters under 16 inches and collapse criteria for pipeline diameters above or equal to 16 inches.

The propagation criteria is more conservative and should be used where optimization of the wall thickness is not required or for pipeline installation methods not compatible with the use of buckle arrestors such as reel and tow methods. It is generally economical to design for propagation pressure for diameters less than 16 inches. For greater diameters, the wall thickness penalty is too high. When a pipeline is designed based on the collapse criteria, buckle arrestors are recommended. The external pressure criteria should be based on nominal wall thickness, as the safety factors included in the following text account for wall variations.

11.2.2.2.2.1 Propagation Criteria

Although a large number of empirical relationships have been published, the recommended formula is the latest given by AGA (1990):

$$P_P = 33S_y\left(\frac{t_{NOM}}{D}\right)^{2.46} \quad \textbf{(11.65)}$$

The nominal wall thickness should be determined such that:

$$P_p > 1.3\,P_e$$

The safety factor of 1.3 is recommended to account for uncertainty in the envelope of data points used to derive Equation (11.65). It can be rewritten as:

$$t_{NOM} \geq D\left(\frac{1.3P_e}{33S_y}\right)^{\frac{1}{2.46}}$$

For the reel barge method, the preferred pipeline grade is below X-60. However, X-65 steel can be used if the ductility is kept high by selecting the proper steel chemistry and micro alloying. For deepwater pipelines, *D/t* ratios of less than 30 are recommended. It has been noted that bending loads have no demonstrated influence on the propagation pressure.

11.2.2.2.2.2 Collapse Criteria

The mode of collapse is a function of *D/t* ratio, pipeline imperfections, and load conditions. The theoretical background is not given in this book. An empirical general formulation is provided that applies to all situations. It corresponds to the transition mode of collapse under external pressure (P_e), axial tension (T_a), and bending strain (ε_b) (Murphey and Langner 1985).

The nominal wall thickness should be determined such that:

$$\frac{1.3P_e}{P_C} + \frac{\varepsilon_b}{\varepsilon_B} \leq g_p \tag{11.66}$$

where 1.3 is the recommended safety factor on collapse, ε_B is the bending strain of buckling failure due to pure bending, and, g_p is an imperfection parameter defined in the following text.

The safety factor on collapse is calculated for *D/t* ratios along with the loads (P_e,ε_b,T_a) and initial pipeline out-of-roundness (δ_o). The equations are:

$$P_C = \frac{P_{el}P_y^{'}}{\sqrt{P_{el}^2 + P_y^{'2}}} \tag{11.67}$$

$$P_y^{'} = P_y\left[\sqrt{1 - 0.75\left(\frac{T_a}{T_y}\right)^2} - \frac{T_a}{2T_y}\right] \tag{11.68}$$

$$P_{el} = \frac{2E}{1-\nu^2}\left(\frac{t}{D}\right)^3 \tag{11.69}$$

$$P_y = 2S_y\left(\frac{t}{D}\right) \quad \textbf{(11.70)}$$

$$T_y = AS_y \quad \textbf{(11.71)}$$

where g_p is based on pipeline imperfections such as initial out-of-roundness (δ_o), eccentricity (usually neglected), and residual stress (usually neglected). Hence,

$$g_p = \sqrt{\frac{1+p^2}{p^2 - \frac{1}{f_p^{\,2}}}} \quad \textbf{(11.72)}$$

with

$$p = \frac{P_y'}{P_{el}} \quad \textbf{(11.73)}$$

$$f_p = \sqrt{1+\left(\delta_o\frac{D}{t}\right)^2} - \delta_o\frac{D}{t} \quad \textbf{(11.74)}$$

$$\varepsilon_B = \frac{t}{2D} \quad \textbf{(11.75)}$$

$$\delta_o = \frac{D_{\max} - D_{\min}}{D_{\max} + D_{\min}} \quad \textbf{(11.76)}$$

When a pipeline is designed using the collapse criterion, a good knowledge of the loading conditions is required (T_a and ε_b). An upper conservative limit is necessary and must often be estimated.

Under high bending loads, care should be taken in estimating ε_b using an appropriate moment-curvature relationship. A Ramberg Osgood relationship can be used:

$$K^* = M^* + AM^{*B} \tag{11.77}$$

where $K^* = K/K_y$ and $M^* = M/M_y$ with $K_y = 2S_y/ED$ is the yield curvature and $M_y = 2IS_y/D$ is the yield moment. The coefficients A and B are calculated from the two data points on stress-strain curve generated during a tensile test.

11.2.2.2.2.3 Corrosion Allowance

Extra wall thickness is added to account for corrosion when water is present in a fluid along with contaminants such as oxygen, hydrogen sulfide (H_2S), and carbon dioxide (CO_2). A review of standards, rules, and codes of practices (Hill and Warwick 1986) shows that wall allowance is only one of several methods available to prevent corrosion, and often the least recommended.

For H_2S and CO_2 contaminants, corrosion is often localized (pitting) and the rate of corrosion allowance ineffective. Corrosion allowance is made to account for damage during fabrication, transportation, and storage. A value of 1/16 inch may be appropriate. A thorough assessment of the internal corrosion mechanism and rate is necessary before any corrosion allowance is taken.

11.2.2.2.2.4 Check for Hydrotest Condition

The minimum hydrotest pressure for gas lines is given in Table 11–6, and is equal to 1.25 times the design pressure for pipelines. Codes do not require that the pipeline be designed for hydrotest conditions, but sometimes give a tensile hoop stress limit 90% the specified minimum yield strength (SMYS) which is always satisfied if credit has not been taken for external pressure. For cases in which the wall thickness is based on $P_d = P_i - P_e$, codes recommend not to overstrain the pipe. Some of the codes are ASME B31.8 (no limit on hoop stress during hydrotest), and DnV [Clause 8.8.4.3].

For design purposes, condition $\sigma_h \leq \sigma_y$ should be confirmed and increasing the wall thickness or reducing the test pressure should be considered in other cases. For pipelines connected to riser sections requiring $P_h = 1.4\ P_i$, it is recommended to consider testing the riser separately (for prefabricated sections), or determining the hydrotest pressure based on the actual internal pressure experienced by the pipeline section. It is important to note that most pressure testing of subsea pipelines is done with water, but on occasion, nitrogen or air has been used. For low *D/t* ratios (less than 20), the actual hoop stress in a pipeline tested from the surface is overestimated when using the thin wall equations provided in this chapter. Credit for this effect is allowed by DnV Clause 4.2.2.2, but is not normally taken into account.

11.3 Transportation of LNG

Liquefied natural gas (LNG) is natural gas that has been converted to liquid form for ease of storage or transport. LNG typically contains more than 90% methane. It also contains small amounts of ethane, propane, butane and some heavier alkanes. LNG takes up about 1/600th the volume of natural gas at a stove burner tip. The density of LNG is roughly 0.41 to 0.5 kg/L, depending on temperature, pressure and composition. The heat value depends on the source of gas and the process that is used to liquefy the gas. The higher heating value of LNG is estimated to be 24 MJ/L at 164 degrees Celsius. It is odorless, colorless, non-toxic and non-corrosive. Hazards include flammability and freezing.

The liquefaction process involves removal of certain components, such as dust, helium, water, and heavy hydrocarbons, which could cause difficulty downstream. The natural gas is then condensed into a liquid at close to atmospheric pressure (maximum transport pressure set around 3.6 psi) by cooling it to approximately 260 °F. The reduction in volume makes it much more cost-efficient to transport over long distances where pipelines do not exist. Where moving natural gas by pipelines is not possible or economical, it can be transported by specially designed cryogenic sea vessels (LNG carriers) or cryogenic road tankers.

The most important infrastructure needed for LNG production and transportation is an LNG plant consisting of one or more LNG trains, each of which is an independent unit for gas liquefaction. The largest LNG train in operation is now in Qatar. Until recently it was the Train 4 of Atlantic LNG in Trinidad and Tobago with a production capacity of 5.2 million metric ton per annum (mmtpa), followed by the SEGAS LNG plant in Egypt with a capacity of 5 mmtpa. The Qatargas II plant has a production capacity of 7.8 mmtpa for each of its two trains. LNG is loaded onto ships and delivered to a regasification terminal, where the LNG is reheated and turned into gas. Regasification terminals are usually connected to a storage and pipeline distribution network to distribute natural gas to local distribution companies (LDCs) or Independent Power Plants (IPPs).

LNG is shipped around the world in specially constructed seagoing vessels. The trade of LNG is completed by signing a sale and purchase agreement (SPA) between a supplier and receiving terminal, and by signing a gas sale agreement (GSA) between a receiving terminal and end-users. Most of the contract terms used to be DES or, which meant the seller was responsible for the transportation. But with low shipbuilding costs, and the buyer preferring to ensure reliable and stable supply, there are more and more contract terms of FOB, under which the buyer is responsible for the transportation, which is realized by the buyer owning the vessel or signing a long-term charter agreement with independent carriers.

The agreements for LNG trade used to be long-term portfolios that were relatively inflexible both in price and volume. If the annual contract quantity is confirmed, the buyer is obliged to take and pay for the product, or pay for it even if not taken, which is called the obligation of TOP. In contrast to LNG imported to North America, where the price is pegged to Henry Hub, most of the LNG imported to Asia is pegged to crude oil prices by a formula consisting of indexation called the Japan Crude Cocktail (JCC).

With new demand from China, India and US increasing dramatically, and crude oil price skyrocketing, the LNG price is on the rise too. Natural gas can be considered as the most environmentally friendly of the fossil fuels, because it has the lowest CO_2 emissions per unit of energy and because it is suitable for use in high efficiency combined cycle power stations. Because of the energy required to liquefy and to transport it, the environmental performance of LNG is inferior to that of natural gas, although in most cases LNG is still superior to alternatives such as fuel oil or coal. This is particularly so in the case where the source gas would otherwise be flared.

Whilst natural gas power plants emit approximately half the carbon dioxide of an equivalent coal power plant, the natural gas combustion required to produce and transport LNG to the plants adds 20 to 40 percent more carbon dioxide than burning natural gas alone. To ensure safe and reliable operation, particular measures are taken in the design, construction and operation of LNG facilities.

In its liquid state, LNG is not explosive and can not burn. For LNG to burn, it must first vaporize, then mix with air in the proper proportions (the flammable range is 5% to 15%), and then be ignited. In the case of a leak, LNG vaporizes rapidly, turning into a gas (methane plus trace gases), and mixing with air. While this mixture is within the flammable range, there is risk of ignition which would create fire and thermal radiation hazards. Note that since 1944, only one serious accident at a regasification facility has taken place.

Modern LNG storage tanks are typically the full containment type, which is a double-wall construction with reinforced concrete outer wall and a high-nickel steel inner tank, with extremely efficient insulation between the walls. Large tanks are low aspect ratio (height to width) and cylindrical in design with a domed roof. Storage pressures in these tanks are very low, less than 7 psig. Sometimes more expensive frozen-earth, underground storage is used. Pre-stressed concrete backed up with suitable thermal insulation, are designed to be both under and above ground to suit sites conditions and local safety regulations and requirements.

Smaller quantities (say 190,000 US gallons and less), may be stored in horizontal or vertical, vacuum-jacketed, pressure vessels. These tanks may be at pressures anywhere from less than 7 psig to over 250 psig. LNG must be kept cold to remain a liquid, independent of pressure. Despite efficient insulation, there will inevitably be some heat leakage into the LNG, resulting in vapourisation of the LNG. This boil-off gas acts to keep the LNG cold. The boil-off gas is typically compressed and exported as natural gas, or is reliquefied and returned to storage.

LNG is transported using both tanker truck, railway tanker, and purpose built ships known as LNG carriers. LNG will be sometimes taken to cryogenic temperatures to increase the tanker capacity. Recently ship-to-ship transfer (STS) transfers have been carried out by the Belgian gas tanker owner in the Gulf of Mexico which involved the transfer of LNG from a conventional LNG carrier to an LNG regasification vessel (LNGRV).

11.4 References

American Gas Association. "Collapse of Offshore Pipelines." Paper presented at the Pipeline Research Committee Seminar, Houston, Texas, February 20, 1990.

American Society of Mechanical Engineers. "Gas Transmission and Distribution Piping Systems." ASME Code for Pressure Piping, B31.8, 1989 Edition and 1990 Addendum.

Bai, Y. *Pipelines and Risers*. Vol. 3. Amsterdam: Elsevier Ocean Engineering Book Series, 2001.

Colebrook, C. F. J. *Inst. Civil Eng*. **11,** 1983: 133.

Det norske Veritas. "Rules for Submarine Pipeline Systems." 1981 Edition.

Drew, T. B., E. C. Koo, and W. H. McAdams. *Trans. Am. Inst. Chem. Eng*. **28**, 1930: 56.

Guo, B. and A. Ghalambor. *Gas Voule Requirements for Underbalanced Drilling Deviated Holes*. Tulsa: PennWell Books, 2002.

Guo, B., S. Song, J. Chacko, and A. Ghalambor. *Offshore and Deepwater Pipelines.* Amsterdam: Elsevier, 2005.

Hill, R. T. and P. C. Warwick. "Internal Corrosion Allowance for Marine Pipelines: A Question of Validity." OTC paper No. 5268, 1986.

Ikoku, C. U. *Natural Gas Production Engineering*. New York: John Wiley & Sons, Inc 1984.

Jain, A. K. "An Accurate Explicit Equation for Friction Factor." *J. Hydraulics Div*. ASCE **102**, HY5, May 1976.

Katz, D. L. and R. L. Lee, *Natural Gas Engineering—Production and Storage*. New York: McGraw-Hill, 1990.

Moody, L. F. "Friction Factor for Pipe Flow," *Trans. ASME* **66** (1944), 671.

Murphey, C. E. and C. G. Langner. "Ultimate Pipe Strength Under Bending, Collapse, and Fatigue." Proceedings of the OMAE Conference, 1985.

Nikuradse, J. *Forschungshelf* (1933): 301.

Weymouth, T. R. "Problems in Natural Gas Engineering." *Trans. ASME* **34** (1912): 185.

Wichert, E. and K. Aziz. "Calculate Zs for Sour Gases." *Hydrocarbon Processing* **51** (May 1972): 119.

11.5 Problems

11-1 For the following data given for a horizontal pipeline in Texas, estimate gas flow rate through the pipeline using the Weymouth, Panhandle A, Panhandle B, and Clinedinst equations.

Pipeline ID: 12.09 in

Pipeline length: 100 mi

Temperature: 70 °F

Gas-specific gravity: 0.65

Delivery pressure: 150 psia

Compressor pressure: 500 psia

11-2 If the pipeline described in Problem 11-1 goes up a hill with an inclination angle of 3°, at what rate can the pipeline deliver natural gas?

11-3 For the following data given for a horizontal pipeline in Oklahoma, predict the minimum compression pressure required to transport the gas using Weymouth, Panhandle A, Panhandle B, and Clinedinst equations.

Pipeline ID: 12.09 in

Pipeline length: 150 mi

Temperature: 75 °F

Gas specific gravity: 0.55

Delivery pressure: 100 psia

Compressor pressure: 500 psia

Gas flow rate: 20 MMscfd

11-4 If the pipeline described in Problem 11-3 goes up a hill with an inclination angle of 4°, at what rate can the pipeline deliver natural gas?

11-5 Consider a 6-in pipeline that is 15 miles long. Assuming that the compression and delivery pressures will remain unchanged, calculate gas capacity increases by taking the following measures of improvement:

(a) Replace 7.5 miles of the 6-in pipeline by a 10-in pipeline segment

(b) Place a 10-in parallel pipeline to share gas transmission

(c) Loop 7.5 miles of the 6-in pipeline with a 10-in pipeline segment

11-6 Develop a spreadsheet program and duplicate the chart in Figure 11–3.

11-7 If a 6-in pipeline is looped by a 12-in pipe segment for 40%, 50%, 60%, 70%, and 80%, how much increase in gas flow rate will be expected, respectively?

Chapter 12

Special Problems

12.1 Introduction

There are some problems in natural gas production operations that need to be paid special attentions. One of them is liquid loading of gas production wells, which reduces deliverability of gas wells. Blockage of gas hydrates of pipelines and equipment is another problem that reduces pipeline efficiency and affects normal operations of gas-processing facilities. Yet cleaning pipelines during operation presents a challenging task for engineers in gas-production operations. This chapter addresses these problems and their solutions.

12.2 Liquid Loading on Gas Wells

As shown in Figure 12–1, high-pressure gas wells produce gas carrying liquid water and/or condensate in the form of mist. As the gas flow velocity in the well drops owing to the reservoir pressure depletion, the carrying capacity of the gas decreases. When the gas velocity drops to a critical level, liquids begin to accumulate in the well and the well flow can undergo annular flow regime followed by a slug flow regime. The accumulation of liquids (liquid loading) increases bottom hole pressure that reduces gas production rate. Low gas production rate will cause gas velocity to drop further. Eventually the well will undergo bubbly flow regime and cease producing.

Liquid loading is not always obvious and recognizing the liquid loading problem is not an easy task. A thorough diagnostic analysis of well data needs to be performed. The symptoms to look for include onset of liquid slugs at the surface of the well, increasing difference between the tubing and casing pressures with time, sharp changes in gradient on a flowing

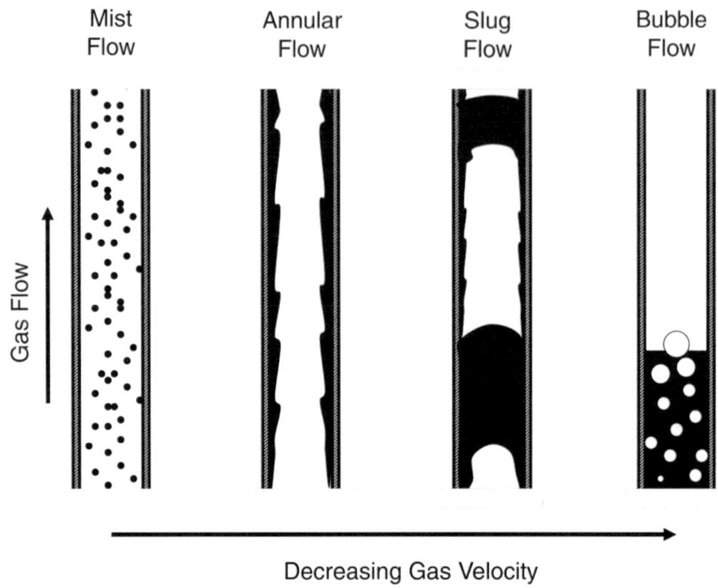

Figure 12–1 Four flow regimes commonly encountered in gas wells.

pressure survey, sharp drops in a production decline curve, and prediction with analytical methods.

Accurate prediction of the problem is vitally important for taking timely measures to solve the problem. Previous investigators have suggested several methods to predict the problem. Results from these methods often show discrepancies. Also some of these methods are not easy to use due to the difficulties with prediction of bottom hole pressure in multi-phase flow. In this section, the method presented by Turner, Hubbard, and Dukler (1969) is summarized and the method by Guo, Ghalambor, and Xu (2005) is described in more detail.

12.2.1 Turner's Method

The method presented by Turner, Hubbard, and Dukler (1969) is referred to as Turner's method. They are the pioneer investigators who analyzed and predicted the minimum gas flow rate capable of removing liquids from the gas production wells. They presented two mathematical models to describe the liquid loading problem: the film movement model and entrained drop movement model. On the basis of analyses on field data they had, they concluded that the film movement model does not represent the controlling liquid transport mechanism.

Turner's entrained drop movement model was derived on the basis of the terminal free settling velocity of liquid drops and the maximum drop diameter corresponding to the critical Weber number of 30. Turner's terminal velocity equation is expressed as

$$v_{sl} = \frac{k_v \sigma^{1/4} \left(\rho_L - \rho_g\right)^{1/4}}{C_d^{1/4} \rho_g^{1/2}} \tag{12.1}$$

where

v_{sl} = terminal settling velocity, ft/s

$k_v = 1.3$

σ = interfacial tension, dynes/cm

ρ_L = liquid density, $\mathrm{lb_m/ft^3}$

ρ_g = gas density, $\mathrm{lb_m/ft^3}$

C_d = drag coefficient, Turner et al. (1969) recommended a value 0.44.

According to Turner et al. (1969), gas will continuously remove liquids from the well until its velocity drops to below the terminal velocity. The minimum gas flow rate for a particular set of conditions (pressure and conduit geometry) can be calculated using equations (12.1) and (12.2).

$$Q_{gm} = \frac{3.06 p v_{sl} A}{Tz} \tag{12.2}$$

where

Q_{gm} = the minimum required gas flow for liquid removal, MMscf/day

p = pressure at depth of interest, psia

A = cross-sectional area of conduit, $\mathrm{ft^2}$

T = temperature, °R

z = gas compressibility factor

Turner et al. (1969) found that this entrained drop movement model underestimates the minimum gas flow rates. They recommended that the equation-derived values be adjusted upward by approximately 20 percent to ensure removal of all drops. Turner et al. believed that the discrepancy was attributed to several facts including the use of drag coefficients for solid spheres, the assumption of stagnation velocity, and the critical Weber number established for drops falling in air, not in compressed gas.

The main problem that hinders the application of Turner et al.'s entrained drop model to gas wells comes from the difficulties of estimating the values of gas density and pressure. Using an average value of gas-specific gravity (0.6) and gas temperature (120 °F), Turner et al. derived an expression for gas density as 0.0031 times the pressure. However, they did not present a method for calculating the gas pressure in a multiphase flow wellbore.

Turner et al.'s entrained drop movement model was later modified by a number of authors. Coleman et al. (1991) suggested the use of Equation (12.1) with a coefficient k_v value of 1.59. Nosseir et al. (2000) expanded Turner et al.'s entrained drop model to more than one flow regime in a well. Lea and Nickens (2004) corrected Turner et al.'s equation with k_v =1.92. However, the original drawbacks (neglected transport velocity and multi-phase flow pressure) with Turner et al.'s approach still remain unsolved by these investigators.

Example Problem 12.1

The following data is for a gas well. Estimate the terminal slip velocity of a liquid droplet and the minimum required gas production rate for liquid removal using Turner's method:

Gas-specific gravity: 0.6

Tubing diameter: 1.995 in

Pressure: 636 psia

Temperature: 127 °F

Liquid density: 67.4 lbm/ft^3

Interfacial tension: 60 dynes/cm

Solution

The spreadsheet program TurnerLoading.xls was written to perform Turner velocity and the minimum unloading gas flow rate calculations. The solution by the spreadsheet is shown in Table 12–1.

It can be shown that if a coefficient value $k_v = 1.92$ (suggested by Lea and Nickens [2004]) is used in the spreadsheet, the minimum gas flow rate will be 0.88 MMscfd.

12.2.2 Guo's Method

Guo's method refers to the method presented by Guo, Ghalambor, and Xu (2005a) Starting from Turner et al.'s entrained drop model, Guo et al. (2005a) determined the minimum kinetic energy of gas that is required to lift liquids. A four-phase (gas, oil, water, and solid particles) mist-flow model was developed. Applying the minimum kinetic energy criteria to the four-phase flow model resulted in a closed form analytical equation for predicting the minimum gas flow rate. Through case studies Guo et al. (2005a) demonstrated that Guo's method is more conservative and accurate.

12.2.2.1 Gas Kinetic Energy

Kinetic energy per unit volume of gas can be expressed as

$$E_k = \frac{\rho_g v_g^{\,2}}{2g_c} \tag{12.3}$$

where

E_k = gas-specific kinetic energy, $lb_f\text{-}ft/ft^3$

v_g = gas velocity, ft/s

g_c = unit conversion factor, $lb_m\text{-}ft/lb_f\text{-}s^2$

Table 12–1 Turner Velocity and the Minimum Unloading Gas Flow Rate Given by TurnerLoading.xls[a]

Instructions: 1) Update input data; 2) View results.		
Input Data		
Coefficient k_V:		1.3
Gas-specific gravity γ_g:		0.6
Tubing diameter d:		1.995 in
Pressure p_{wf}:		636 psia
Temperature T_{wf}:		127 °F
Heavy-liquid density ρ_l:		67.4 lbm/ft^3
Interfacial tension σ:		60 dynes/cm
Solution		
T	=	587.00 °R
ρ_g	=	1.76 lbm/ft^3
A	=	0.0217 ft^2
p_{pc}	=	672.50 psia
T_{pc}	=	358.50 °R
T_{av}	=	587.00 °R
p_{av}	=	636.00 psia
p_{pr}	=	0.95
T_{pr}	=	1.64
Z	=	0.94
v_{gm}	=	9.54 ft/s
Q_{gm}	=	0.732 MMscf/d

a. This spreadsheet calculates the minimum required unloaded gas production rate.

The gas kinetic energy has been used in the well drilling industry to determine the minimum required gas flow rate for effectively transporting drill cuttings in boreholes (Guo and Ghalambor 2002). Substituting Equation (12.1) into Equation (12.3) gives an expression for the minimum kinetic energy required to keep liquid droplets from falling:

$$E_{ksl} = 0.026\sqrt{\frac{\sigma\left(\rho_L - \rho_g\right)}{C_d}} \tag{12.4}$$

where E_{ksl} = kinetic energy required to hold liquid drops stationary, $lb_f\text{-}ft/ft^3$.

If $C_d = 0.44$ is used, and the effect of gas density is neglected (a conservative assumption), Equation (12.4) becomes:

$$E_{ksl} = 0.04\sqrt{\sigma\rho_L} \tag{12.5}$$

In gas wells producing water, typical values for water/gas interfacial tension and water density are 60 dynes/cm and 65 lb_m/ft^3, respectively. This yields the minimum kinetic energy value of 2.5 $lb_f\text{-}ft/ft^3$. In gas wells producing condensate, typical values for condensate/gas interfacial tension and condensate density are 20 dynes/cm and 45 lb_m/ft^3, respectively. This yields the minimum kinetic energy value of 1.2 $lb_f\text{-}ft/ft^3$.

The minimum gas velocity required for transporting the liquid droplets upward is equal to the minimum gas velocity required for floating the liquid droplets (keeping the droplets from falling) plus the net transport velocity of the droplets, that is,

$$v_{gm} = v_{sl} + v_{tr} \tag{12.6}$$

where

v_{gm} = minimum gas velocity required to transport liquid drops, ft/s

v_{tr} = liquid transport velocity, ft/s

The liquid transport velocity v_{tr} may be calculated on the basis of liquid production rate, geometry of the conduit, and liquid volume fraction, which is difficult to quantify. Instead of formulating an expression for the transport velocity v_{tr}, this study uses v_{tr} as an empirical constant to lump the effects of nonstagnation velocity, drag coefficients for solid spheres, and the critical Weber number established for drops falling in air. On the basis of Turner et al.'s (1969) work, the value of v_{tr} was taken by Guo et al. (2005a) as 20 percent of v_{sl}. Use of this value results in

$$v_{gm} \approx 1.2v_{sl} \quad \textbf{(12.7)}$$

Substituting equations (12.1) and (12.7) into Equation (12.3) results in the expression for the minimum kinetic energy required for transporting the liquid droplets as:

$$E_{km} = 0.0576\sqrt{\sigma\rho_L} \quad \textbf{(12.8)}$$

where E_{km} = minimum required kinetic energy, lb_f-ft/ft^3.

For typical gas wells producing water, this equation yields the minimum kinetic energy value of 3.6 lb_f-ft/ft^3. For typical gas wells producing condensate, this equation gives the minimum kinetic energy value of 1.73 lb_f-ft/ft^3.

In order to evaluate the actual kinetic energy E_k of a given gas stream with Equation (12.3) and compare it with the minimum required kinetic energy E_{km} given by Equation (12.8), the values of gas density ρ_g and gas velocity v_g need to be determined. Expressions for ρ_g and v_g can be obtained from gas law assuming ideal gas:

$$\rho_g = \frac{S_g P}{53.34T} \quad \textbf{(12.9)}$$

and

$$v_g = 4.71 \times 10^{-5} \frac{TQ_g}{AP} \quad \textbf{(12.10)}$$

where

S_g = specific gravity of gas, air =1

P = pressure, lb$_f$/ft^2

Q_g = gas production rate, scf/day

Substituting Equation (12.9) and Equation (12.10)) into Equation (12.3) yields:

$$E_k = 6.46 \times 10^{-13} \frac{S_g T Q_g{}^2}{A^2 P} \tag{12.11}$$

Equation (12.11) indicates that the gas kinetic energy decreases with increased pressure, which means that the controlling conditions are bottom hole conditions where gas has higher pressure and lower kinetic energy. This analysis is consistent with the observations from air-drilling operations where solid particles accumulate at bottom hole rather than top hole (Guo and Ghalambor 2002). However, this analysis is in contradiction with Turner et al.'s results that indicated that the wellhead conditions are, in most instances, controlling.

12.2.2.2 Minimum Required Gas Flow Rate

A logical procedure for predicting the minimum required gas flow rate Q_{gm} involves calculating gas density ρ_g, gas velocity v_g, and gas kinetic energy E_k at bottom hole condition using an assumed gas flow rate Q_g, and compare the E_k with E_{km}. If the E_k is greater than E_{km}, the Q_g is higher than the Q_{gm}. The value of Q_g should be reduced and the calculation should be repeated until the E_k is very close to E_{km}. Because this procedure is tedious, a simple equation is derived by Guo et al. (2005a) for predicting the minimum required gas flow rate in this section. Under the minimum unloaded condition (the last point of the mist flow regime), Equation (12.11) becomes:

$$E_{km} = 6.46 \times 10^{-13} \frac{S_g T Q_{gm}{}^2}{A^2 P} \tag{12.12}$$

where Q_{gm} = minimum gas flow rate required to transport liquid drops, scf/day.

Equation (12.12) gives:

$$P = 6.46 \times 10^{-13} \frac{S_g T Q_{gm}^2}{A^2 E_{km}} \tag{12.13}$$

which is a P-Q_{gm} relation established based on the minimum kinetic energy theory. Another P-Q_{gm} relation is determined by the gas-oil-water-solid four-phase mist-flow model presented in Chapter 4:

$$b\left(P - P_{hf}\right) + \frac{1-2bm}{2}\ln\left|\frac{(P+m)^2+n}{\left(P_{hf}+m\right)^2+n}\right| - \frac{m + \frac{b}{c}n - bm^2}{\sqrt{n}}\left[\tan^{-1}\left(\frac{P+m}{\sqrt{n}}\right) - \tan^{-1}\left(\frac{P_{hf}+m}{\sqrt{n}}\right)\right] = a\left(1+d^2e\right)L \tag{12.14}$$

Substituting Equation (12.13) into Equation (12.14) results in:

$$b\left(6.46 \times 10^{-13} \frac{S_g T Q_{gm}^2}{A^2 E_{km}} - P_{hf}\right) + \frac{1-2bm}{2}\ln\left|\frac{\left(6.46 \times 10^{-13} \frac{S_g T Q_{gm}^2}{A^2 E_{km}} + m\right)^2 + n}{\left(P_{hf}+m\right)^2+n}\right| - \frac{m + \frac{b}{c}n - bm^2}{\sqrt{n}}\left[\tan^{-1}\left(\frac{6.46 \times 10^{-13} \frac{S_g T Q_{gm}^2}{A^2 E_{km}} + m}{\sqrt{n}}\right) - \tan^{-1}\left(\frac{P_{hf}+m}{\sqrt{n}}\right)\right] = a\left(1+d^2e\right)L \tag{12.15}$$

where all the parameters should be evaluated at Q_{gm}. The minimum required gas flow rate Q_{gm} can be solved from Equation (12.15) with a numerical method such as the Newton-Raphson iteration technique. The spreadsheet program GasWellLoading.xls is for this purpose.

Example Problem 12.2

For the data given in Example Problem 12.1, estimate the minimum required gas production rate for liquid removal using Guo's method.

Solution

The spreadsheet program GasWellLoading.xls was written to calculate the minimum unloading gas flow rate. The solution by the spreadsheet is shown in Table 12–2.

12.2.3 Comparison of Methods

Field data containing 106 test points used by Turner et al. (1969) was employed in this section to compare Guo's method with Turner's method. The wells produced gas and condensate and/or water from either tubing or annulus. A gas-specific gravity of 0.6 was assumed for all the wells. Water-specific gravity was assumed to be 1.08. Gas/condensate and gas/water interfacial tensions were assumed to be 20 dynes/cm and 60 dynes/cm, respectively. Wellhead temperature and geothermal gradient were assumed to be 60 °F and 0.01 °F/ft, respectively. The roughness of the conduit wall was assumed to be 0.000015 inches All other required data were given by Turner et al. (1969).

Figure 12–2 illustrates Turner et al.'s calculated minimum flow rates mapped against the test flow rates. The diagonal line (boundary) separates the predicted unloaded and loaded regions with the upper being the unloaded and lower being loaded regions. The map shows only two unloaded points in the loaded region. The one with a question mark might be reported with error, that is, it might not be an unloaded well. The other point is just below the boundary. This seems to indicate that Turner et al.'s method is accurate in predicting unloaded wells. However, this map shows nine loaded points in the unloaded region with one point being very close to the boundary. This means Turner et al.'s method underestimates the minimum gas flow rates.

Table 12–2 Input Data and Solution Given by GasWellLoading.xls[a]

Instructions: 1) Update input data; 2) Run Macro Solution and view result.	
Input Data	
Gas-specific gravity:	0.6 air = 1
Hole inclination:	0°
Producing depth:	6,700 ft
Wellhead pressure:	500 psi
Wellhead temperature:	60 °F
Producing zone temperature:	127 °F
Condensate gravity:	70.8 API
Condensate make:	0 bbl/MMscf
Water-specific gravity:	1.08 water = 1
Water make:	10.5 bbl/MMscf
Solid-specific gravity:	2.65 water = 1
Solid make:	0 ft^3/MMscf
Conduit OD:	1.995 in
Conduit ID:	0 in
Conduit wall roughness:	0.000015 in
Major liquid (1 = water; –1 = condensate):	1
Heavy liquid-gas interfacial tension:	60 dyne/cm
Heavy liquid density:	67.39 lb/ft^3
Calculated Parameters	
Hydraulic diameter:	0.1663 ft
Conduit cross-sectional area:	0.0217 ft^2
Average temperature:	553.5 °R
Minimum kinetic energy:	3.6627 lbf-ft/ft^3
a =	2.21319E–05
b =	2.61599E–08
c =	993,082
d =	0.025982888
e =	0.000721177
f =	0.007714093
m =	18.60862817
n =	711,232,734
Solution	
Gas production rate (Q_{gm}) =	826 Mscf/day
Pressure (P) =	590 psia
Objective function f(Q_{gm}) =	–0.000191102

a. This spreadsheet calculates the minimum unloading gas production rate.

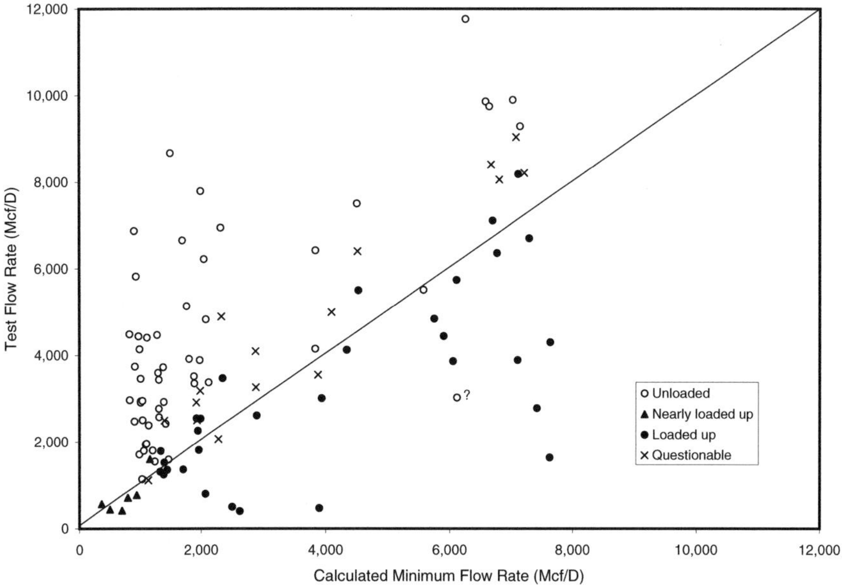

Figure 12–2 Turner et al.'s model-calculated minimum flow rates mapped against the test flow rates.

Figure 12–3 illustrates Guo's method-calculated minimum flow rates mapped against the test flow rates. The map shows the same two unloaded points in the loaded region. Again the one with a question mark might be reported with error. The other point is just below the boundary. This map shows six loaded points in the unloaded region but they are very close to the boundary. This means Guo's method is more accurate than Turner et al.'s method in estimating the minimum flow rates.

The minimum gas flow rates for liquid removal in wells of various geometries were generated with Guo's method. The results are shown in Appendix D and Appendix E.

12.2.4 Solutions to the Liquid Loading Problem

Several measures can be taken to reduce the liquid loading problem in gas production wells. Foaming the liquid water can enable the gas to lift water from the well. This is not only observed to be an effective solution but also theoretically evidenced by Equation (12.8). Using smaller tubing or creating a lower wellhead pressure sometimes can keep mist flow longer.

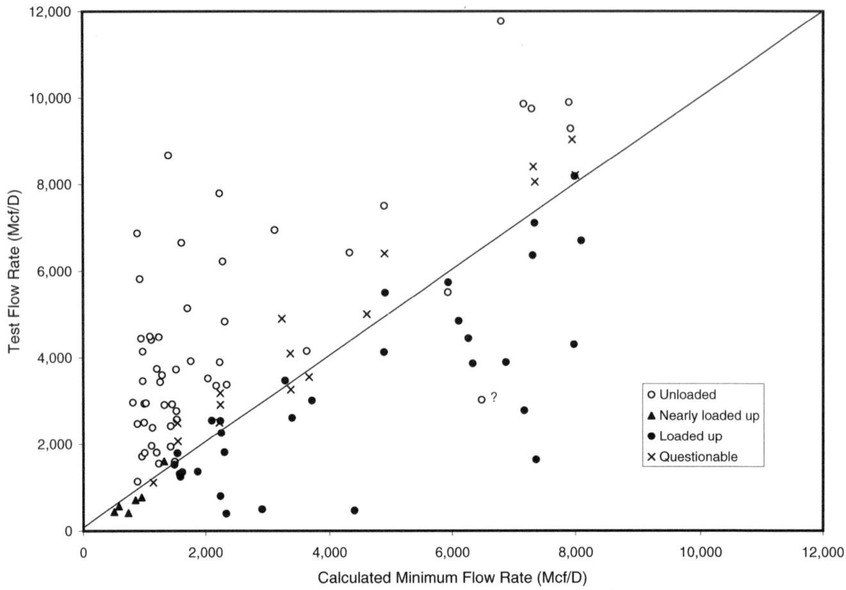

Figure 12–3 The minimum flow rates given by Guo's method mapped against the test flow rates.

The loaded gas wells can be unloaded by gas-lifting or pumping the liquids out of the wellbore. Heating the wellbore can prevent condensation. Down hole injection of water into an underlying disposal zone is another option. Lea and Nickens (2004) give a detailed description of solving the liquid-loading problem.

12.3 Hydrate Control

Natural gas hydrates are solid crystalline compounds formed by the chemical combination of natural gas and water under pressure at temperatures considerably above the freezing point of water. They have often been found responsible for operating difficulties at wellheads, pipelines, and other processing equipment. In the presence of free water, hydrates form when the temperature is below a certain degree (hydrate temperature). The hydrate temperature would be less than or equal to the dew point temperature of the hydrate forming gas.

The chemical formulas of some natural gas hydrates are:

Methane hydrates: $CH_4 \cdot 7H_2O$

Ethane hydrates: $C_2H_6 \cdot 8H_2O$

Propane hydrates: $C_3H_8 \cdot 18H_2O$

CO_2 hydrates: $CO_2 \cdot 7H_2O$

The crystals of gas hydrate resemble ice or wet snow in appearance but do not have ice's solid structure, are much less dense, and exhibit properties that are generally associated with chemical compounds. The main framework of their structure is water; the hydrocarbon molecule occupies the void space in a crystalline network held together by chemically weak bonds with the water. The water framework is ice like; unlike ice, however, it has void space and a network structure.

During the flow of natural gas, it becomes necessary to define, and thereby avoid, conditions that promote the formation of hydrates. This is essential because hydrates may choke the flow string, surface lines, and other equipment, resulting in lower flow rates of gas. The conditions that tend to promote the formation of natural gas hydrates are:

- presence of liquid water
- low temperature
- high pressure
- high velocity or agitation
- presence of "seed" crystals of hydrate
- presence of highly soluble gas in water, such as H_2S or CO_2

12.3.1 Hydrate-Forming Conditions

A hydrate-forming condition depends on the composition of natural gas. For normal natural gases with compositions defined by their specific gravities, Table 12–3 and Figure 12–4 present some critical pressure and temperature values for hydrate-forming conditions.

Figure 12–4 shows hydrate-forming conditions for natural gases with various specific gravities. A rigorous technique for predicting conditions for

Table 12–3 Hydrate-Forming Conditions of Natural Gases (Courtesy of SPE-AIME)

Temperature °F	Gas-Specific Gravity (air = 1)					
	Methane	0.6	0.7	0.8	0.9	1
32	381	150	95	76	60	50
40	600	243	165	132	105	90
45	800	335	230	188	150	136
50	1,100	470	335	267	220	200
55	1,500	658	462	381	335	300
60	2,150	900	650	550	500	450
65	3,000	1,380	1,000	851	780	700
70	4,122	2,463	1,860	1,500	1,400	1,300
75		4,000	3,400	2,980	2,670	2,460

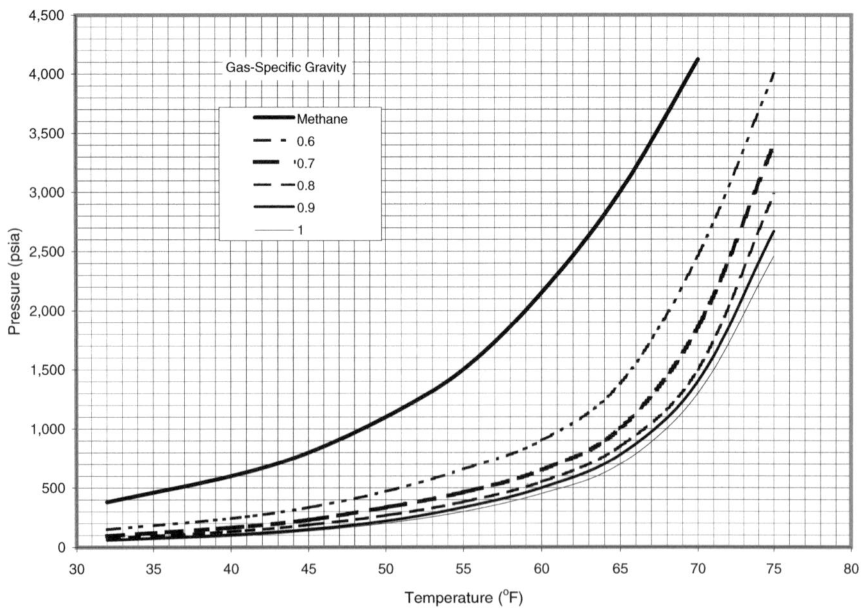

Figure 12–4 Hydrate-forming conditions of natural gases (Courtesy of SPE-AIME).

hydrate formation involves the use of vapor/solid equilibrium constants such as that given by Katz et al. (1945). The calculations are analogous to a dew point calculation for multicomponent mixtures. This method of hydrate prediction has proved to be rather reliable.

It is convenient to divide hydrate formation into two categories: (a) hydrate formation due to a decrease in temperature with no sudden pressure drop, such as in the flow string or pipelines, and (b) hydrate formation where a sudden expansions occurs, such as in the flow provers, orifices, back-pressure regulators, or chokes. For problems in category (a), Figure 12–4 gives approximate values of the hydrate temperature as a function of pressure and specific gravity. Hydrates will form whenever temperature and pressure plot to the left of the hydrate formation line for the gas in question. For problems in category (b), Figure 12–5 through Figure 12–8 may be used to approximate the conditions for hydrate formation. These figures are strictly applicable to sweet natural gases. They may be used for sour gases, keeping in mind that the presence of H_2S and CO_2 will increase the hydrate temperature and reduce the pressure above which hydrates will form. In other words, the presence of H_2S and CO_2 increases the possibility of hydrate formation. A number of computer packages are available in the natural gas industry for prediction of hydrate-forming conditions.

Example Problem 12.3

A gas of specific gravity 0.7 is at a pressure of 1,000 psia. Assuming presence of free water, (a) to what extent can the temperature be lowered without hydrate formation? (b) how far can the gas be expanded without hydrate formation, if the initial gas temperature is 80 °F? (c) how far can the gas be expanded without hydrate formation, if the initial gas temperature is 100 °F?

Solution

(a) From Figure 12–4 at a specific gravity of 0.7 and a pressure of 1,000 psia, hydrate temperature is 65 °F. Thus, hydrates may form at or below 65 °F.

(b) From Figure 12–6, the intersection of the 1,000 psia initial pressure line with the 80 °F initial temperature curve gives a final pressure of 620 psia. Hence, this gas may be expanded to a final pressure of 620 psia without a possibility of hydrate formation.

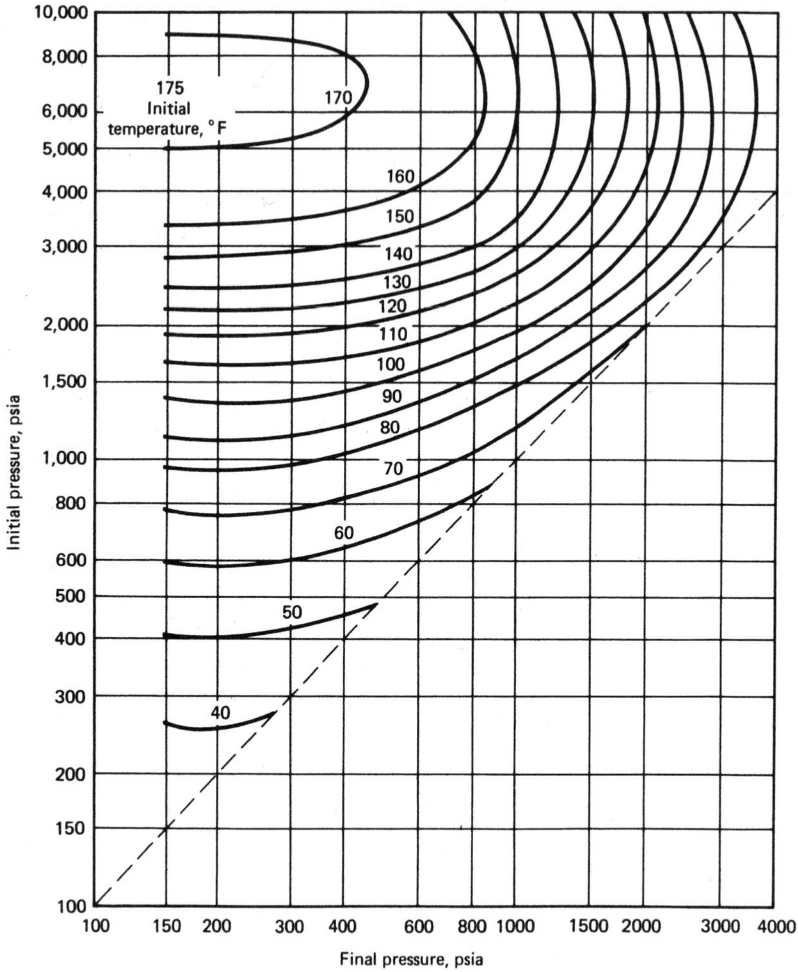

Figure 12–5 Permissible expansion of a 0.60-specific gravity natural gas without hydrate formation (Courtesy of Gas Processors Suppliers Association).

(c) From Figure 12–6, the 100 °F initial temperature curve does not intersect the 1,000 psia initial pressure line. Hence, this gas may be expanded to atmospheric pressure without hydrate formation.

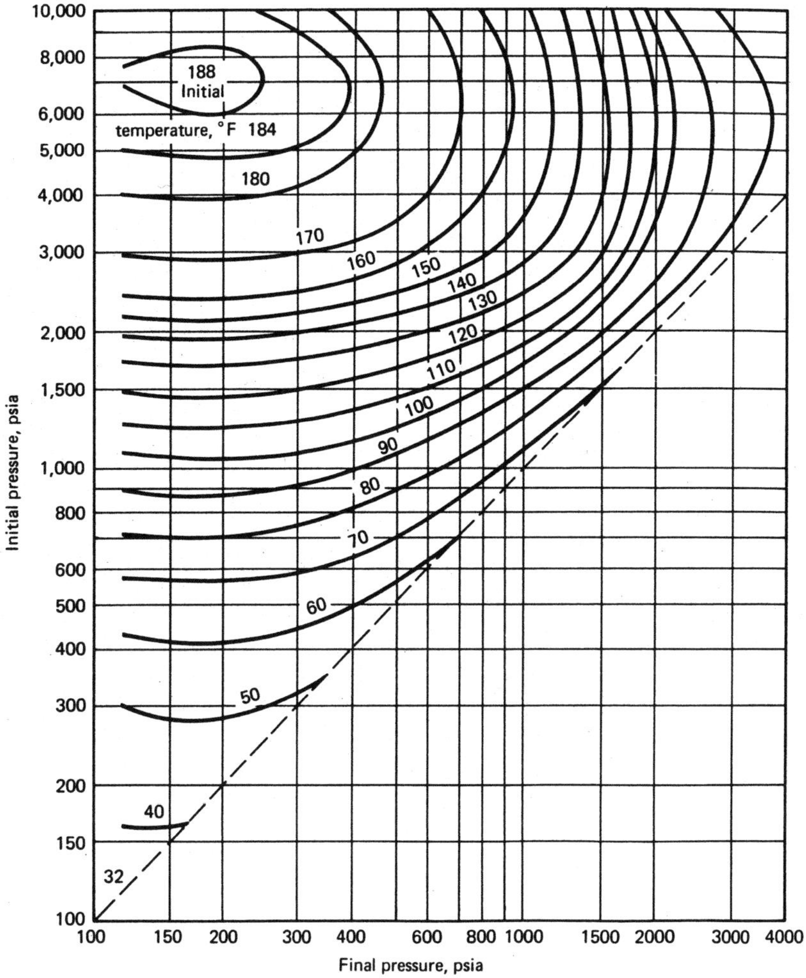

Figure 12–6 Permissible expansion of a 0.70-specific gravity natural gas without hydrate formation (Courtesy of Gas Processors Suppliers Association).

12.3.2 Preventing Hydrate Formation

The hydrate formation lines in Figure 12–4 indicate that hydrate formation can be prevented by heating if the pressure and water content of the gas remain constant. If the pipelines and equipment must be maintained below the hydrate temperature, the water must be inhibited for trouble-

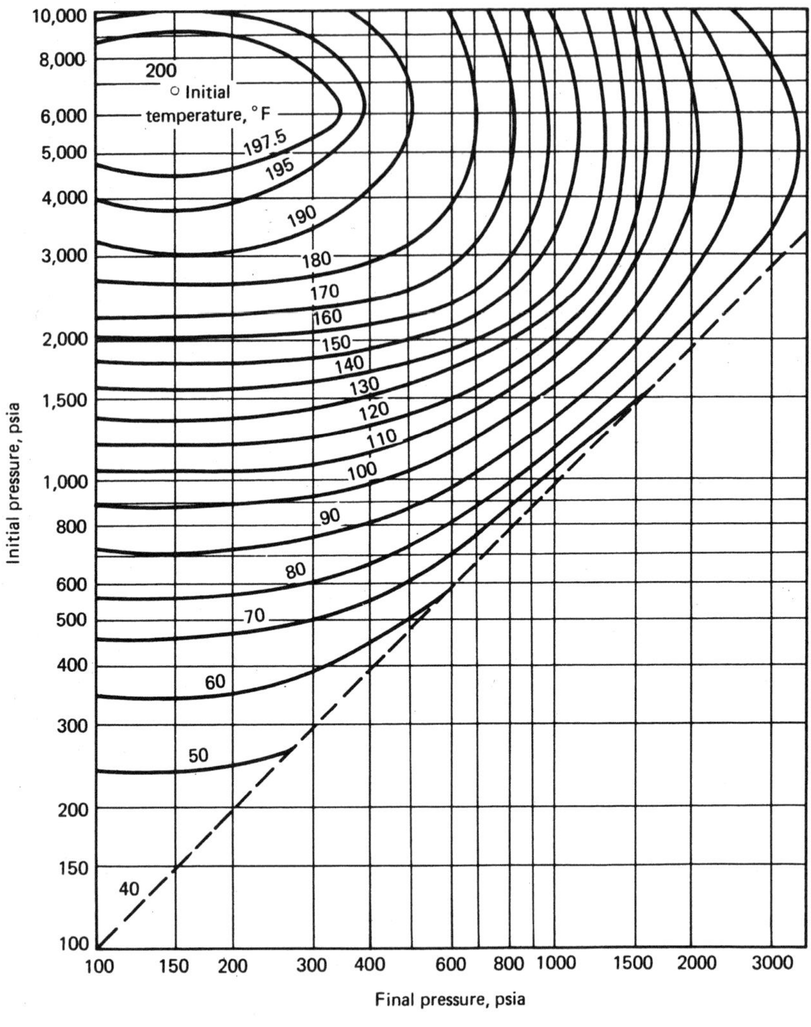

Figure 12–7 Permissible expansion of a 0.80-specific gravity natural gas without hydrate formation (Courtesy of Gas Processors Suppliers Association).

free operation. Guo et al. (2005b) discuss hydrate mitigation strategies for gas pipelines.

Methanol, Ethylene Glycol (EG), and Diethylene Glycol (DEG) are commonly injected into gas streams to depress the freezing point. All of these inhibitors can be recovered and recycled; however, the recovery of methanol is often uneconomical. Hydrate inhibitor injection does not always

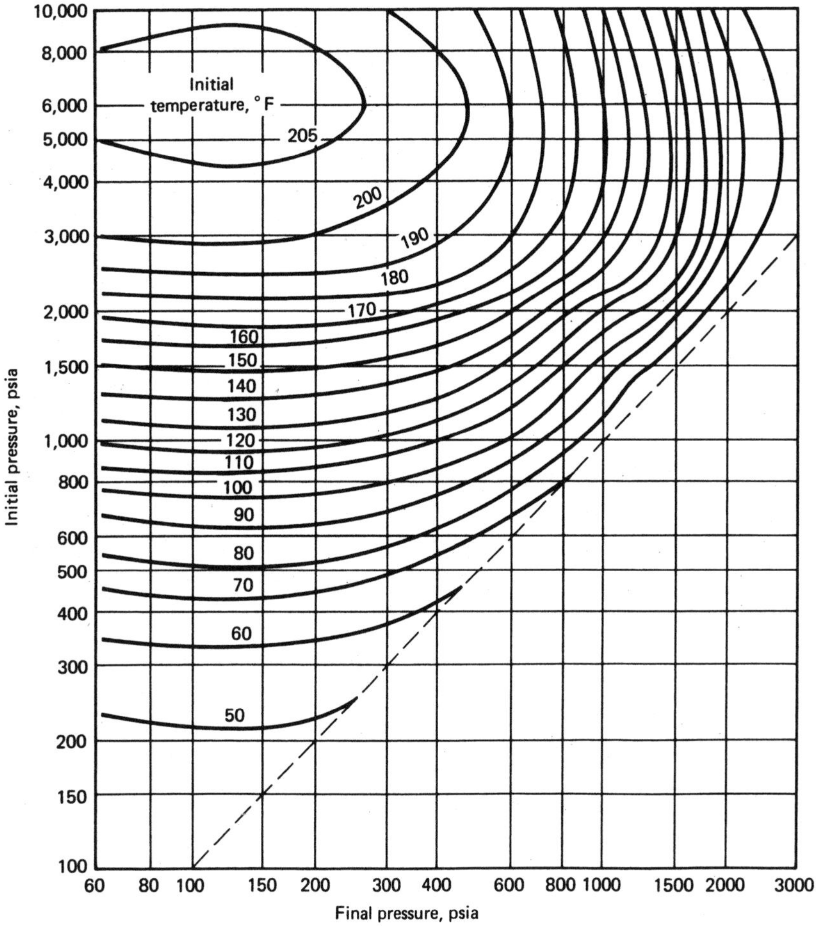

Figure 12–8 Permissible expansion of a 0.90-specific gravity natural gas without hydrate formation (Courtesy of Gas Processors Suppliers Association).

provide the ultimate degree of dehydration specified by the purchaser or required by the process conditions.

Methanol injection systems are frequently installed at facilities where low gas volumes prohibit dehydration. They are also temporarily used for situations where hydrate inhibition requires high capital investment equipment before a decision regarding a permanent facility is made. These systems have been utilized in fields where hydrate problems are relatively mild, infrequent, seasonal, or expected during start-up. The EG and DEG are injected primarily at low-temperature processing plants for extracting natural gas liquids. The glycol prevents freezing in these plants during the condensation of water and hydrocarbons. The water phase of the process liquid contains the EG or DEG, which is always recovered and regenerated.

The minimum amount of hydrate inhibitor required can be calculated using Hammerschmidt's (1939) method:

$$W_h = \frac{(MW)_{inh}\Delta t_h}{(MW)_{inh}\Delta t_h + K_H} \times 100 \tag{12.16}$$

where

W_h = weight of pure inhibitor in liquid water phase, %

$(MW)_{inh}$ = molecular weight of inhibitor

Δt_h = depression of hydrate formation temperature, °F

K_H = Hammerschmidt constant for inhibitor, 2,335 for methanol, 4,000 for EG and DEG

If glycol is used as the inhibitor at an operating temperature of below 20 °F, the freezing point of the glycol must be considered. It is a common practice to keep the glycol concentrations (W_h) between 60 and 80 wt % to avoid "mushy" glycol in the system (Kohl and Riesenfeld 1985). If the calculated W_h value from Equation (12.16) is less than 60 percent, the quantity of inhibitor required should be calculated by a material balance:

$$W_G = I_{100}\left(\frac{100}{W_{out}} - \frac{100}{W_{in}}\right) \tag{12.17}$$

or

$$I_{100} = \frac{W_G}{\left(\frac{100}{W_{out}} - \frac{100}{W_{in}}\right)} \tag{12.18}$$

where

W_G = water removed from gas stream, lb_m/MMscf

I_{100} = pure inhibitor required, lb_m/MMscf

W_{out} = concentration of inhibitor in outlet inhibitor stream, wt %

W_{in} = concentration of inhibitor in inlet inhibitor stream, wt %

Combining equations (12.16) and (12.18) to eliminate W_{out} gives:

$$I_{100} = \frac{W_G}{1 + \frac{K_H}{(MW)_{inh} \Delta t_h} - \frac{100}{W_{in}}} \tag{12.19}$$

If methanol is used as the inhibitor, the vapor-phase inhibitor losses should also be considered. The following correlation has been obtained based on Jacoby's (1955) chart:

$$\frac{\text{lb}_\text{m}\ \text{MeOH/MMscf}}{\text{WT \% MeOH in Water}} = f(P,t) \tag{12.20}$$

$$f(P,t) = a_0 + a_1 p + a_2 p^2 + a_3 p^3 + a_4 p^4 + a_5 p^5 \tag{12.21}$$

where

p = pressure, psia

t = temperature, °F

Temp. (F)	a_0	a_1	a_2	a_3	a_4	a_5
25	3.9078	−0.0151	2.00E−05	3.00E−09	−4.00E−11	3.00E−14
30	5.409	−0.03	9.00E−05	−2.00E−07	1.00E−10	−4.00E−14
35	4.8559	−0.0192	4.00E−05	−5.00E−08	3.00E−11	−6.00E−15
40	3.8309	−0.0086	1.00E−05	−6.00E−09	2.00E−12	−2.00E−16
45	3.1025	−0.0041	3.00E−06	−1.00E−09	2.00E−13	−2.00E−17
50	3.3181	−0.004	3.00E−06	−1.00E−09	2.00E−13	−2.00E−17
55	3.5711	−0.0038	3.00E−06	−1.00E−09	2.00E−13	−1.00E−17
60	2.4814	−0.0009	2.00E−07	−2.00E−11	0	0
65	2.3502	−0.0006	1.00E−07	−1.00E−11	0	0

Example Problem 12.4

Ten MMscfd of a 0.7 specific gravity natural gas cools down to 40 °F in a buried pipeline. The minimum pipeline pressure is 900 psia. Concentrations of commercially available glycol and methanol inhibitors are 75% and 100% by weight, respectively. What volume of inhibitor solution must be added daily if the gas enters a line saturated at 90 °F? Consider both DEG and MeOH inhibitors.

Solution

This problem can be solved quickly with two spreadsheet programs. Spreadsheet GlycolInjection.xls is for glycol injection calculations; while MethanolInjection.xls is for methanol injection calculations. Solutions to this example problem are shown in Tables 12.4, 12.5, and 12.6 for glycol and methanol inhibitions, respectively.

Table 12–4 Glycol Inhibition Input Data Given by GlycolInjection.xls[a]

Instructions: 1) Update input data; 2) View result.				
Input Data				
Gas flow rate:			10 MMscfd	
Gas-specific gravity (γ_g):			0.7 air = 1	
Minimum pressure (p_m):			900 psia	
Inlet gas temperature (t_i):			90 °F	
Hydrate temperature t_m:			64 °F	
Operating temperature (t_o):			40 °F	
Water content at t_i and p_m:			48 lb/MMscf	
Water content at t_o and p_m:			9.6 lb/MMscf	
Inhibitor data:	**Used (1=Yes; 0=No)**	W_{in}	W_{out}	**Density**
EG	1	75%	65%	9.08 ppg
DEG	0	75%	65%	9.08 ppg
Concentration of inhibitor in its commercial solution:			75 Wt %	

a. This spreadsheet calculates the minimum amount of required inhibitor usage.

Table 12-5 Glycol Inhibition Calculations and Results Given by GlycolInjection.xls

Intermediate Calculations		
Water to be removed from gas (w_G):		38.4 lb/MMscf
Molecular weight of inhibitor $(MW)_{inh}$:		62
Density of inhibitor (ρ_{inh}):		9.08 ppg
Hammerschmidt constant for inhibitor (K_H):		4,000
Depression of hydrate formation temperature (Δt_h):		24 °F
W_{out} for the selected inhibitor:		65 Wt %
W_{in} for the selected inhibitor:		75 Wt %
Inhibitor requirement based on Hammerschmidt's method:		
$W_h = \dfrac{(MW)_{inh}\Delta t_h}{(MW)_{inh}\Delta t_h + K_H} \times 100$	=	27.11 Wt %
Inlet concentration to be used		75 Wt %
$I_{100} = \dfrac{W_G}{\left(\dfrac{100}{W_{out}} - \dfrac{100}{W_{in}}\right)}$	=	187.20 lb/MMscf
Result		
Required weight of commercial inhibitor solution		249.6 lb/MMscf
Required volume of commercial inhibitor solution		27.49 gal/MMscf
Hourly inhibitor injection rate		11.45 gal/hour
Daily inhibitor injection rate		274.89 gal/day

12.4 Pipeline Cleaning

Gas pipelines are cleaned with tools called pigs. The term pig was originally used to refer to Go-Devil scrapers driven through the pipeline by the

Table 12-6 Methanol Inhibition Input Data and Calculations Given by MethanolInjection.xls[a]

Instructions: 1) Update input data; 2) View result.		
Input Data		
Gas flow rate:		10 MMscfd
Gas-specific gravity (γ_g):		0.7 air = 1
Minimum pressure (p_m):		900 psia
Inlet gas temperature (t_i):		90 °F
Hydrate temperature at t_m:		64 °F
Operating temperature (t_o):		40 °F
Water content at t_i and p_m:		48 lb/MMscf
Water content at t_o and p_m:		9.6 lb/MMscf
Inhibitor solution density:		6.64 ppg
Inhibitor concentration (W_{in}):		100 Wt %
Intermediate Calculations		
Water to be removed from gas (w_G):		38.4 lb/MMscf
Molecular weight of inhibitor $(MW)_{inh}$:		32
Hammerschmidt constant for inhibitor (K_H):		2,335
Depression of hydrate formation temperature (Δt_h):		24 °F
Inhibitor requirement based on Hammerschmidt's method:		
$W_h = \dfrac{(MW)_{inh}\Delta t_h}{(MW)_{inh}\Delta t_h + K_H} \times 100$	=	24.75 Wt %
Liquid phase:		
$I_{100} = \dfrac{W_G}{\left(\dfrac{100}{W_{out}} - \dfrac{100}{W_{in}}\right)}$	=	12.63 lb/MMscf

Table 12–6 Methanol Inhibition Input Data and Calculations Given by MethanolInjection.xls[a] (Continued)

Gas phase:		
$\frac{lb_m \text{ MeOH/MMscf}}{\text{WT \% MeOH in Water}} = f(P,t)$	=	1.05
Inhibitor requirement:		25.99 lb/MMscf
System total:		
Inhibitor requirement:		38.62 lb/MMscf
Daily injection rate:		386.18 lb/day
		58.16 gal/day
Hourly injection rate:		2.42 gal/hour

a. This spreadsheet calculates the minimum amount of required inhibitor usage

flowing fluid trailing spring-loaded rakes to scrape wax off the internal walls. One of the tales about the origin of the name pig is that the rakes made a characteristic loud squealing noise. Pipeline operators now use the word pig to describe any device made to pass through a pipeline for cleaning and other purposes. The process of driving the pig through a pipeline by fluid is called a pigging operation.

Although pigs were originally developed to remove deposits, which could obstruct or retard flow through a pipeline, today pigs are used during all phases in the life of a pipeline and for many different reasons. During pipeline construction, pigging is used for removing debris, gauging, cleaning, flooding, and dewatering. During fluid production operations, pigging is utilized for removing wax in oil pipelines, removing liquids in gas pipelines, and meter proving. Pigging is widely employed for pipeline inspection purposes such as wall thickness measurement and detection of spanning and burial. Pigging is also run for coating the inside surface of a pipeline with inhibitors and providing pressure resistance during other pipeline maintenance operations. Figure 12–9 shows pipeline deposits displaced by a pig. This chapter describes how to apply different pigging techniques to solve various problems in the pipeline operations.

Figure 12–9 Pipeline deposits that could obstruct or retard flow through a pipeline (Courtesy of Pigging Products & Services Association).

12.4.1 Pigging System

A pigging system includes pigs, a launcher, and a receiver. It also includes pumps and compressors, which are not discussed here because they have to be available for transporting the product fluids anyway. Obviously pigs are the most essential equipment. Although each pipeline has its own set of characteristics that affects how and why pigging is utilized, there are basically three reasons to pig a pipeline: (1) to batch or separate dissimilar products; (2) to displace undesirable materials; and (3) to perform internal inspections. The pigs used to accomplish these tasks fall into three categories:

- Utility pigs: used to perform functions such as cleaning, separating, or dewatering
- In-line inspection tools: provide information on the condition of the line, as well as the extent and location of any problems
- Gel pigs: used in conjunction with conventional pigs to optimize pipeline dewatering, cleaning, and drying tasks

12.4.1.1 Utility Pigs

Utility pigs can be divided into two groups based upon their fundamental purpose: (1) cleaning pigs used to remove solid or semisolid deposits or debris from the pipeline, and (2) sealing pigs used to provide a good seal in order to either sweep liquids from the line, or provide an interface between two dissimilar products within the pipeline. Within these two groups, a further subdivision can be made to differentiate among the various types or forms of pigs: spherical pigs, foam pigs, mandrel pigs, and solid cast pigs.

Spherical pigs, or spheres, are of either a solid composition or inflated to their optimum diameter with glycol and/or water. Figure 12–10 shows some spheres. Spheres have been used for many years as a sealing pig. There are four basic types of spheres: inflatable, solid, foam, and soluble. Soluble spheres are usually used in crude oil pipelines containing microcrystalline wax and paraffin inhibitor. Spheres normally dissolve in a few hours. The dissolving rate depends on fluid temperature, fluid movement, friction, and absorbability of the crude. If the line has never been pigged, it is a good idea to run the soluble pig. If it hangs up in the line, it will not obstruct the flow. Inflatable spheres are manufactured of various elastomers (polyurethane, neoprene, nitrile, and Viton) depending on their applications. An inflatable sphere has a hollow center with filling valves that are used to inflate the sphere with liquid. Spheres are filled with water, or water and glycol and inflated to the desired size. Spheres should never be inflated with air. Depending on the application and material, the sphere is inflated 1 to 2 percent more than the pipe's inside diameter. As the sphere wears from service, it is resized, extending its life. In small sizes the sphere can be manufactured solid, eliminating the need to inflate it. The solid sphere does not have the life of an inflatable sphere because it cannot be resized. Spheres can also be manufactured from open cell polyurethane foam. They can be coated with a polyurethane material to give better wear. For cleaning purposes they can have wire brushes on the surface. The advantages of the foam sphere are that they are lightweight, economical, and do not need to be inflated. Spheres in general are easy to handle, and negotiate short radius 90° irregular turns and bends. They go from smaller lateral lines to larger main lines and are easier to automate than other styles of pigs. Spheres are commonly used to remove liquids from wet gas systems, serve to prove fluid meters, control paraffin in crude oil pipelines, flood pipelines to conduct hydrostatic tests, and dewater after pipeline rehabilitation or new construction. Special design considerations for the pipeline should be considered when using spheres. They should never be run in lines that do not have special flow tees installed.

Figure 12–10 Some spheres used in the pipeline pigging operations (Courtesy of Girard Industries, Inc.).

Foam pigs, also known as Polly-Pigs, are molded from polyurethane foam with various configurations of solid polyurethane strips and/or abrasive materials permanently bonded to them. Figure 12–11 demonstrates a foam pig and how it works.

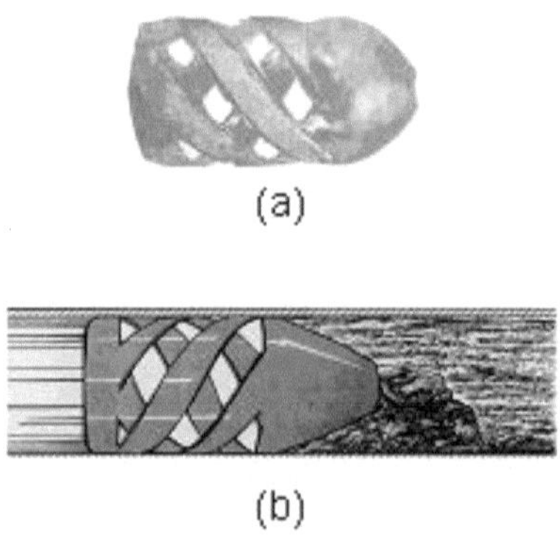

Figure 12–11 (a) A foam pig; (b) An ideal foam pig cleaning the pipeline (Courtesy of Montauk Service, Inc.).

Foam pigs are molded from open cell polyurethane foams of various densities ranging from light density (2 lbs/ft^3), medium density (5 to 8 lbs/ft^3), to heavy density (9 to 10 lbs/ft^3). They are normally manufactured in a bullet shape. They can be bare foam or coated with a 90-durometer polyurethane material. The coated pigs may have a spiral coating of polyurethane, various brush materials, or silicon carbide coating. If the pig is of bare foam, it will have the base coated. The standard foam pig length is twice the diameter. Foam pigs are compressible, expandable, lightweight, and flexible. They travel through multiple diameter pipelines, go around mitered bends, and short radius 90° bends. They make abrupt turns in tees so laterals can be cleaned. They also go through valves with as little as a 65 percent opening. The disadvantages of foam pigs are that they are one-time use products; shorter length of runs and high concentrations of some acids will shorten life. Foam pigs are also inexpensive. Foam pigs are used for pipeline proving, drying, and wiping; removal of thick soft deposits; condensate removal in wet gas pipelines; and pigging multiple diameter lines. Foam pigs coated with a wire brush or silicon carbide are used for scraping and mild abrasion of the pipeline.

A mandrel pig has a central body tube, or mandrel, and various components can be assembled onto the mandrel to configure a pig for a specific duty. Figure 12–12 demonstrates some mandrel pigs.

Figure 12–12 Some mandrel pigs used in pipeline pigging operations (Courtesy of Girard Industries, Inc.).

The pig is equipped with wire brushes or polyurethane blades for cleaning the line. The mandrel pig can be either a cleaning pig, sealing pig, or a combination of both. The seals and brushes can be replaced to make the pig reusable. Cleaning pigs are designed for heavy scraping and can be equipped with wire brushes or polyurethane blades. These pigs are designed for long runs. Bypass holes in the nose of the pig control the speed or act as jet ports to keep debris suspended in front of the pig. The cost of redressing the pig is high, and larger pigs require special handling equipment to load and unload the pig. Occasionally the wire brush bristles break off and get into instrumentation and other unwanted places. Smaller size mandrel pigs do not negotiate 1.5D bends.

Solid cast pigs are usually molded in one piece from polyurethane, however, neoprene, nitrile, Viton, and other rubber elastomers are available in smaller size pigs. Figure 12–13 demonstrates some solid cast pigs. Solid

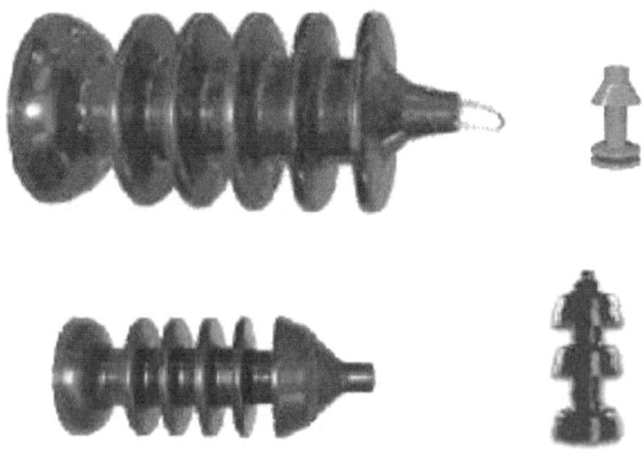

Figure 12–13 Some solid cast pigs used in pipeline pigging operations (Courtesy of *Apache Pipeline Products, Inc.*).

cast pigs are considered sealing pigs although some solid cast pigs are available with wraparound brushes and can be used for cleaning purposes. The solid cast pig is available in cup, disc, or a combination cup/disc design. Most of the pigs are of one-piece construction but several manufacturers have all urethane pigs with replaceable sealing elements. Because of the cost to redress a mandrel pig, many companies use the solid cast pig up through 14 or 16 inches. Some solid cast designs are available in sizes of up to 36 inches. Solid cast pigs are extremely effective in removing liquids from product pipelines, removing condensate and water from wet gas systems, and controlling paraffin buildup in crude oil systems.

12.4.1.2 In-Line Inspection Tools

In-line inspection tools are used to carry out various types of tasks including:

- measuring pipe diameter/geometry
- monitoring pipeline curvature
- determining pipeline profile
- recording temperature/pressure
- measuring bend
- detecting metal-loss/corrosion
- performing photographic inspection
- detecting cracks
- measuring wax deposition
- detecting leaks
- taking product samples
- mapping

A typical in-line inspection tool is an ultrasonic tool shown in Figure 12–14. Ultrasonic in-line inspection tools are used for measuring metal loss and detecting crack of pipelines. Ultrasonic tools are especially suitable if there are high requirements regarding sensitivity and accuracy, which is especially relevant in offshore pipelines. Ultrasonic tools are also well suited with regard to the range of wall thickness usually experienced in offshore lines.

12.4.1.3 Gel Pigs

Gel pigs have been developed for use in pipeline operations, either during initial commissioning, or as part of a continuing maintenance program.

Figure 12–14 An ultrasonic inspection tool (Courtesy of Pigging Products and Services Association).

Figure 12–15 shows how gel pigs work The principle pipeline applications for gel pigs are as follows:

- product separation
- debris removal
- line filling/hydrotesting
- dewatering and drying
- condensate removal from gas lines
- inhibitor and biocide laydown
- special chemical treatment
- removal of stuck pigs

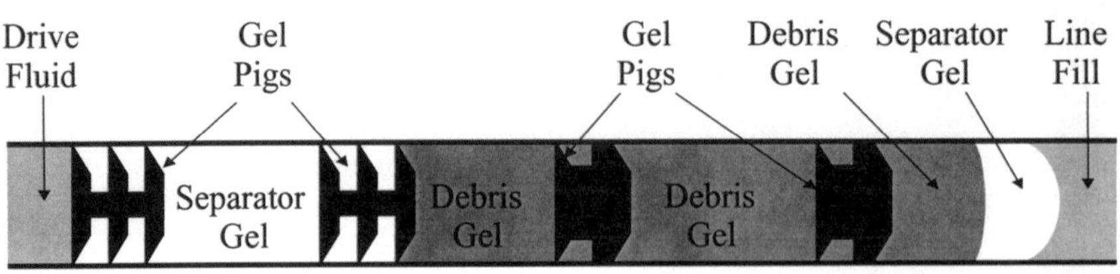

Figure 12–15 Application of gel pigs in pipeline pigging operations.

Most pipeline gels are water-based, but a range of chemicals, solvents, and even acids can be gelled. Some chemicals can be gelled as the bulk liquid and others only diluted in a carrier. Gelled diesel is commonly used as a carrier of corrosion inhibitor in gas lines. The four main types of gel used in pipeline applications are batching or separator gel, debris pickup gel, hydrocarbon gel, and dehydrating gel. The gel can be pumped through any line accepting liquids. Gel pigs can be used alone (in liquid lines), in place of batching pigs, or in conjunction with various types of conventional pigs. When used with conventional pigs, gelled pigs can improve overall performance while almost eliminating the risk of sticking a pig. Gel pigs do not wear out in service like conventional pigs. They can, however, be susceptible to dilution and gas cutting. Care must be taken when designing a pig train that incorporates gel pigs to minimize fluid bypass of the pigs, and to place a conventional pig at the back of the train when displacing with gas. Specially formulated gels have also been used to seal gate valves during hydrostatic testing. Gels have been developed with a controlled gellation time and a controlled viscosity for temporary pipeline isolation purposes.

12.4.1.4 Launchers and Receivers

Pigs generally need specially designed launching and receiving vessels (launcher and receiver) to introduce them into the pipeline. The launcher and receiver are installed at the upstream and downstream of the pipeline section being pigged, respectively. The distance between the launcher and receiver depends on the service, location of pump (liquid product) or compressor (gas product) stations, operating procedures, and the materials used in the pig. In crude oil pipeline systems, the distance between launcher and receiver can be as long as 500 miles for spheres and 300 miles for pigs. The amount of sand, wax, and other materials carried along the pig can affect the proper distance. In gas transmission service, the distance between the launcher and receiver can be as long as 200 miles for spheres and 100 miles for pigs, depending on the amount of lubrication used.

The launcher and receiver consist of a quick opening closure for access, an oversized barrel, a reducer and a neck pipe for connection to the pipeline. Pigs can be located using fixed signalers along the pipe or electronic tracking systems mounted inside the pig.

Typical configurations of pig launchers and receivers for gas service are depicted in Figure 12–16 and Figure 12–17, respectively. The inclined barrels should be long enough to hold ten or more spheres. In large-diameter gas pipelines, the barrel diameter can be one inch larger than the pipeline.

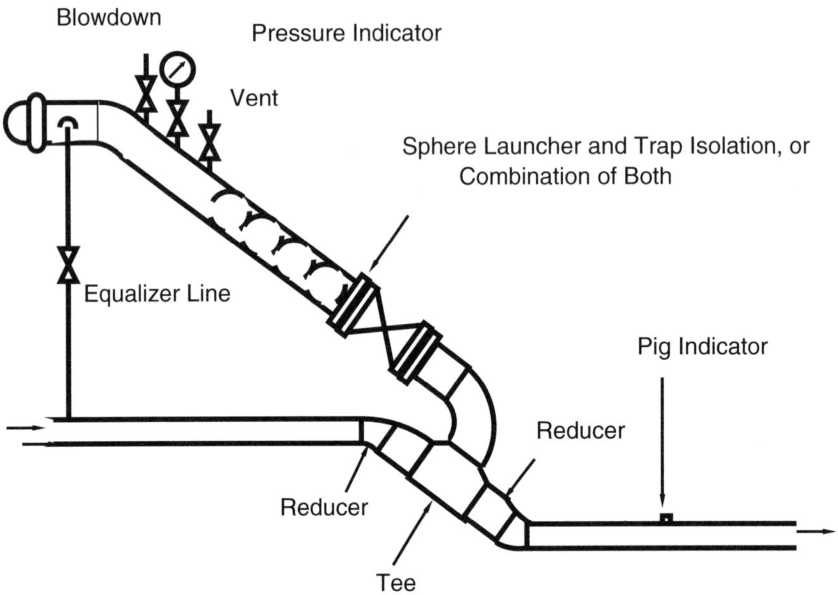

Figure 12–16 A typical configuration of a pig launcher for gas services.

12.4.2 Selection of Pigs

The purpose of operational pigging is to obtain and maintain efficiency of the pipeline to be pigged. The pipeline's efficiency depends on two things: first, it must operate continuously, and second, the required throughput must be obtained at the lowest operating cost. The type of pig

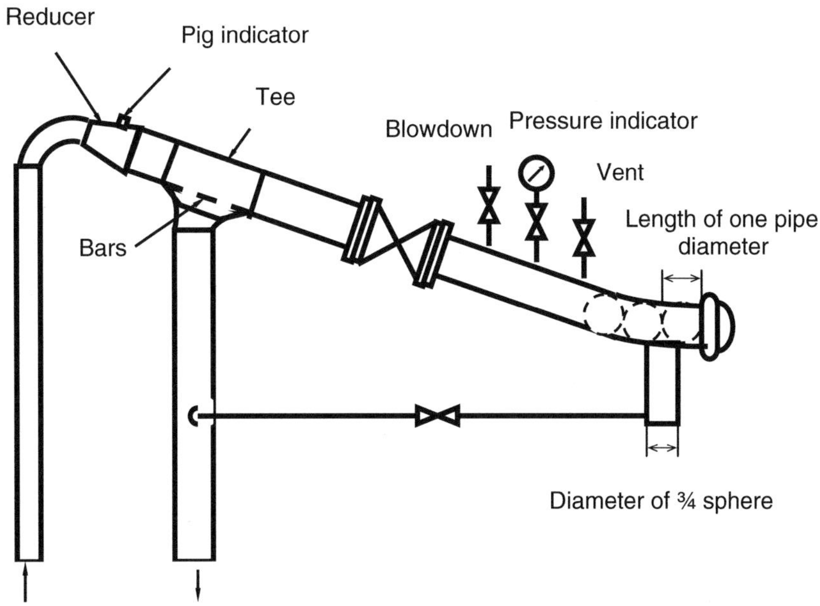

Figure 12–17 A typical configuration of a pig receiver for gas services.

to be used and its optimum configuration for a particular task in a particular pipeline should be determined based upon several criteria including:

Purpose of pigging

- type, location, and volume of the substance to be removed or displaced
- type of information to be gathered from an intelligent pig run
- objectives and goals for the pig run

Line contents

- contents of the line while pigging
- available versus required driving pressure
- velocity of the pig

Characteristics of the pipeline

- the minimum and maximum internal line sizes
- the maximum distance the pig must travel
- the minimum bend radius and bend angles
- additional features such as valve types, branch connections, and the elevation profile

12.4.2.1 Cleaning Pigs

Cleaning pigs are designed to remove solids or accumulated debris in the pipeline. This increases the efficiency and lowers the operating cost. They have wire brushes to scrape the walls of the pipe to remove the solids. Pigs 14 inches and smaller normally use rotary wire wheel brushes. These brushes are easy to replace and inexpensive. Special rotary brushes are used on some larger pigs. Larger pigs have wear-compensating brushes. These brushes can be individually replaced as needed and are mounted on either leaf springs, cantilever springs, or coil springs. The springs push the brushes against the pipe wall. As the wire brushes wear, the force of the spring keeps it in contact with the pipe wall compensating for the brush wear. There are many different brush materials available. The standard brushes are made of fine or coarse carbon steel wire. Prostran is the material of choice for pipelines with internal coatings. Some service requires a stainless steel brush. Special brush designs such as the pit cleaning brush are also available. When soft deposits of paraffin, mud, and so forth, need to be removed, the urethane blade is an excellent choice. The blade design is interchangeable with the brushes. Bypass ports are installed in the nose of the pig or on the body. These ports are used to control fluid bypass. If the ports are on the body of the pig, the flow will also flow through the brushes and keep them clean. As the fluid passes through the ports on the nose of the pig, it helps keep the debris in front of the pig stirred up and moving. Plugs are used to regulate the bypass. The sealing elements are either elastomer cups or discs. They are used as a combination cleaning and sealing element to remove soft deposits. Cups are of standard or conical design. Specialty cups are available for some applications. The cup and disc material is normally manufactured from a polyurethane material, which gives outstanding abrasion, and tear resistance but is limited in temperature range. Neoprene, nitrile, EPDM, and Viton are available for higher temperature applications.

The best choice for cleaning applications are normally pigs with discs, conical cups, spring mounted brushes, and bypass ports. Figure 12–18 shows details of two pigs of this type. Discs are effective for pushing out solids and providing good support for the pig. Conical cups provide sealing characteristics, good support, and long wear. Spring-mounted brushes provide continuous forceful scraping for removal of rust, scale, and other buildups on the pipe wall. Instead of brushes, polyurethane scraper blades can also be selected for cleaning waxy crude oil lines because the scraper blades are easier to clean than brushes. Bypass ports allow some of the flow to bypass through the pig. This can help minimizing solids buildup in front of the pig. For a new pipeline construction, it is a good practice to include a magnetic cleaning assembly in the pig.

Figure 12–18 Some mandrel pigs used in pipeline pigging operations (Courtesy of T. D. Williamson, Inc.).

12.4.2.2 Gauging Pigs

Gauging pigs are used after constructing the pipeline to determine if there are any obstructions in the pipeline. These obstructions can be caused by partially closed valves, wrinkle bends, ovality caused by overburden, dents caused by rocks underneath the pipe, third-party damage, and buckles caused by flooding, earthquakes, and so forth. A gauging pig assures that the ovality of the line is within accepted tolerance. The gauging plate may be mounted on the front or rear of the pig and is made of a mild steel or aluminum. The plate may be slotted or solid. The outside diameter of the plate is 90 to 95 percent of the pipe's inside diameter. Gauging runs are normally done during new construction and prior to running a corrosion inspection pig. The best practice is to choose inspection tools that can provide critical information about the line such as determining the location (distance), o'clock position, and severity of a reduction.

12.4.2.3 Caliper Pigs

Caliper pigs are used to measure pipe internal geometry. They have an array of levers mounted in one of the pig cups. The levers are connected to a recording device in the pig body. The body is normally compact, about 60 percent of the internal diameter, which, combined with flexible cups, allows the pig to pass constrictions up to 15 percent of bore. Caliper pigs can be used as gauging pigs. The ability of a caliper pig to pass constrictions means minimum risk of jamming. This is very important for subsea pipelines where it would be very difficult and expensive to locate a stuck pig.

12.4.2.4 Displacement Pigs

Displacement pigs displace one fluid with another based on a sealing mechanism. They can be bidirectional or unidirectional in design. They are used in the testing and commissioning phase of the pipeline, that is, hydrostatic testing, line fills, and dewatering, and so forth. Line evacuation and abandonment is another application for the displacement pig. Bidirectional (Figure 12–19) pigs can be sent back to the launch site by reversing flow should they encounter an obstacle. They are also used in filling and dewatering associated with hydrostatic testing when the water used to fill the line has to be pushed back to its source after completion of a test. The best choice of displacement pigs are normally pigs with multilipped conical cups (Figure 12–20). Conical cups can maintain contact with the pipe wall even in out-of-round pipe, which is more common in large-diameter pipelines. Conventional cups and discs usually cannot maintain a seal in the out-of-round pipe. Multilipped cups have numerous, independent sealing lips on each cup which significantly improve their ability to maintain a seal.

12.4.2.5 Profile Pigs

A profile pig is a gauging pig with multiple gauging plates, usually three plates. One plate is mounted on the front, one in the middle, and one on the rear of the pig. It is normally used before running an ILI (In-Line Inspection) tool to assure the tool's passage around bends and through the pipeline.

12.4.2.6 Transmitter Pigs

Occasionally pigs will get stuck in a line. The location of the stuck pig can be found by using a detector pig with a transmitter in its body. The transmitter emits a signal so it can be located with a receiver. Transmitters are normally mounted into a mandrel, solid cast, or Polly-Pig.

12.4.2.7 Special Pigs

Many applications require special pigs. Manufacturers in the pigging industry have made special pigs for many applications. Figure 12–21 illustrates that a special pig can be used for spraying corrosion inhibitor to the upper side of pipe interior. Dual diameter pigs are designed for pigging dual diameter pipelines. They are usually mandrel pigs fitted with solid discs for the smaller line and slotted discs for the larger line. If it is a cleaning pig, the brushes will support it in the line and keep the pig centered. The Polly-Pig is also widely used in this application.

Other special pigs include pinwheel pigs, which use steel pins with hardened tips. They were developed to remove wax and scale from a pipeline. Magnetic cleaning pigs were developed to pick up ferrous debris left in the pipeline.

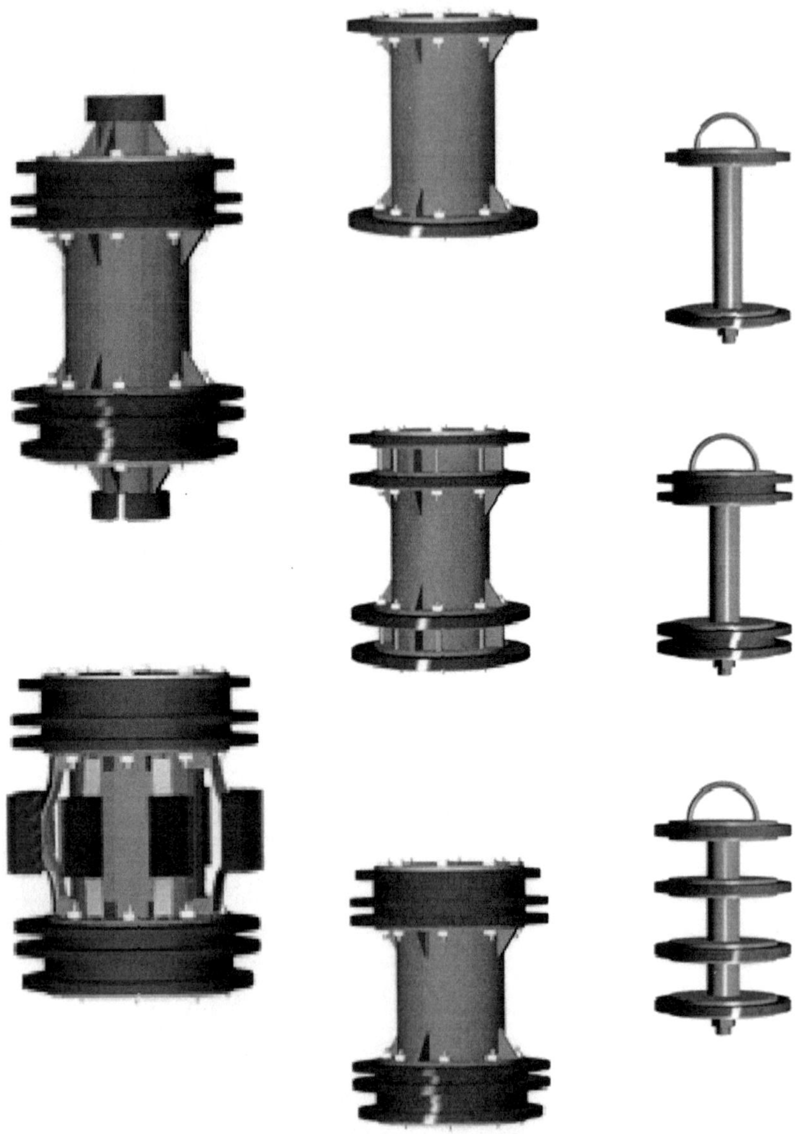

Figure 12–19 Some bidirectional pigs used in pipeline pigging operations (Courtesy of Apache Pipeline Products, Inc.).

There are many pig configurations to choose from but some configurations will not work in some pipelines. It is very important to compare

Figure 12–20 Pig with multilipped conical cups (Courtesy of T. D. Williamson, Inc.).

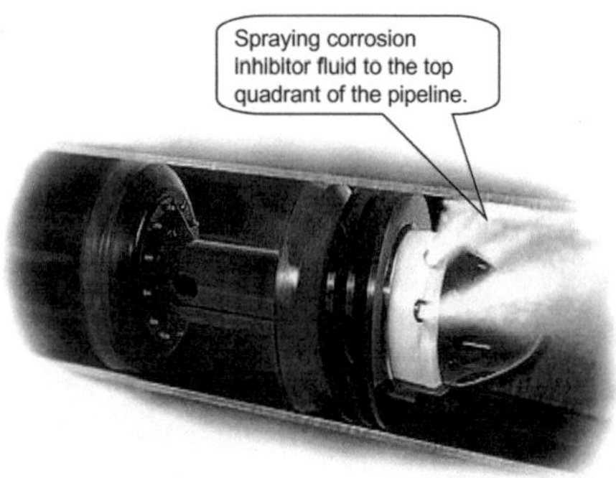

Figure 12–21 A special pig for spraying corrosion inhibitor (Courtesy of T. D. Williamson, Inc.).

pipeline information to the pig specifications. The best way to stay out of trouble is to provide the pipeline specifications to the pig manufacturer and ask them to recommend a pig.

12.4.3 Major Applications

Major applications of pigging are found during pipeline construction, operation, inspection, and maintenance. Depending on application type and pipeline conditions, different kinds of pigs are chosen to minimize the cost of pigging operations. Tiratsoo (1992) presents a comprehensive description of applications of pigging in the pipeline industry.

12.4.3.1 Construction

During pipeline construction, it is quite possible for construction debris to get inside the pipeline. The debris could harm downstream equipment such as filters and pumps. The only way to remove possible debris is to run a pig through the pipeline. Typically, debris removal is done section by section as the lay barge moves forward. An air-driven cleaning pig is usually sent through the pipeline section to sweep out the debris. Features of the cleaning pig should be selected based on anticipated pipeline conditions. The most effective way to clear debris is with the use of a magnetic cleaning assembly, which can be mounted on conventional pigs. The removal of this type of debris is a must before attempting to run corrosion inspection pigs

Pipelines subjected to subsea conditions may buckle in certain sections. The place for a buckle to occur during pipe laying is most likely the sag bend just before the touchdown on the seabed. To detect the buckle, a gauging pig is pulled along behind the touchdown point. If the pig encounters a buckle, the towing line goes taut indicating that it is necessary to retrieve and replace the buckled section of the pipeline. Features of the gauging pig should be selected based on anticipated pipeline conditions. Caliper pigs can be used as gauging pigs after completion of construction. The ability of the caliper pig to pass constrictions can reduce the possibility of jamming, which is vitally important for subsea pipelines where it would be very difficult and expensive to locate a stuck pig.

Upon completion of construction, the pipeline should be cleaned to remove the rust, dirt, and millscale that contaminate product fluids. These contaminates also reduce the effectiveness of corrosion inhibitor. A typical cleaning operation would consist of sending through a train of displacement pigs with different features suitable to the pipeline conditions. Gel slugs are used to pick up the debris in suspension, clearing the pipeline more efficiently. Corrosion inhibitor can also be added to the interior of the pipeline in the trip of cleaning pigging.

After cleaning, the pipeline is flooded with water for hydrotest. Air must be completely removed so that the pipeline can be pressurized efficiently. Pigging with displacement pigs is normally the best solution for flooding a pipeline. Use of bidirectional batching pigs is favorable for the afterward-dewatering operation.

Upon a successful hydrotest, water is usually displaced with air, nitrogen, or the product fluid. Because dewatering is the reverse process of flooding, a bidirectional batching pig used to flood the pipeline, left during the hydrotest, can be used to dewater the pipeline. In cases of gas service pipelines, it is necessary to dry the pipeline to prevent formation of hydrates and waxy solids. For this purpose, methanol or glycol slugs can be sent through the pipeline between batching pigs. An alternative means of drying the pipeline is to vacuum the pipeline with vacuum pumps.

12.4.3.2 Operation

During fluid production operations, pigging is utilized to maintain efficiency of pipelines by removing wax in oil pipelines and liquids (water and condensate) in gas pipelines. Sometimes pigging operations are for meter proving. Pipeline wax is characterized as long-chain paraffin formed and deposited in pipelines due to changes in pressure and temperature. Accumulation of wax in a pipeline reduces the effective pipeline hydraulic diameter and hence efficiency of the pipeline. A variety of cleaning pigs are available to remove wax. Most of them work on the principle of causing a bypass flow through the body of the pig over the brushes or scrapers and out to the front. The pigs used for removing wax should be selected to have features that induce the bypass flow. The action of the pig also polishes wax remaining on the pipe wall, leaving a surface for low flow resistance of product fluids. To remove hard scale deposits, aggressive and progressive pigs are the best choice. They can be used with cleaning fluids that help to attack the deposits and/or help to keep the deposits in suspension while being pushed out of the line. This is a very special application that would normally be provided by a pipeline cleaning service company. Samples of the deposits are usually required for chemical analysis and to see what cleaning fluids are best suited.

Sometimes chemical cleaning is used for removal of specific types of pipe deposits. Chemical cleaning is a process of using pigs in conjunction with environmentally friendly detergent-based cleaning fluids and is almost always done by pipeline cleaning service companies. The detergents help to suspend solid particles and keep them in a slurry thus allowing removal of large volumes of solids in one pig run. Samples of the material to be removed from the line are required in order to select the best cleaning fluid. The cleaning fluids are captured between batching and cleaning pigs and normally a slug of the fluid is introduced in front of the first pig.

In gas service pipelines, liquid water and/or gas condensate can form and accumulate on the bottom of the pipeline. The liquid accumulation reduces flow efficiency of the pipeline. It can also develop slug flow causing problems with the processing facilities. Different types of displacement and cleaning pigs are available to remove the liquids. Because gas is the drive fluid, the pigs used for removing the liquids in the gas pipeline should be selected to have features of good sealing. Spheres are usually the preferred choice for liquid removal from wet gas systems. Most of these systems are designed to automatically (remotely) launch and receive spheres. A large number of spheres can be loaded into the automatic launcher and launched at predetermined frequencies. At the receiving end of the line is a slug catcher to capture all the liquid pushed in by a sphere. If more liquid is brought in than the slug catcher can handle, the plant normally shuts down. Thus spheres are launched at a frequency that prevents exceeding the capacity of the slug catcher. Pipeline systems are normally designed for use with spheres or pigs but not both. Pipelines designed for spheres may require modifications of launchers and receivers in order to run conventional pigs.

To clean pipelines with known internal corrosion, special pigs are available that have independent scraping wires that will go into a pit to break up and remove deposits that prevent corrosion inhibitors from getting to the corroding area. Brushes on conventional pigs will not extend into a pit. To clean internally coated pipelines, the preferred choice is a pig with discs and cups as these will normally remove deposits from the coating due to the "teflon-like" characteristics of epoxy coatings. Conventional cleaning pigs with "prostran" brushes or polyurethane blades can also be used on internally coated pipelines.

12.4.3.3 Inspection

A variety of intelligent pigs have been employed for pipeline inspection purposes including detection of not only dents and buckles but also corrosion pitting, cracks, spanning and burial, and measurement of wall thickness. The information obtained from the pigging operations is used for assessment of pipeline safety and integrity.

Magnetic-flux leakage pigs have been used for detection of dents and buckles, and measurement of pipe ovality and wall thickness over the entire pipe surface. The principle of magnetic-flux leakage detection relies on measurement of metal loss, and hence the size of the defect. Usually a series of survey runs over years are required to establish trends. Magnetic-flux leakage pigging can be utilized in liquid and gas pipelines.

An ultrasonic intelligent pig is used to make direct measurement of wall thickness of the entire pipe surface. They are better suited to liquid pipelines and cannot be used in gas pipelines without a liquid couplant.

Pipeline spans have traditionally been found by external inspection using side-scan sonar. In recent years, neutron-scatter pigs have been employed to detect spanning and burial in subsea pipelines with lower cost and better accuracy.

12.4.3.4 Maintenance

Pigging is also run for maintenance of pipelines for coating the inside surface of the pipelines, providing pressure resistance, and installing barrier valves. Traditionally the internal surfaces of pipe joints are precoated with a smooth epoxy liner and leave the welds uncoated. Recently a pigging system has been developed to coat the entire pipeline internal surface by first cleaning the surface and then pushing through a number of slugs of epoxy paint.

Shutting down offshore, especially deepwater, pipelines for maintenance is very expensive. With advanced technology, it is possible to carry out some maintenance jobs without shutting down the pipeline. In cases where there are not enough isolation valves, a pressure-resisting plug may be pigged into the pipeline to seal off downstream operation.

Corrosion inhibitors are normally injected into the line on a continuous basis and carried through the line with the product flow. Sometimes inhibitors are batched between two pigs but there is no way to guarantee the effectiveness of this method especially at the twelve o'clock position. Special pigs have been developed that spray inhibitor to the top of the pipe as they travel through the pipe. This is done using a siphoning effect created by bypass flow through an orifice specifically designed to pick up inhibitor from the bottom of the pipe.

12.4.4 Pigging Procedure

12.4.4.1 Pressure and Flow Rate

Any pigging operation should follow a safe procedure that is suitable to the given pipeline conditions. Operating pigging pressures and fluid flow rates should be controlled carefully. Velocity of driving fluid is usually between three and five feet per second during pigging. Recommended ranges of operating pressures and flow rates are presented in Table 12–7.

12.4.4.2 Prerun Inspection

The pig must be in good condition if it is to do the job it was selected to do. If the pig has been run before, it should be inspected to assure it will run again without stopping in the pipeline. Measure the outside diameter of the pig's sealing surface. This diameter must be larger than the inside pipe diameter to maintain a good seal. Inspect the sealing surfaces to assure there are no cuts, tears, punctures, or other damage that will affect the pig's ability to run in the pipeline. The unrestrained diameter of brush pigs should also be measured to assure that the brushes will maintain contact with the pipe wall during the complete run. When using brush-type mandrel cleaning pigs, the brushes should be inspected for corrosion or breakage. Every precaution should be taken to prevent these brushes from breaking in the pipeline. Loose bristles can damage valves, instrumentation, and other pipeline equipment. All components of brush-type mandrel pigs should be checked to be certain that they are tight and in good condition.

Table 12–7 Recommended Pigging Pressures and Flow Rates

Pipe Inner Diameter (in)	Typical Pigging Pressure (psig)		Liquid Flow Rate (GPM)		Gas Flow Rate (SCFM)	
	Launching	Running	3 FPS	5 FPS	5 FPS	10 FPS
2	100 to 200	40 to 100	20	40	30	60
3	100 to 150	35 to 85	60	100	70	140
4	75 to 125	30 to 80	110	190	120	240
6	50 to 100	30 to 75	260	430	270	540
8	30 to 80	25 to 70	460	770	440	880
10	30 to 60	25 to 50	720	1,200	580	1,200
12	30 to 50	20 to 45	1,040	1,700	760	1,500
14	20 to 50	15 to 40	1,400	2,300	930	1,900
16	15 to 45	10 to 40	1,800	3,100	1,100	2,200
18	15 to 40	10 to 30	2,300	3,900	1,200	2,400
20	10 to 25	5 to 20	2,900	4,800	1,200	2,400
24	10 to 25	5 to 20	4,100	6,900	1,700	3,400
30	10 to 20	5 to 15	6,500	10,900	2,400	4,800
36	10 to 20	5 to 10	9,400	15,700	3,200	6,400
40	10 to 20	5 to 10	11,600	19,400	4,000	8,000
42	10 to 20	5 to 10	12,800	21,400	4,400	8,800
48	10 to 20	5 to 10	16,700	27,900	5,800	11,600
54	10 to 20	5 to 10	21,200	35,300	7,300	14,600
60	10 to 20	5 to 10	26,200	43,600	9,000	18,000
72	10 to 20	5 to 10	37,700	62,900	13,000	26,000

12.4.4.3 Pig Launching and Receiving

Pig launchers are used to launch the pig into the pipeline, and pig receivers are used to receive the pigs after they have made a successful run. The design of these pig traps will depend on the type of pig to be run and pipeline design conditions. Provisions in the station design should include handling equipment for pigs 20 inches and larger. Caution should be taken for liquid spillage from the pig traps.

The following pig-launching procedures can be used as a guideline for developing operating procedures. Because company policies vary regarding whether the pig launcher is left on stream or isolated from the pipeline after the pig is launched, the operator should verify that the trap is isolated from the pipeline and depressurized before commencing any part of the launch procedure.

To launch pigs, make sure that the isolation valve and the kicker valves are closed. In liquid systems, open the drain valve and allow air to displace the liquid by opening the vent valve. In natural gas systems, open the vent and vent the launcher to atmospheric pressure. When the pig launcher is completely drained (no pressure left), with the vent and drain valves still open, open the trap (closure) door. Install the pig with the nose firmly in contact with the reducer between the barrel and the nominal bore section of the launcher. Clean the closure seal and other sealing surfaces, lubricate if necessary, and close and secure the closure door. Close the drain valve. Slowly fill the trap by gradually opening the kicker valve and venting through the vent valve. When the filling is completed, close the vent valve to allow pressure to equalize across the isolation valve. Open the isolation valve. The pig is ready for launching. Partially close the main line valve. This will increase the flow through the kicker valve and behind the pig. Continue to close the main line valve until the pig leaves the trap and moves into the main line as indicated by the pig signaler. After the pig leaves the trap and enters the main line, fully open the main line valve. Close the isolation valve and the kicker valve. The pig launching is complete.

To receive pigs, make sure the receiver is pressurized. Fully open the bypass valve. Fully open the isolation valve and partially close the main line valve. Monitor the pig signaler for pig arrival. Close the isolation valve and bypass valve. Open the drain valve and the vent valve. Check the pressure gauge on the receiver to assure the trap is completely depressurized. Open the trap closure and remove the pig from the receiver.

Clean the closure seal and other sealing surfaces, lubricate if necessary, and close and secure the trap (closure) door. Return the receiver to the original condition.

12.4.4.4 Freeing a "Stuck" Pig

The goals of pigging a pipeline include not only running pigs to remove a product or to clean the line, but to do the work without sticking the pig. Getting the pig stuck rarely happens in a pipeline that is pigged routinely, but can happen when pigging a pipeline that has been neglected or has never been pigged before. It is a good practice to run a low density (2 lb/ft^3) foam pig in any "suspect" pipeline and examine the foam pig for wear patterns, tears, gouges, and so forth. The pigging project should be continued only after feeling comfortable that the line is piggable. If a pig becomes stuck, it is important to identify the cause. Retrieving the pig is the first priority. When bidirectional pigs are used, the stuck pigs may be recovered with reverse flow.

Pig tracking is normally done on critical projects and when attempting to locate stuck pigs. A pig tracking system consists of a transmitter mounted on the pig, an antenna and a receiver that records and stores each pig passage. In addition, the operator can see and hear the signal of the pig passing under the antenna. The antenna and receiver are simply laid on the ground above and in line with the pipe and then the passage of the pig is heard, seen, and recorded. Inexpensive audible pig tracking systems are also available, however, they cannot be used to find a stuck pig as they rely on the noise the pig makes as it travels through the line. Sometimes a pig without a transmitter fails to come in to the receiver because it gets stuck somewhere in the line. When this happens, the pig cups usually flip forward and flow continues around the stuck pig. In order to find the stuck pig, another pig with a transmitter is launched and tracked closely at all points that are readily accessible. When the transmitter pig passes one tracking point but never reaches the next point, it is assumed the transmitter pig has reached the stuck pig and they are both stuck. The line is then walked carrying the antenna and receiver until the transmitter pig is pinpointed. Both pigs and the debris ahead of the pigs are then removed by cutting the pipe behind and well ahead of the stuck pig.

12.5 References

Coleman, S. B., H. B. Clay, D. G. McCurdy, and L. H. Norris III. "A New Look at Predicting Gas Well Loading-Up." *Journal of Petroleum Technology* Trans. AIME **291** (March 1991): 329.

Guo, B. and A. Ghalambor. *Gas Volume Requirements for Underbalanced Drilling Deviate Holes*. Tulsa: PennWell Books, 2002.

Guo, B., A. Ghalambor, and C. Xu. "A Systematic Approach to Predicting Liquid Loading in Gas Wells." paper SPE 94081 presented at the SPE Production Operations Symposium, Oklahoma City, April 17–19, 2005(a).

Guo, B., S. Song, J. Chacko, and A. Ghalambor. *Offshore and Deepwater Pipelines*. Amsterdam: Elsevier, 2005(b).

Hammerschmidt, E. G. "Preventing and Removing Gas Hydrate Formations in Natural Gas Pipelines." *Oil & Gas Journal* **37**, No. 52 (May 11, 1939): 66–72.

Jacoby, R. S. "Calculation of Methanol Requirements to Prevent Formation of Gas Hydrates." *Gas* **31**, No. 2 (February 1955): 114.

Katz, D. L. "Prediction of Conditions for Hydrate Formation in Natural Gas" *Trans. AIME* **160** (1945): 140.

Kennedy, J. L.: *Oil and Gas Pipeline Fundamentals*. Tulsa: PennWell Books, 1993.

Kohl, A. L. and F. C. Riesenfield. "Gas Purification." 4th ed. Houston: Gulf Publishing Co., 1985.

Lea, J. F. and H. V. Nickens. "Solving Gas-Well Liquid-Loading Problems." *SPE Prod. & Facilities* (April 2004): 30.

Mare, R. F. D. *Advances in Offshore Oil & Gas Pipeline Technology*. Boston: Gulf Publishing Company, 1985.

McAllister, E. W. *Pipeline Rules of Thumb Handbook*. Boston: Gulf Publishing Company, 2002.

Muhlbauer, W. K. *Pipeline Risk Management Manual*. Houston: Gulf Publishing Company, 1992.

Nosseir, M. A., T. A. Darwich, M. H. Sayyouh, and M. E. Sallaly. "A New Approach for Accurate Prediction of Loading in Gas Wells Under Different Flowing Conditions." *SPE Prod. & Facilities* **15**, No. 4 (November 2000): 245.

Tiratsoo, J. N. H. *Pipeline Pigging Technology.* 2nd ed. Accrington, United Kingdom: Gulf Publishing Company, 1992.

Turner, R. G., M. G. Hubbard, and A, E. Dukler. "Analysis and Prediction of Minimum Flow Rate for the Continuous Removal of Liquids from Gas Wells." *Journal of Petroleum Technology* Trans. AIME 246 (November 1969): 1475.

12.6 Problems

12-1 The following data is for a gas well. Predict the minimum required gas production rate for liquid removal using both Turner's method and Guo's method:

Gas-specific gravity: 0.7

Tubing diameter: 1.99 in

Bottom hole pressure: 800 psia

Bottom hole temperature: 130 °F

Liquid density: 66 lbm/ft^3

Interfacial tension: 55 dyne/cm

12-2 The following data is for a gas well. Predict the minimum required gas production rate for liquid removal using both Turner's method and Guo's method:

Gas-specific gravity: 0.7 air = 1

Hole inclination: 0°

Producing depth: 8,000 ft

Wellhead pressure: 560 psi

Wellhead temperature: 70 °F

Producing zone temperature: 140 °F

Condensate gravity: 65 API

Condensate make: 10 bbl/MMscf

Tubing ID: 1.995 in

12-3 A gas of specific gravity 0.65 is at a pressure of 1,200 psia. Assuming presence of free water, (a) to what extent can the temperature be lowered without hydrate formation? (b) How far can the gas be expanded without hydrate formation, if the initial gas temperature is 120 °F? (c) How far can the gas be expanded without hydrate formation, if the initial gas temperature is 100 °F?

12-4 A gas of specific gravity 0.75 is at a temperature of 100 °F. Assuming presence of free water, (a) to what extent can the pressure be increased without hydrate formation? (b) How far can the gas be expanded from 1,000 psia without hydrate formation?

12-5 Twenty MMscfd of a 0.65-specific gravity natural gas cools down to 35 °F in a deepwater pipeline. The minimum pipeline pressure is 1,200 psia. What volume of inhibitor solution must be added daily if the gas enters a line saturated at 90 °F? Consider both DEG and MeOH inhibitors.

Appendix A

Pseudopressures of Sweet Natural Gases

Real gas pseudopressure *m(p)* is defined as

$$m(p) = \int_{p_b}^{p} \frac{2p}{\mu z} dp \qquad \textbf{(A.1)}$$

where p_b is the base pressure that is about 14.7 psia. The pseudopressure curves presented in this appendix were generated using gas viscosity given by the correlation of Carr, Kobayashi, and Burrows. Gas deviation factors were calculated with the correlation of Brill and Beggs. The following correlations were used for computing pseudocritical properties with impurities:

$$p_{pc} = 678 - 50(\gamma_g - 0.5) - 206.7 y_{N_2} + 440 y_{CO_2} + 606.7 y_{H_sS} \qquad \textbf{(A.2)}$$

$$T_{pc} = 326 + 315.7(\gamma_g - 0.5) - 240 y_{N_2} - 83.3 y_{CO_2} + 133.3 y_{H_sS}$$

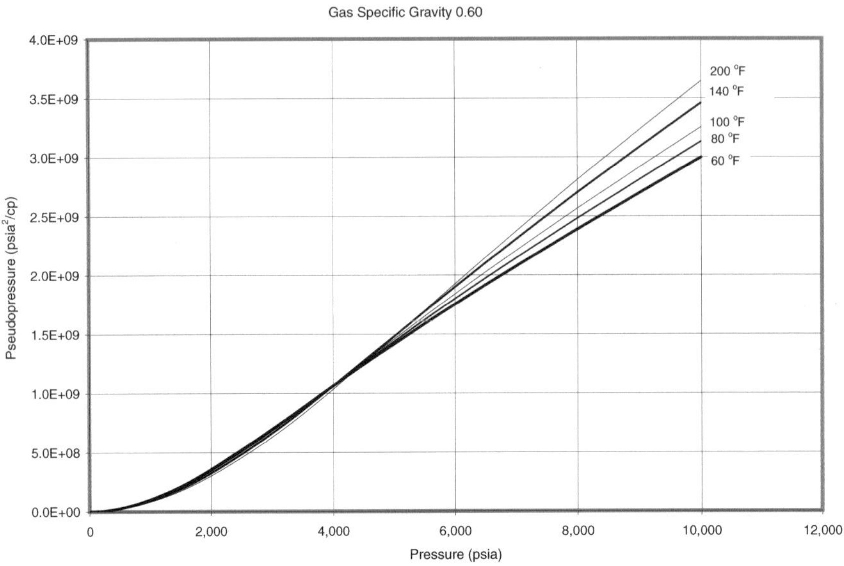

Figure A–1 Real gas pseudopressure of a sweet natural gas, specific gravity 0.60.

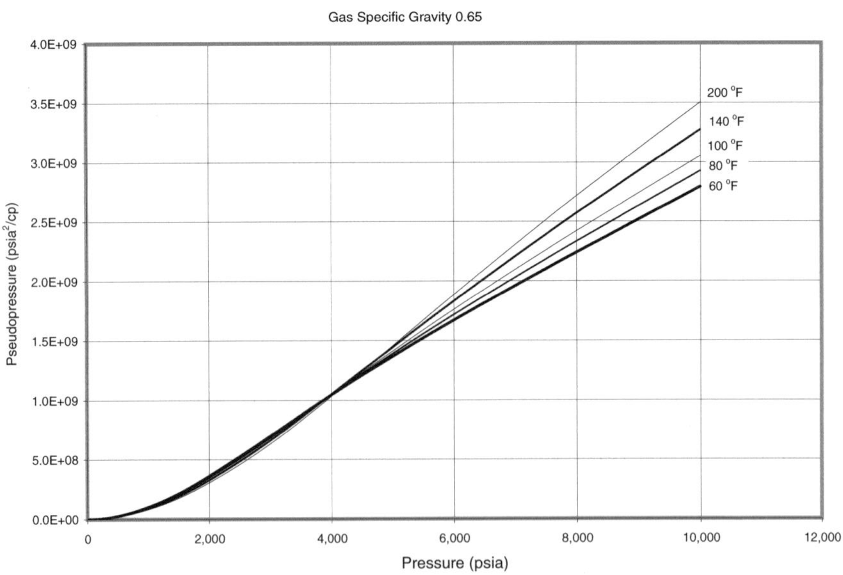

Figure A–2 Real gas pseudopressure of a sweet natural gas, specific gravity 0.65.

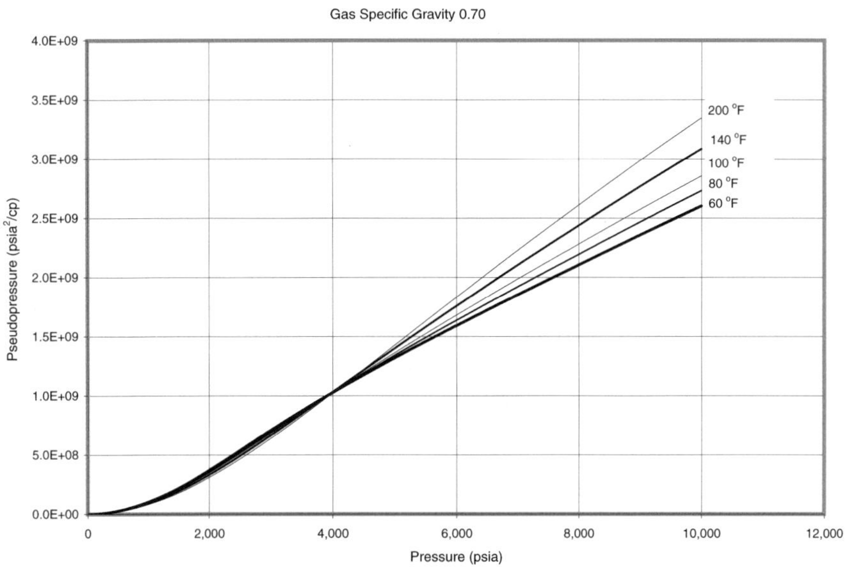

Figure A–3 Real gas pseudopressure of a sweet natural gas, specific gravity 0.70.

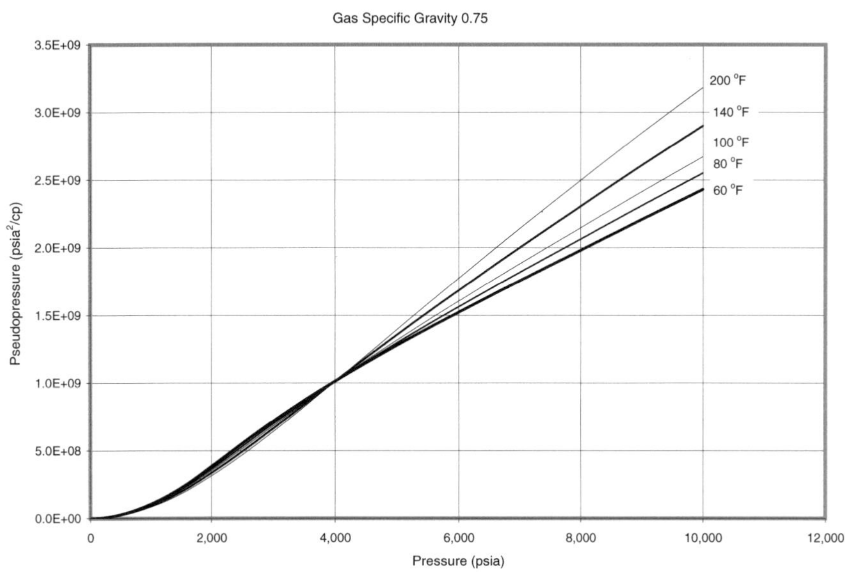

Figure A–4 Real gas pseudopressure of a sweet natural gas, specific gravity 0.75.

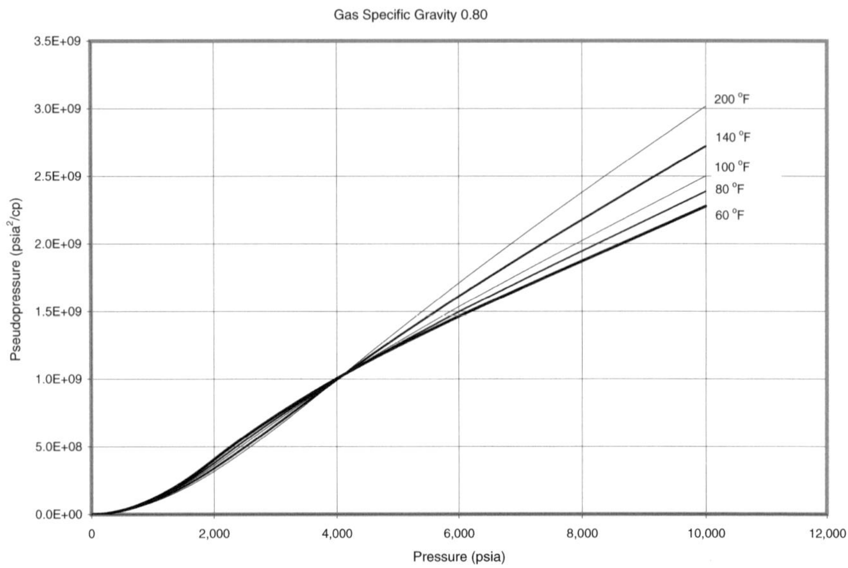

Figure A–5 Real gas pseudopressure of a sweet natural gas, specific gravity 0.80.

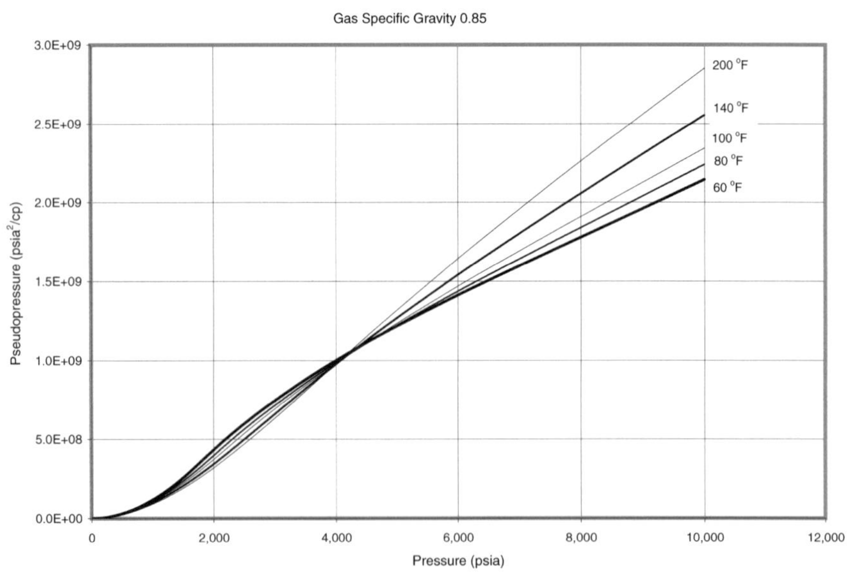

Figure A–6 Real gas pseudopressure of a sweet natural gas, specific gravity 0.85.

Normalized Pressures of Sweet Natural Gases

Real gas normalized pressure *n(p)* is defined as

$$n(p) = \int_0^{p_r} \frac{p_r}{z} dp_r \qquad \textbf{(B.1)}$$

where p_r is the pseudoreduced pressure. The normalized pressure curves presented in this appendix were generated using gas deviation factors calculated with the correlation of Brill and Beggs. The following correlations were used for computing pseudocritical properties with impurities:

$$p_{pc} = 678 - 50(\gamma_g - 0.5) - 206.7 y_{N_2} + 440 y_{CO_2} + 606.7 y_{H_2S} \qquad \textbf{(B.2)}$$

$$T_{pc} = 326 + 315.7(\gamma_g - 0.5) - 240 y_{N_2} - 83.3 y_{CO_2} + 133.3 y_{H_2S} \qquad \textbf{(B.3)}$$

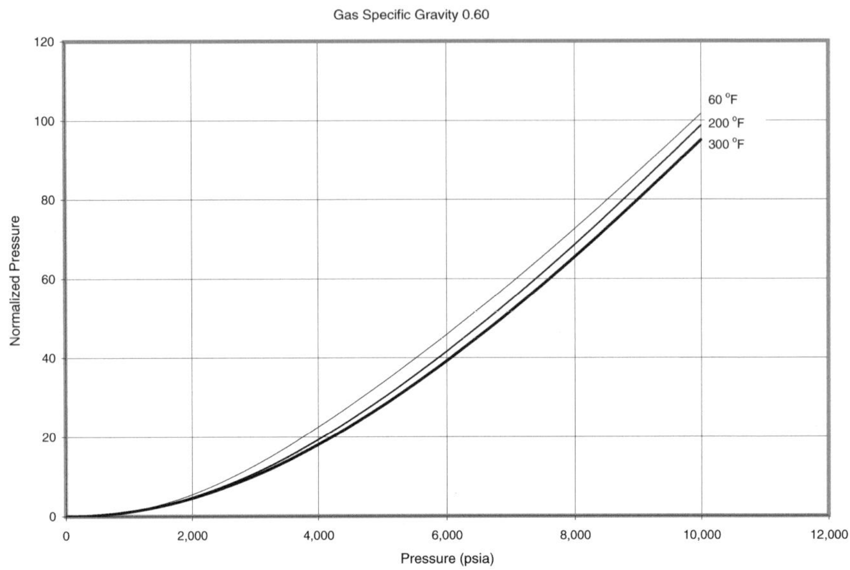

Figure B–1 Normalized pressure of a sweet natural gas, specific gravity 0.60.

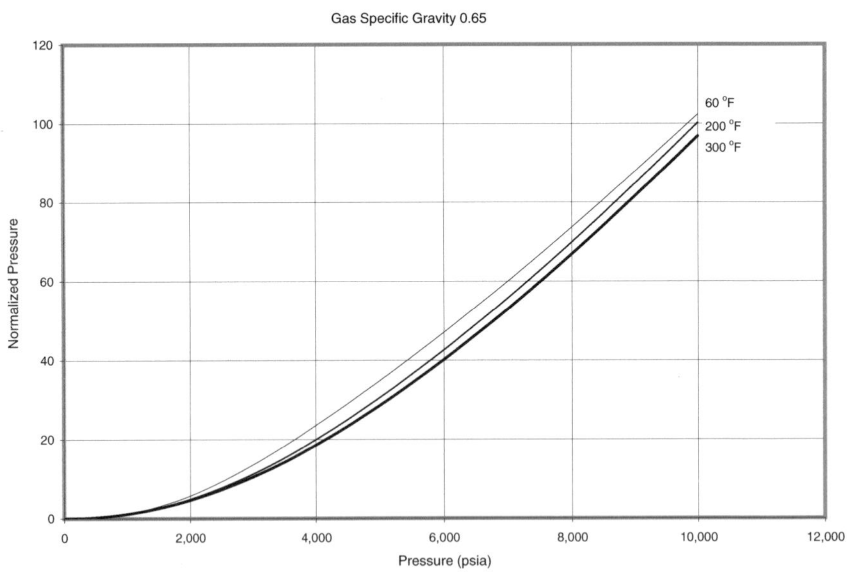

Figure B–2 Normalized pressure of a sweet natural gas, specific gravity 0.65.

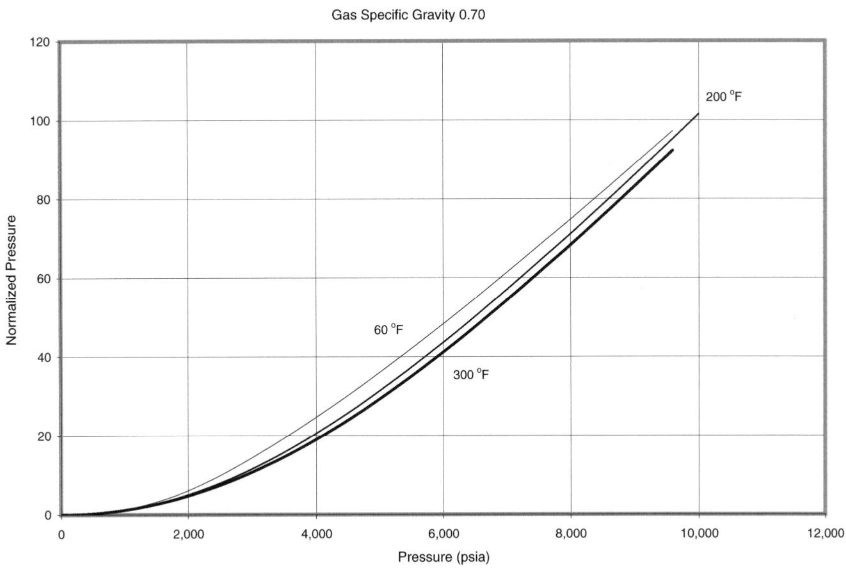

Figure B–3 Normalized pressure of a sweet natural gas, specific gravity 0.70.

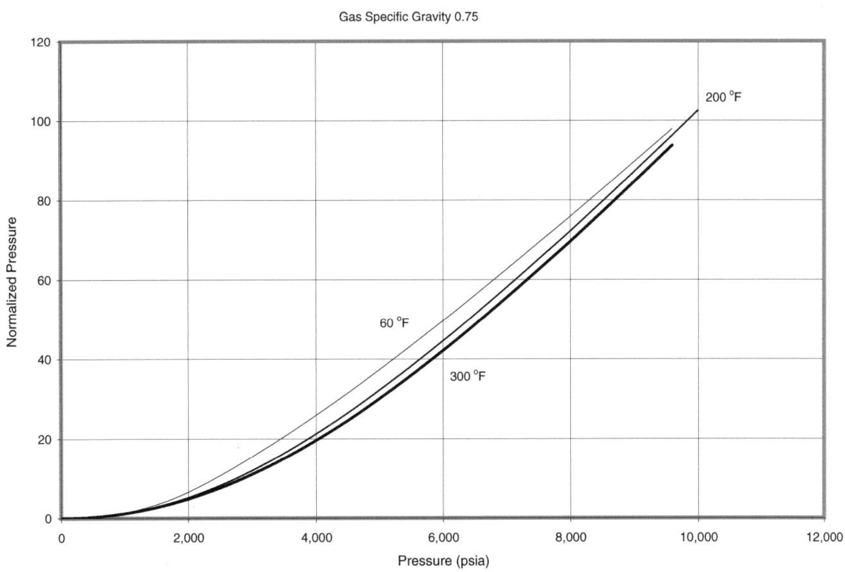

Figure B–4 Normalized pressure of a sweet natural gas, specific gravity 0.75.

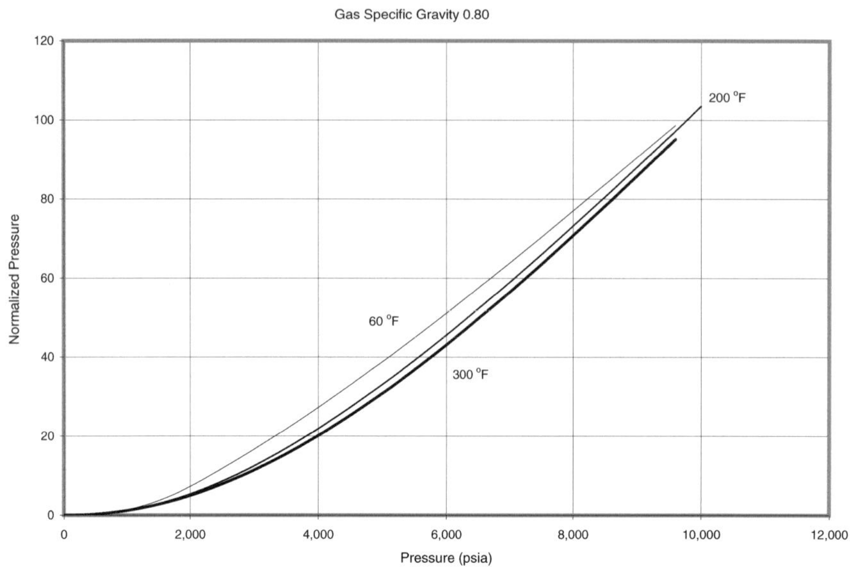

Figure B–5 Normalized pressure of a sweet natural gas, specific gravity 0.80.

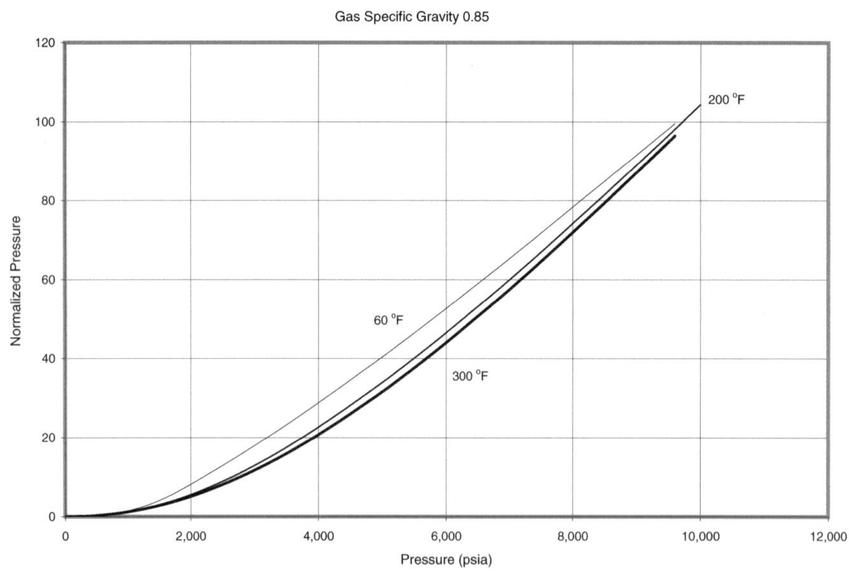

Figure B–6 Normalized pressure of a sweet natural gas, specific gravity 0.85.

Orifice Meter Tables for Natural Gas

Table C-1 F_b Basic Orifice Factors—Flange Taps

Base temperature = 60 °F
Flowing temperature = 60 °F
Base pressure = 14.73 psia
Specific gravity = 1.0

Pipe Sizes—Nominal and Published Inside Diameters, in

Orifice Diameter (in)	2			3			4		
	1.689	1.939	2.067	2.300	2.626	2.900	3.068	3.152	3.438
0.250	12.695	12.707	12.711	12.714	12.712	12.708	12.705	12.703	12.697
0.375	28.474	28.439	28.428	28.411	28.393	28.382	28.376	28.373	28.364
0.500	50.777	50.587	50.521	50.435	50.356	50.313	50.292	50.284	50.258
0.625	80.090	79.509	79.311	79.052	78.818	78.686	78.625	78.598	78.523
0.750	117.09	115.62	115.14	114.52	113.99	113.70	113.56	113.50	113.33
0.875	162.95	159.56	158.47	157.12	156.00	155.41	155.14	155.03	154.71
1.000	219.77	212.47	210.22	207.44	205.18	204.04	203.54	203.33	202.75
1.125	290.99	276.20	271.70	266.35	262.06	259.95	259.04	258.65	257.63
1.250	385.78	353.58	345.13	335.12	327.39	323.63	322.03	321.37	319.61
1.375		448.57	433.50	415.75	402.18	395.80	393.09	391.97	389.03
1.500			542.26	510.86	487.98	477.36	472.96	471.14	466.39
1.625				623.91	586.82	569.65	562.58	559.72	552.31
1.750					701.27	674.44	663.42	658.96	647.54
1.875					834.88	793.88	777.18	770.44	753.17
2.000						930.65	906.01	896.06	870.59
2.125						1,091.2	1,052.5	1,038.1	1,001.4
2.250							1,223.2	1,199.9	1,147.7
2.375									1,311.7
2.500									1,498.4

Orifice Diameter (in)	4		6				8		
	3.826	**4.026**	**4.897**	**5.182**	**5.761**	**6.065**	**7.625**	**7.981**	**8.071**
0.250	12.687	12.683							
0.375	28.353	28.348							
0.500	50.234	50.224	50.197	50.191	50.182	50.178			
0.625	78.450	78.421	78.338	78.321	78.296	78.287			
0.750	113.15	113.08	112.87	112.82	112.75	112.72			
0.875	154.40	154.27	153.88	153.78	153.63	153.56	153.34	153.31	153.31
1.000	202.20	201.99	201.34	201.19	200.96	200.85	200.46	200.39	200.38
1.125	256.69	256.33	255.31	255.08	254.72	254.56	253.99	253.69	253.87
1.250	318.03	317.45	315.83	315.48	314.95	314.72	313.91	313.78	313.74
1.375	386.45	385.51	382.99	382.47	381.70	381.37	380.25	380.06	380.02
1.500	462.27	460.79	456.93	456.16	455.03	454.57	453.02	452.78	452.72
1.625	545.89	543.61	537.77	536.64	535.03	534.38	532.27	531.95	531.87
1.750	637.84	634.39	625.73	624.09	621.79	620.88	618.02	617.60	617.50
1.875	738.75	733.68	721.03	718.69	715.44	714.19	710.32	709.77	709.64
2.000	849.41	842.12	823.99	820.68	816.13	814.41	809.22	808.50	808.34
2.125	970.95	960.48	934.97	930.35	924.07	921.71	914.79	913.86	913.64
2.250	1,104.7	1,089.9	1,054.4	1,048.1	1,039.5	1,036.3	1,027.1	1,025.9	1,025.6
2.375	1,252.1	1,231.7	1,182.9	1,174.2	1,162.6	1,158.3	1,146.2	1,144.7	1,144.3
2.500	1,415.0	1,387.2	1,320.9	1,309.3	1,293.8	1,288.2	1,272.3	1,270.3	1,269.8
2.625	1,595.6	1,558.2	1,469.2	1,453.9	1,433.5	1,426.0	1,405.4	1,402.9	1,402.3
2.750	1,797.1	1,746.7	1,628.9	1,608.7	1,582.1	1,572.3	1,545.7	1,542.5	1,541.8
2.875		1,955.5	1,801.0	1,774.5	1,740.0	1,727.5	1,693.4	1,689.3	1,688.4
3.000		2,194.9	1,986.6	1,952.4	1,907.8	1,891.9	1,848.6	1,843.5	1,842.3
3.125			2,187.2	2,143.4	2,086.4	2,066.1	2,011.6	2,005.2	2,003.8
3.250			2,404.2	2,348.8	2,276.5	2,250.8	2,182.6	2,174.6	2,172.9
3.375			2,639.5	2,569.8	2,479.1	2,446.8	2,361.8	2,352.0	2,349.9
3.500			2,895.5	2,808.1	2,695.1	2,654.9	2,654.9	2,537.7	2,535.0
3.625			3,180.8	3,065.3	2,925.7	2,876.0	2,746.5	2,731.8	2,728.6
3.750				3,345.5	3,172.1	3,111.2	2,952.6	2,934.8	2,930.8
3.875				3,657.7	3,435.7	3,361.5	3,168.3	3,146.9	3,142.1
4.000					3,718.2	3,628.2	3,394.3	3,368.5	3,362.9
4.250					4,354.8	4,216.6	3,879.4	3,842.3	3,834.2
4.500						4,900.9	4,412.8	4,360.5	4,349.0
4.750							5,000.7	4,928.1	4,912.2
5.000							5,650.0	5,551.1	5,529.5
5.250							6,369.3	6,236.4	6,207.3
5.500							7,170.9	6,992.0	6,953.6
5.750								7,830.0	7,777.8
6.000									8,706.9

Orifice Diameter (in)	10			12			16		
	9.564	10.020	10.136	11.376	11.938	12.090	14.688	15.000	15.250
1.000	200.20								
1.125	253.55	253.48	253.47						
1.250	313.31	313.20	313.18	312.94	312.85	312.83			
1.375	379.44	379.29	379.26	378.94	378.82	378.79			
1.500	451.95	451.76	451.72	451.30	451.14	451.10	450.53	450.48	
1.625	530.87	530.63	530.57	530.04	529.83	529.78	529.06	528.99	528.94
1.750	616.21	615.90	615.83	615.16	614.90	614.84	613.94	613.85	613.78
1.875	707.99	707.61	707.51	706.68	706.36	706.28	705.18	705.07	704.99
2.000	806.23	805.76	805.65	804.61	804.23	804.13	802.78	802.65	802.55
2.125	910.97	910.38	910.24	908.98	908.51	908.39	906.77	906.61	906.49
2.250	1,022.2	1,021.5	1,021.3	1,019.8	1,019.2	1,019.1	1,017.1	1,017.0	1,016.8
2.375	1,140.1	1,139.2	1,139.0	1,137.1	1,136.4	1,136.2	1,133.9	1,133.7	1,133.5
2.500	1,264.5	1,263.4	1,263.1	1,260.8	1,260.0	1,259.8	1,257.1	1,256.8	1,256.6
2.625	1,395.6	1,394.2	1,393.9	1,391.1	1,390.1	1,389.9	1,386.7	1,386.4	1,386.1
2.750	1,533.4	1,531.7	1,531.3	1,528.0	1,526.8	1,526.5	1,522.7	1,522.4	1,522.1
2.875	1,678.0	1,675.9	1,675.4	1,671.4	1,670.0	1,669.6	1,665.2	1,664.8	1,664.5
3.000	1,829.4	1,826.9	1,826.3	1,821.4	1,819.7	1,819.3	1,814.1	1,813.7	1,813.3
3.125	1,987.8	1,984.7	1,984.0	1,978.1	1,976.1	1,975.6	1,969.6	1,969.0	1,966.6
3.250	2,153.2	2,149.5	2,148.6	2,141.5	2,139.2	2,130.6	2,131.5	2,130.9	2,130.4
3.375	2,325.7	2,321.2	2,320.2	2,311.7	2,308.9	2,308.2	2,299.9	2,299.2	2,293.7
3.500	2,505.6	2,500.1	2,498.9	2,488.7	2,485.4	2,484.6	2,474.9	2,474.1	2,473.5
3.625	2,692.8	2,686.2	2,684.7	2,672.6	2,668.7	2,667.7	2,656.4	2,655.5	2,654.8
3.750	2,887.6	2,879.7	2,877.9	2,863.5	2,858.8	2,857.7	2,844.6	2,843.5	2,842.7
3.875	3,090.1	3,080.7	3,078.5	3,061.4	3,055.9	3,054.6	3,039.4	3,038.1	3,037.2
4.000	3,300.6	3,289.3	3,286.8	3,266.4	3,260.0	3,258.5	3,240.8	3,239.4	3,238.3
4.250	3,746.1	3,730.2	3,726.7	3,698.4	3,689.6	3,687.5	3,663.8	3,661.9	3,660.5
4.500	4,226.0	4,204.1	4,199.2	4,160.4	4,148.4	4,145.5	4,113.9	4,111.5	4,109.7
4.750	4,742.7	4,712.8	4,706.2	4,653.4	4,637.2	4,633.4	4,591.5	4,508.4	4,586.0
5.000	5,298.6	5,258.5	5,249.6	5,179.0	5,157.4	5,152.3	5,097.2	5,093.1	5,090.1
5.250	5,897.4	5,843.6	5,831.8	5,738.5	5,710.0	5,703.3	5,631.4	5,626.1	5,622.3
5.500	6,543.1	6,471.9	6,456.3	6,333.8	6,296.6	6,287.9	6,194.8	6,180.1	6,183.1

Orifice Diameter (in)	10			12			16		
	9.564	**10.020**	**10.136**	**11.376**	**11.938**	**12.090**	**14.688**	**15.000**	**15.250**
5.750	7,240.0	7,146.9	7,126.5	6,966.9	6,919.0	6,907.8	6,788.1	6,779.5	6,773.3
6.000	7,993.3	7,873.0	7,846.6	7,640.4	7,579.0	7,564.7	7,412.3	7,401.5	7,393.6
6.250	8,808.9	8,654.8	8,621.1	8,357.3	8,278.9	8,260.7	8,060.4	8,054.8	8,044.8
6.500	9,693.3	9,498.1	9,455.3	9,121.0	9,021.7	8,998.7	8,757.3	8,740.3	8,727.9
6.750	10,654	10,409	10,355	9,935.2	9,810.5	9,781.6	9,480.4	9,459.4	9,444.0
7.000	11,711	11,394	11,327	10,804	10,649	10,613	10,239	10,213	10,194
7.250		12,467	12,381	11,732	11,540	11,496	11,035	11,003	10,980
7.500		13,656	13,541	12,725	12,489	12,434	11,869	11,831	11,803
7.750				13,787	13,500	13,433	12,745	12,698	12,664
8.000				14,927	14,578	14,498	13,664	13,607	13,566
8.250				16,156	15,730	15,633	14,628	14,560	14,511
8.500				17,505	16,962	16,845	15,642	15,560	15,501
8.750					18,296	18,148	16,706	16,609	16,539
9.000						19,565	17,826	17,711	17,628
9.250							19,004	18,868	18,770
9.500							20,245	20,085	19,969
9.750							21,552	21,365	21,230
10.000							22,930	22,712	22,555
10.250							24,385	24,132	23,948
10.500							25,924	25,628	25,416
10.750							27,567	27,210	26,962
11.000							29,331	28,899	28,600
11.250								30,710	30,348

Orifice Diameter (in)	20			24			30		
	18.814	19.000	19.250	22.626	23.000	23.250	28.628	29.000	29.250
2.000	801.40	801.35	801.29						
2.125	905.11	905.06	904.98						
2.250	1,015.2	1,015.1	1,015.0						
2.375	1,131.6	1,131.5	1,131.4	1,130.2	1,130.1	1,130.0			
2.500	1,254.4	1,254.3	1,254.2	1,252.8	1,252.6	1,252.6			
2.625	1,383.6	1,383.5	1,383.3	1,381.7	1,381.5	1,381.4			
2.750	1,519.1	1,519.0	1,518.8	1,517.0	1,516.8	1,516.7			
2.875	1,661.0	1,660.9	1,660.7	1,658.6	1,658.4	1,658.3	1,656.0		
3.000	1,809.4	1,809.2	1,809.0	1,806.6	1,806.4	1,806.2	1,803.7	1,803.5	1,803.4
3.125	1,964.1	1,963.9	1,963.7	1,961.0	1,960.7	1,960.6	1,957.7	1,957.5	1,957.4
3.250	2,125.3	2,125.1	2,124.8	2,121.7	2,121.5	2,121.3	2,118.0	2,117.9	2,117.7
3.375	2,292.9	2,292.6	2,292.3	2,280.9	2,288.6	2,288.4	2,284.8	2,284.5	2,284.4
3.500	2,466.9	2,466.6	2,466.3	2,462.4	2,462.1	2,461.8	2,457.8	2,457.6	2,457.5
3.625	2,647.3	2,647.0	2,646.6	2,642.4	2,642.0	2,641.7	2,637.3	2,637.0	2,636.8
3.750	2,834.2	2,833.9	2,833.5	2,828.7	2,828.3	2,828.0	2,823.1	2,822.8	2,822.6
3.875	3,027.5	3,027.3	3,026.8	3,021.5	3,021.0	3,020.7	3,015.2	3,014.9	3,014.7
4.000	3,227.5	3,227.1	3,226.5	3,220.6	3,220.1	3,219.8	3,213.8	3,213.5	3,213.2
4.250	3,646.7	3,646.2	3,645.6	3,638.3	3,637.7	3,637.2	3,630.1	3,629.7	3,629.4
4.500	4,092.1	4,091.5	4,090.6	4,081.8	4,081.0	4,080.5	4,071.9	4,071.4	4,071.1
4.750	4,563.7	4,562.9	4,561.9	4,551.1	4,550.1	4,549.5	4,539.4	4,538.8	4,538.4
5.000	5,061.8	5,060.8	5,050.6	5,046.4	5,045.2	5,044.5	5,032.5	5,031.8	5,031.4
5.250	5,586.6	5,585.4	5,583.8	5,567.7	5,566.4	5,565.5	5,551.3	5,550.5	5,550.0
5.500	6,138.2	6,136.7	6,134.8	6,115.3	6,113.6	6,112.6	6,095.8	6,094.9	6,094.4
5.750	6,717.0	6,715.2	6,712.3	6,689.1	6,687.2	6,685.9	6,666.2	6,665.2	6,664.5
6.000	7,323.4	7,321.1	7,318.2	7,289.4	7,287.1	7,285.6	7,262.5	7,261.3	7,260.5
6.250	7,957.5	7,954.7	7,951.2	7,916.4	7,913.6	7,911.9	7,864.7	7,883.4	7,882.5
6.500	8,620.0	8,616.5	8,612.2	8,570.2	8,566.9	8,564.8	8,533.0	8,531.4	8,530.4
6.750	9,311.1	9,306.9	9,301.6	9,251.1	9,247.2	9,244.7	9,207.4	9,205.6	9,204.4
7.000	10,031	10,026	10,020	9,959.3	9,954.6	9,951.7	9,908.0	9,905.9	9,904.6
7.250	10,782	10,776	10,768	10,695	10,669	10,686	10,635	10,633	10,631
7.500	11,562	11,555	11,546	11,459	11,452	11,448	11,388	11,386	11,384
7.750	12,374	12,365	12,354	12,250	12,243	12,238	12,168	12,165	12,163
8.000	13,218	13,207	13,194	13,071	13,062	13,056	12,975	12,971	12,969
8.250	14,095	14,082	14,066	13,920	13,910	13,903	13,809	13,805	13,802
8.500	15,005	14,990	14,971	14,799	14,787	14,779	14,669	14,665	14,661

Orifice Diameter (in)	20			24			30		
	18.814	**19.000**	**19.250**	**22.626**	**23.000**	**23.250**	**28.628**	**29.000**	**29.250**
8.750	15,950	15,933	15,911	15,708	15,693	15,684	15,557	15,552	15,548
9.000	16,932	16,911	16,885	16,648	16,630	16,620	16,473	16,466	16,462
9.250	17,950	17,926	17,895	17,618	17,596	17,585	17,416	17,409	17,404
9.500	19,007	18,979	18,943	18,620	18,597	18,582	18,387	18,379	18,373
9.750	20,104	20,071	20,030	19,655	19,628	19,611	19,386	19,377	19,371
10.000	21,243	21,205	21,157	20,723	20,692	20,672	20,414	20,403	20,396
10.250	22,426	22,332	22,326	21,825	21,789	21,767	21,471	21,458	21,450
10.500	23,654	23,603	23,538	22,926	22,921	22,895	22,556	22,542	22,533
10.750	24,931	24,672	24,797	24,134	24,087	24,058	23,672	23,656	23,646
11.000	26,257	26,190	26,104	25,344	25,290	25,257	24,817	24,799	24,787
11.250	27,636	27,559	27,460	26,592	26,531	26,492	25,992	25,972	25,959
11.500	29,070	28,982	28,870	27,878	27,809	27,766	27,199	27,176	27,161
11.750	30,562	30,462	30,334	29,205	29,126	29,077	28,437	28,411	28,394
12.000	32,116	32,001	31,856	30,574	30,485	30,429	29,706	29,677	29,659
12.500	35,417	35,270	35,084	33,444	33,330	33,259	32,343	32,306	32,283
13.000	39,003	38,817	38,581	36,502	36,357	36,267	35,114	35,068	35,039
13.500	42,913	42,673	42,375	39,762	39,581	39,467	38,025	37,968	37,932
14.000	47,244	46,921	46,523	43,241	43,015	42,874	41,082	41,012	40,968
14.500				46,958	46,679	46,505	44,291	44,206	44,151
15.000				50,934	50,591	50,378	47,622	47,557	47,490
15.500				55,192	54,774	54,513	51,202	51,075	50,993
16.000				59,759	59,251	58,935	54,923	54,769	54,671
16.500				64,701	64,060	63,670	58,835	58,649	58,531
17.000					69,288	68,792	62,950	62,728	62,586
17.500							67,282	67,017	66,848
18,000							71,844	71,530	71,330
18.500							76,653	76,282	76,046
19.000							81,725	81,289	81,012
19.500							87,079	86,568	86,244
20.000							92,734	92,140	91,761
20.500							98,728	98,025	97,564
21.000							105,130	104,280	103,750
21.500								110,980	110,340

Table C–2 "*b*" Values for Reynolds Number Factor F_r Determination—Flange Taps

Orifice Diameter (in)	Pipe Sizes—Nominal and Published Inside Diameters, in								
		2			3				4
	1.689	1.939	2.067	2.3	2.626	2.9	3.068	3.152	3.438
0.250	0.0879	0.0911	0.0926	0.0950	0.0979	0.0999	0.1010	0.1014	0.1030
0.375	0.0677	0.0709	0.0726	0.0755	0.0792	0.0820	0.0836	0.0844	0.0867
0.500	0.0562	0.0576	0.0588	0.0612	0.0648	0.0677	0.0695	0.0703	0.0728
0.625	0.0520	0.0505	0.0506	0.0516	0.0541	0.0566	0.0583	0.0591	0.0618
0.750	0.0536	0.0485	0.0471	0.0462	0.0470	0.0486	0.0498	0.0504	0.0528
0.875	0.0595	0.0506	0.0478	0.0445	0.0429	0.0433	0.0438	0.0442	0.0460
1.000	0.0677	0.0559	0.0515	0.0458	0.0416	0.0403	0.0402	0.0403	0.0411
1.125	0.0762	0.0630	0.0574	0.0495	0.0427	0.0396	0.0386	0.0383	0.0380
1.250	0.0824	0.0707	0.0646	0.0550	0.0456	0.0408	0.0388	0.0381	0.0365
1.375		0.0772	0.0715	0.0614	0.0501	0.0435	0.0406	0.0394	0.0365
1.500			0.0773	0.0679	0.0554	0.0474	0.0436	0.0420	0.0378
1.625				0.0735	0.0613	0.0522	0.0477	0.0457	0.0402
1.750					0.0669	0.0575	0.0524	0.0500	0.0434
1.875					0.0717	0.0628	0.0574	0.0549	0.0473
2.000						0.0676	0.0624	0.0598	0.0517
2.125						0.0715	0.0669	0.0642	0.0563
2.250							0.0706	0.0685	0.0607
2.375									0.0648
2.500									0.0683

Orifice Diameter (in)	4			6				8	
	3.826	**4.026**	**4.897**	**5.189**	**5.761**	**6.065**	**7.625**	**7.981**	**8.071**
0.250	0.1047	0.1054							
0.375	0.0894	0.0907							
0.500	0.0763	0.0779	0.0836	0.0852	0.0880	0.0892			
0.625	0.0653	0.0670	0.0734	0.0753	0.0785	0.0801			
0.750	0.0561	0.0578	0.0645	0.0665	0.0701	0.0718			
0.875	0.0487	0.0502	0.0567	0.0587	0.0625	0.0643	0.0723	0.0738	0.0742
1.000	0.0430	0.0442	0.0500	0.0520	0.0557	0.0576	0.0660	0.0676	0.0680
1.125	0.0388	0.0396	0.0444	0.0462	0.0498	0.0517	0.0602	0.0517	0.0623
1.250	0.0361	0.0364	0.0399	0.0414	0.0447	0.0464	0.0549	0.0566	0.0571
1.375	0.0347	0.0344	0.0363	0.0375	0.0403	0.0419	0.0518	0.0518	0.0523
1.500	0.0345	0.0336	0.0336	0.0344	0.0367	0.0381	0.0457	0.0474	0.0479
1.625	0.0354	0.0338	0.0318	0.0322	0.0337	0.0348	0.0418	0.0435	0.0439
1.750	0.0372	0.0350	0.0307	0.0306	0.0314	0.0322	0.0383	0.0399	0.0403
1.875	0.0398	0.0370	0.0305	0.0298	0.0298	0.0303	0.0353	0.0366	0.0371
2.000	0.0430	0.0395	0.0308	0.0296	0.0287	0.0288	0.0327	0.0340	0.0343
2.125	0.0467	0.0427	0.0318	0.0300	0.0281	0.0278	0.0304	0.0315	0.0318
2.250	0.0507	0.0462	0.0334	0.0310	0.0281	0.0274	0.0286	0.0295	0.0297
2.375	0.0548	0.0501	0.0354	0.0324	0.0286	0.0274	0.0271	0.0278	0.0280
2.500	0.0589	0.0540	0.0378	0.0342	0.0295	0.0279	0.0259	0.0264	0.0265
2.625	0.0626	0.0579	0.0406	0.0365	0.0308	0.0287	0.0251	0.0253	0.0254
2.750	0.0659	0.0615	0.0436	0.0391	0.0324	0.0300	0.0246	0.0245	0.0245
2.875		0.0647	0.0468	0.0418	0.0343	0.0314	0.0244	0.0240	0.0240
3.000		0.0673	0.0500	0.0448	0.0366	0.0332	0.0245	0.0238	0.0237
3.125			0.0533	0.0479	0.0389	0.0353	0.0248	0.0239	0.0237
3.250			0.0564	0.0510	0.0416	0.0375	0.0254	0.0242	0.0240
3.375			0.0594	0.0541	0.0443	0.0400	0.0263	0.0248	0.0244
3.500			0.0620	0.0569	0.0472	0.0426	0.0273	0.0255	0.0251
3.625			0.0643	0.0597	0.0500	0.0452	0.0286	0.0265	0.0260
3.750				0.0621	0.0527	0.0479	0.0300	0.0274	0.0271
3.875				0.0640	0.0553	0.0505	0.0316	0.0289	0.0283
4.000					0.0578	0.0531	0.0334	0.0304	0.0297
4.250					0.0620	0.0579	0.0372	0.0338	0.0330
4.500						0.0618	0.0414	0.0386	0.0366

Orifice Diameter (in)	4			6				8	
	3.826	4.026	4.897	5.189	5.761	6.065	7.625	7.981	8.071
4.750							0.0457	0.0416	0.0405
5.000							0.0500	0.0457	0.0446
5.250							0.0539	0.0497	0.0487
5.500							0.0574	0.0535	0.0524
5.750								0.0569	0.0559
6.000									0.0588

Orifice Diameter (in)		10			12			16	
	9.564	10.02	10.136	11.376	11.938	12.09	14.688	15	15.25
1	0.0738								
1.125	0.0685	0.0701	0.0705						
1.25	0.0635	0.0652	0.0656	0.0698	0.0714	0.0718			
1.375	0.0588	0.0606	0.061	0.0654	0.0671	0.676			
1.5	0.0545	0.0563	0.0568	0.0612	0.0631	0.0635	0.0706	0.0713	
1.625	0.0504	0.0523	0.0527	0.0573	0.0592	0.0597	0.067	0.0678	0.0684
1.75	0.0467	0.0485	0.049	0.0536	0.0555	0.056	0.0636	0.0644	0.065
1.875	0.0433	0.0451	0.0455	0.0501	0.0521	0.0526	0.0604	0.0612	0.0618
2	0.0401	0.0419	0.0414	0.0469	0.0488	0.0492	0.0572	0.0581	0.0587
2.125	0.0372	0.0389	0.0383	0.0438	0.0458	0.0463	0.0542	0.0551	0.0558
2.25	0.0346	0.0362	0.0356	0.041	0.0429	0.0434	0.0514	0.0523	0.0529
2.375	0.0322	0.0337	0.033	0.0383	0.0402	0.0407	0.0467	0.0496	0.0502
2.5	0.0302	0.0315	0.0308	0.0359	0.0377	0.0382	0.0461	0.047	0.0476
2.625	0.0283	0.0296	0.0287	0.0336	0.0354	0.0358	0.0436	0.0445	0.0452
2.75	0.0267	0.0278	0.0269	0.0316	0.0332	0.0336	0.0413	0.0422	0.0428
2.875	0.0254	0.0263	0.0253	0.0297	0.0312	0.0317	0.0391	0.0399	0.0406
3	0.0243	0.025	0.0252	0.0278	0.0294	0.0298	0.037	0.0378	0.0385
3.125	0.0234	0.0239	0.0241	0.0264	0.0278	0.0282	0.035	0.0358	0.0365
3.25	0.0226	0.023	0.0231	0.0251	0.0263	0.0266	0.0331	0.0339	0.0346
3.375	0.0221	0.0223	0.0224	0.0239	0.025	0.0253	0.0314	0.0321	0.0328
3.5	0.0219	0.0218	0.0218	0.0229	0.0238	0.0241	0.0298	0.0305	0.0311
3.625	0.0218	0.0214	0.0214	0.0221	0.0226	0.023	0.0282	0.029	0.0295
3.75	0.0218	0.0213	0.0212	0.0214	0.0219	0.0221	0.0268	0.0275	0.0281
3.875	0.0221	0.213	0.0211	0.0208	0.0212	0.0213	0.0255	0.0262	0.0267
4	0.0225	0.0214	0.0212	0.0204	0.0206	0.0207	0.0243	0.0249	0.0254

Orifice Diameter (in)	10			12			16		
	9.564	**10.02**	**10.136**	**11.376**	**11.938**	**12.09**	**14.688**	**15**	**15.25**
4.25	0.0238	0.0222	0.0219	0.02	0.0198	0.0198	0.0223	0.0228	0.0232
4.5	0.0256	0.0236	0.0231	0.0201	0.0195	0.0194	0.0206	0.021	0.0213
4.75	0.0279	0.0254	0.0249	0.0207	0.0196	0.0194	0.0193	0.0196	0.0198
5	0.0307	0.0277	0.027	0.0217	0.0202	0.0199	0.0184	0.0185	0.0187
5.25	0.0337	0.0303	0.0295	0.0231	0.0212	0.0208	0.0178	0.0178	0.0179
5.5	0.037	0.0332	0.0323	0.0249	0.0226	0.0221	0.0176	0.0174	0.0174
5.75	0.0404	0.0363	0.0354	0.027	0.0243	0.0237	0.0176	0.0174	0.0172
6	0.0438	0.396	0.0386	0.0294	0.0263	0.0255	0.018	0.0176	0.0173
6.25	0.0473	0.0437	0.0418	0.032	0.0285	0.0277	0.0186	0.016	0.0177
6.5	0.0505	0.0462	0.0451	0.0347	0.0309	0.03	0.0195	0.0188	0.0183
6.75	0.0536	0.0493	0.0483	0.0376	0.0335	0.0325	0.0206	0.0198	0.0192
7	0.0562	0.0523	0.0513	0.0406	0.0362	0.0351	0.022	0.021	0.0202
7.25		0.055	0.054	0.0435	0.039	0.0379	0.0235	0.0224	0.0216
7.5		0.0572	0.0564	0.0463	0.0418	0.0407	0.0252	0.024	0.023
7.75				0.0491	0.0446	0.0434	0.0271	0.0257	0.0246
8				0.0517	0.0473	0.0461	0.0291	0.0276	0.0264
8.25				0.054	0.0498	0.0487	0.0312	0.0296	0.0288
8.5				0.056	0.0522	0.0511	0.0334	0.0317	0.0303
8.75					0.0543	0.0534	0.0357	0.0338	0.0324
9						0.0533	0.038	0.0361	0.0346
9.25							0.0402	0.0383	0.0368
9.5							0.0425	0.0406	0.039
9.75							0.0447	0.0427	0.0412
10							0.0469	0.0449	0.0434
10.25							0.0489	0.047	0.0455
10.5							0.0508	0.049	0.0475
10.75							0.0526	0.0509	0.0495
11							0.0541	0.0526	0.0513
11.25								0.0541	0.0528

Orifice Diameter (in)	20			24			30		
	18.814	19	19.25	22.626	23	23.25	28.628	29	29.25
2	0.0667	0.0671	0.0676						
2.125	0.064	0.0644	0.0649						
2.25	0.0614	0.0618	0.0622						
2.375	0.0588	0.0592	0.0597	0.0659	0.0665	0.0669			
2.5	0.0563	0.0568	0.0573	0.0636	0.0642	0.0646			
2.625	0.054	0.0544	0.0549	0.0614	0.062	0.0624			
2.75	0.0517	0.0521	0.0526	0.0592	0.0599	0.0603			
2.875	0.0494	0.0499	0.0504	0.0571	0.0578	0.0582	0.0662		
3	0.0473	0.0477	0.0483	0.0551	0.0557	0.0562	0.0644	0.0649	0.0652
3.125	0.0452	0.0457	0.0462	0.0531	0.0538	0.0542	0.0626	0.0631	0.0634
3.25	0.0433	0.0437	0.0442	0.0511	0.052	0.0523	0.0608	0.0613	0.0616
3.375	0.0414	0.0418	0.0423	0.0493	0.05	0.0504	0.059	0.0596	0.0599
3.5	0.0395	0.0399	0.0405	0.0474	0.0481	0.0486	0.0574	0.0579	0.0582
3.625	0.0378	0.0382	0.0387	0.0457	0.0464	0.0468	0.0557	0.0562	0.0566
3.75	0.0361	0.0365	0.037	0.044	0.0447	0.0451	0.0541	0.0546	0.055
3.875	0.0345	0.0349	0.0354	0.0423	0.043	0.0435	0.0525	0.53	0.0534
4	0.0329	0.0333	0.0339	0.0407	0.0414	0.0419	0.0509	0.0515	0.0518
4.25	0.0301	0.0304	0.031	0.0376	0.0384	0.0388	0.0479	0.0485	0.0488
4.5	0.0275	0.0279	0.0283	0.0348	0.0355	0.036	0.045	0.0456	0.046
4.75	0.0252	0.0256	0.026	0.0322	0.0328	0.0333	0.0423	0.0429	0.0433
5	0.0232	0.0235	0.0239	0.0297	0.0304	0.0308	0.0397	0.0403	0.0407
5.25	0.0214	0.0217	0.022	0.0275	0.0281	0.0285	0.0373	0.0378	0.0382
5.5	0.0199	0.0201	0.0204	0.0254	0.026	0.0264	0.0349	0.0355	0.0359
5.75	0.0186	0.0188	0.0191	0.0236	0.0241	0.0245	0.0327	0.0333	0.0337
6	0.0176	0.0177	0.0179	0.0219	0.0224	0.0228	0.0306	0.0312	0.0316
6.25	0.0167	0.0168	0.017	0.0204	0.0208	0.0212	0.0287	0.0292	0.0296
6.5	0.0161	0.0162	0.0163	0.0191	0.0195	0.0198	0.0269	0.0274	0.0277
6.75	0.0157	0.0157	0.0157	0.0179	0.0183	0.0185	0.0252	0.0257	0.026
7	0.0155	0.0155	0.0154	0.0169	0.0172	0.0174	0.0236	0.024	0.0244
7.25	0.0155	0.0154	0.0153	0.0161	0.0163	0.0165	0.0221	0.0226	0.0229
7.5	0.0157	0.0155	0.0154	0.0154	0.0156	0.0157	0.0208	0.0212	0.0215
7.75	0.016	0.0158	0.0156	0.0148	0.015	0.0151	0.0195	0.019	0.0202
8	0.0166	0.0163	0.016	0.0144	0.0145	0.0146	0.0184	0.0187	0.019

Orifice Diameter (in)	20			24			30		
	18.814	**19**	**19.25**	**22.626**	**23**	**23.25**	**28.628**	**29**	**29.25**
8.25	0.0172	0.0169	0.0165	0.0142	0.0142	0.0142	0.0174	0.0177	0.0179
8.5	0.018	0.0177	0.0172	0.0141	0.014	0.014	0.0164	0.0168	0.017
8.75	0.019	0.0186	0.018	0.0141	0.014	0.0139	0.0156	0.0159	0.0161
9	0.0201	0.0196	0.019	0.0143	0.0141	0.014	0.0149	0.0152	0.0153
9.25	0.0213	0.0208	0.0201	0.0146	0.0143	0.0141	0.0143	0.0145	0.0146
9.5	0.0226	0.022	0.0213	0.015	0.0146	0.0144	0.0138	0.0139	0.0141
9.75	0.024	0.0234	0.0226	0.0155	0.015	0.0147	0.0133	0.0135	0.0136
10	0.0256	0.0249	0.024	0.0161	0.0155	0.0152	0.013	0.0131	0.0132
10.25	0.0271	0.0264	0.0255	0.0168	0.0162	0.0158	0.0128	0.0128	0.0128
10.5	0.0288	0.028	0.027	0.0176	0.0169	0.0164	0.0126	0.0126	0.0126
10.75	0.0305	0.0297	0.0286	0.0185	0.0176	0.0172	0.0125	0.0125	0.0125
11	0.0322	0.0314	0.0303	0.0194	0.0186	0.0181	0.0125	0.0124	0.0124
11.25	0.034	0.0332	0.032	0.0205	0.0196	0.019	0.0126	0.0125	0.0124
11.5	0.0358	0.0349	0.0338	0.0216	0.0207	0.02	0.0128	0.0126	0.0125
11.75	0.0376	0.0367	0.0355	0.0228	0.0218	0.0211	0.013	0.0128	0.0127
12	0.0394	0.0385	0.0373	0.0241	0.023	0.0223	0.0134	0.0131	0.0129
12.5	0.0429	0.042	0.0408	0.0267	0.0255	0.0248	0.0142	0.0138	0.0136
13	0.0463	0.0454	0.0442	0.0296	0.0282	0.0274	0.0153	0.0148	0.0145
13.5	0.0494	0.0485	0.0474	0.0326	0.0311	0.0302	0.0166	0.016	0.0157
14	0.052	0.0512	0.0502	0.0356	0.0341	0.0331	0.0182	0.0175	0.0171
14.5				0.0386	0.037	0.036	0.0199	0.0192	0.0187
15				0.0415	0.04	0.039	0.0218	0.0209	0.0204
15.5				0.0443	0.0426	0.0418	0.0239	0.023	0.0224
16				0.047	0.0455	0.0446	0.026	0.025	0.0244
16.5				0.0494	0.048	0.0471	0.0283	0.0273	0.0266
17					0.0503	0.0494	0.0307	0.0296	0.0288
17.5							0.0331	0.0319	0.0312
18							0.0355	0.0343	0.0335
18.5							0.0379	0.0366	0.0358
19							0.0402	0.039	0.0382
19.5							0.0424	0.0412	0.0404
20							0.0446	0.0434	0.0426
20.5							0.0466	0.0455	0.0448
21							0.0485	0.0475	0.0467
21.5								0.0492	0.0485

Table C-3 Y_1 Expansion Factors—Flange Taps (Static Pressure Taken from Upstream Taps)

$$\beta = \frac{d}{D}$$

$\frac{h_w}{P_{f1}}$	0.1	0.2	0.3	0.4	0.45	0.50	0.52	0.54	0.56	0.58	0.60	0.61	0.62	0.63	0.64	0.65	0.66	0.67	0.68	0.69	0.7	0.71	0.72	0.73	0.74	0.75
0.0	1.0000	1.0000	1.0000	1.0000	1.0000	1.0000	1.0000	1.0000	1.0000	1.0000	1.0000	1.0000	1.0000	1.0000	1.0000	1.0000	1.0000	1.0000	1.0000	1.0000	1.0000	1.0000	1.0000	1.0000	1.0000	1.0000
0.1	0.9989	0.9989	0.9989	0.9988	0.9988	0.9988	0.9988	0.9988	0.9988	0.9988	0.9987	0.9987	0.9987	0.9987	0.9987	0.9987	0.9987	0.9987	0.9987	0.9986	0.9986	0.9986	0.9986	0.9986	0.9986	0.9986
0.2	0.9977	0.9977	0.9977	0.9977	0.9976	0.9976	0.9976	0.9976	0.9975	0.9975	0.9975	0.9975	0.9974	0.9974	0.9974	0.9974	0.9974	0.9973	0.9973	0.9973	0.9973	0.9972	0.9972	0.9972	0.9971	0.9971
0.3	0.9966	0.9966	0.9966	0.9965	0.9965	0.9964	0.9964	0.9963	0.9963	0.9963	0.9962	0.9962	0.9962	0.9961	0.9961	0.9961	0.996	0.996	0.996	0.9959	0.9959	0.9958	0.9958	0.9958	0.9957	0.9957
0.4	0.9954	0.9954	0.9954	0.9953	0.9953	0.9952	0.9952	0.9951	0.9951	0.9950	0.9949	0.9949	0.9949	0.9948	0.9948	0.9948	0.9947	0.9947	0.9946	0.9946	0.9945	0.9945	0.9944	0.9943	0.9943	0.9942
0.5	0.9943	0.9943	0.9943	0.9942	0.9941	0.9940	0.9940	0.9939	0.9938	0.9938	0.9937	0.9936	0.9936	0.9935	0.9935	0.9934	0.9934	0.9933	0.9933	0.9932	0.9931	0.9931	0.993	0.9929	0.9929	0.9928
0.6	0.9932	0.9932	0.9931	0.9930	0.9929	0.9928	0.9927	0.9927	0.9926	0.9925	0.9924	0.9924	0.9923	0.9923	0.9922	0.9921	0.9921	0.992	0.9919	0.9918	0.9918	0.9917	0.9916	0.9915	0.9914	0.9913
0.7	0.9920	0.9920	0.9920	0.9919	0.9918	0.9916	0.9915	0.9915	0.9914	0.9913	0.9912	0.9911	0.9910	0.9910	0.9909	0.9908	0.9907	0.9907	0.9906	0.9905	0.9904	0.9903	0.9902	0.9901	0.99	0.9899
0.8	0.9909	0.9909	0.9908	0.9907	0.9906	0.9904	0.9903	0.9902	0.9901	0.9900	0.9899	0.9898	0.9897	0.9897	0.9896	0.9895	0.9894	0.9893	0.9892	0.9891	0.989	0.9889	0.9888	0.9887	0.9886	0.9884
0.9	0.9898	0.9897	0.9897	0.9895	0.9894	0.9892	0.9891	0.9890	0.9889	0.9888	0.9886	0.9885	0.9885	0.9884	0.9883	0.9882	0.9881	0.988	0.9879	0.9878	0.9877	0.9875	0.9874	0.9873	0.9871	0.987
1.0	0.9886	0.9886	0.9885	0.9884	0.9882	0.9880	0.9879	0.9878	0.9877	0.9875	0.9874	0.9873	0.9872	0.9871	0.987	0.9869	0.9868	0.9867	0.9865	0.9864	0.9863	0.9861	0.986	0.9859	0.9857	0.9855
1.1	0.9875	0.9875	0.9874	0.9872	0.9870	0.9868	0.9867	0.9866	0.9864	0.9863	0.9861	0.9860	0.9859	0.9858	0.9857	0.9856	0.9854	0.9853	0.9852	0.9851	0.9849	0.9848	0.9846	0.9844	0.9843	0.9841
1.2	0.9863	0.9863	0.9862	0.9860	0.9859	0.9856	0.9855	0.9853	0.9852	0.9850	0.9848	0.9847	0.9846	0.9845	0.9844	0.9843	0.9841	0.984	0.9838	0.9837	0.9835	0.9834	0.9832	0.983	0.9828	0.9826
1.3	0.9852	0.9852	0.9851	0.9849	0.9847	0.9844	0.9843	0.9841	0.9840	0.9838	0.9836	0.9835	0.9833	0.9832	0.9831	0.9829	0.9828	0.9827	0.9825	0.9823	0.9822	0.982	0.9818	0.9816	0.9814	0.9812
1.4	0.9841	0.9840	0.9840	0.9837	0.9835	0.9832	0.9831	0.9829	0.9827	0.9825	0.9823	0.9822	0.9821	0.9819	0.9818	0.9816	0.9815	0.9813	0.9812	0.981	0.9808	0.9806	0.9804	0.9802	0.98	0.9798
1.5	0.9829	0.9829	0.9828	0.9826	0.9823	0.9820	0.9819	0.9817	0.9815	0.9813	0.9810	0.9809	0.9808	0.9806	0.9805	0.9803	0.9802	0.98	0.9798	0.9796	0.9794	0.9792	0.979	0.9788	0.9786	0.9783
1.6	0.9818	0.9818	0.9817	0.9814	0.9811	0.9808	0.9806	0.9805	0.9803	0.9800	0.9798	0.9796	0.9795	0.9793	0.9792	0.979	0.9788	0.9787	0.9785	0.9783	0.9781	0.9778	0.9776	0.9774	0.9771	0.9769
1.7	0.9806	0.9806	0.9805	0.9802	0.9800	0.9796	0.9794	0.9792	0.9790	0.9788	0.9785	0.9784	0.9782	0.9780	0.9779	0.9777	0.9775	0.9773	0.9771	0.9769	0.9767	0.9764	0.9762	0.976	0.9757	0.9754
1.8	0.9795	0.9795	0.9794	0.9791	0.9788	0.9784	0.9782	0.9780	0.9778	0.9775	0.9772	0.9771	0.9769	0.9768	0.9766	0.9764	0.9762	0.976	0.9758	0.9755	0.9753	0.9751	0.9748	0.9745	0.9743	0.974

$\frac{h_w}{P_{f1}}$	0.1	0.2	0.3	0.4	0.45	0.50	0.52	0.54	0.56	0.58	0.60	0.61	0.62	0.63	0.64	0.65	0.66	0.67	0.68	0.69	0.7	0.71	0.72	0.73	0.74	0.75
1.9	0.9784	0.9783	0.9782	0.9779	0.9776	0.9772	0.9770	0.9768	0.9766	0.9763	0.9760	0.9758	0.9756	0.9755	0.9753	0.9751	0.9749	0.9747	0.9744	0.9742	0.9739	0.9737	0.9734	0.9731	0.9728	0.9725
2.0	0.9772	0.9772	0.9771	0.9767	0.9764	0.9760	0.9758	0.9756	0.9753	0.9750	0.9747	0.9745	0.9744	0.9742	0.974	0.9738	0.9735	0.9733	0.9731	0.9728	0.9726	0.9723	0.972	0.9717	0.9714	0.9711
2.1	0.9761	0.9761	0.9759	0.9756	0.9753	0.9748	0.9746	0.9744	0.9741	0.9738	0.9734	0.9733	0.9731	0.9729	0.9727	0.9725	0.9722	0.972	0.9717	0.9715	0.9712	0.9709	0.9706	0.9703	0.97	0.9696
2.2	0.9750	0.9749	0.9748	0.9744	0.9741	0.9736	0.9734	0.9731	0.9729	0.9725	0.9722	0.9720	0.9718	0.9716	0.9714	0.9711	0.9709	0.9706	0.9704	0.9701	0.9698	0.9659	0.9692	0.9689	0.9685	0.9682
2.3	0.9738	0.9738	0.9736	0.9732	0.9729	0.9724	0.9722	0.9719	0.9716	0.9713	0.9709	0.9707	0.9705	0.9703	0.9701	0.9698	0.9696	0.9693	0.969	0.9688	0.9685	0.9681	0.9678	0.9675	0.9671	0.9667
2.4	0.9727	0.9726	0.9725	0.9721	0.9717	0.9712	0.9710	0.9707	0.9704	0.9700	0.9697	0.9694	0.9692	0.9690	0.9688	0.9685	0.9683	0.968	0.9677	0.9674	0.9671	0.9668	0.9664	0.9661	0.9657	0.9653
2.5	0.9715	0.9715	0.9713	0.9709	0.9705	0.9700	0.9698	0.9695	0.9692	0.9688	0.9684	0.9682	0.9680	0.9677	0.9675	0.9672	0.9669	0.9666	0.9663	0.966	0.9657	0.9654	0.965	0.9646	0.9643	0.9639
2.6	0.9704	0.9704	0.9702	0.9698	0.9694	0.9688	0.9686	0.9683	0.9679	0.9675	0.9671	0.9669	0.9667	0.9664	0.9662	0.9659	0.9656	0.9653	0.965	0.9647	0.9643	0.964	0.9636	0.9632	0.9628	0.9624
2.7	0.9693	0.9692	0.9691	0.9686	0.9682	0.9676	0.9673	0.9670	0.9667	0.9663	0.9659	0.9656	0.9654	0.9651	0.9649	0.9646	0.9643	0.964	0.9637	0.9633	0.963	0.9626	0.9622	0.9618	0.9614	0.961
2.8	0.9681	0.9681	0.9679	0.9674	0.9670	0.9664	0.9661	0.9658	0.9654	0.9650	0.9646	0.9644	0.9641	0.9638	0.9636	0.9633	0.963	0.9626	0.9623	0.962	0.9616	0.9612	0.9608	0.9604	0.96	0.9595
2.9	0.9670	0.9669	0.9668	0.9663	0.9658	0.9652	0.9649	0.9646	0.9642	0.9638	0.9633	0.9631	0.9628	0.9625	0.9623	0.962	0.9616	0.9613	0.961	0.9606	0.9602	0.9598	0.9594	0.959	0.9585	0.9581
3.0	0.9658	0.9658	0.9656	0.9651	0.9647	0.9640	0.9637	0.9634	0.9630	0.9626	0.9621	0.9618	0.9615	0.9613	0.961	0.9606	0.9603	0.96	0.9596	0.9592	0.9588	0.9584	0.958	0.9576	0.9571	0.9566
3.1	0.9647	0.9647	0.9645	0.9639	0.9635	0.9628	0.9625	0.9622	0.9617	0.9613	0.9608	0.9605	0.9603	0.96	0.9597	0.9593	0.959	0.9586	0.9583	0.9579	0.9575	0.9571	0.9566	0.9562	0.9557	0.9552
3.2	0.9636	0.9635	0.9633	0.9628	0.9623	0.9616	0.9613	0.9609	0.9605	0.9601	0.9595	0.9593	0.959	0.9587	0.9584	0.958	0.9577	0.9573	0.9569	0.9565	0.9561	0.9557	0.9552	0.9547	0.9542	0.9537
3.3	0.9624	0.9624	0.9622	0.9616	0.9611	0.9604	0.9601	0.9597	0.9593	0.9588	0.9583	0.958	0.9577	0.9574	0.9571	0.9567	0.9564	0.956	0.9556	0.9552	0.9547	0.9543	0.9538	0.9533	0.9528	0.9523
3.4	0.9613	0.9612	0.961	0.9604	0.9599	0.9592	0.9589	0.9585	0.958	0.9576	0.957	0.9567	0.9564	0.9561	0.9558	0.9554	0.955	0.9546	0.9542	0.9538	0.9534	0.9529	0.9524	0.9519	0.9514	0.9508
3.5	0.9602	0.9601	0.9599	0.9593	0.9588	0.958	0.9577	0.9573	0.9568	0.9563	0.9558	0.9554	0.9551	0.9548	0.9545	0.9541	0.9537	0.9533	0.9529	0.9524	0.952	0.9515	0.951	0.9505	0.95	0.9494
3.6	0.959	0.959	0.9587	0.9581	0.9576	0.9568	0.9565	0.956	0.9556	0.9551	0.9545	0.9542	0.9538	0.9535	0.9532	0.9528	0.9524	0.952	0.9515	0.9511	0.9506	0.9501	0.9496	0.9491	0.9485	0.948
3.7	0.9579	0.9578	0.9576	0.957	0.9561	0.9556	0.9553	0.9548	0.9543	0.9538	0.9532	0.9529	0.9526	0.9522	0.9518	0.9515	0.9511	0.9506	0.9502	0.9497	0.9492	0.9487	0.9482	0.9477	0.9471	0.9465
3.8	0.9567	0.9567	0.9564	0.9558	0.9552	0.9544	0.954	0.9536	0.9531	0.9526	0.952	0.9516	0.9513	0.9509	0.9505	0.9502	0.9497	0.9493	0.9488	0.9484	0.9479	0.9474	0.9468	0.9463	0.9457	0.9451
3.9	0.9556	0.9555	0.9533	0.9546	0.954	0.9532	0.9528	0.9524	0.9519	0.9513	0.9507	0.9504	0.95	0.9496	0.9492	0.9488	0.9484	0.948	0.9475	0.947	0.9465	0.946	0.9454	0.9448	0.9442	0.9436
4	0.9545	0.9544	0.9542	0.9535	0.9529	0.952	0.9516	0.9512	0.9506	0.9501	0.9494	0.9491	0.9487	0.9483	0.9479	0.9475	0.9471	0.9465	0.9462	0.9457	0.9451	0.9446	0.944	0.9434	0.9428	0.9422

Table C–4 Y_2 Expansion Factors—Flange Taps (Static Pressure Taken from Downstream Taps)

$$\beta = \frac{d}{D}$$

$\frac{h_w}{P_{f2}}$	0.1	0.2	0.3	0.40	0.45	0.50	0.52	0.54	0.56	0.58	0.60	0.61	0.62	0.63	0.64	0.65	0.66	0.67	0.68	0.69	0.7	0.71	0.72	0.73	0.74	0.75
0.0	1.0000	1.0000	1.0000	1.0000	1.0000	1.0000	1.0000	1.0000	1.0000	1.0000	1.0000	1.0000	1.0000	1.0000	1.0000	1.0000	1.0000	1.0000	1.0000	1.0000	1.0000	1.0000	1.0000	1.0000	1.0000	1.0000
0.1	1.0007	1.0007	1.0006	1.0006	1.0006	1.0006	1.0006	1.0006	1.0006	1.0006	1.0005	1.0005	1.0005	1.0005	1.0005	1.0005	1.0005	1.0005	1.0004	1.0004	1.0004	1.0004	1.0004	1.0004	1.0004	1.0004
0.2	1.0013	1.0013	1.0013	1.0013	1.0012	1.0012	1.0012	1.0012	1.0011	1.0011	1.0011	1.0011	1.0010	1.0010	1.001	1.001	1.001	1.0009	1.0009	1.0009	1.0009	1.0008	1.0008	1.0008	1.0008	1.0007
0.3	1.0020	1.0020	1.0020	1.0019	1.0019	1.0018	1.0018	1.0018	1.0017	1.0017	1.0016	1.0016	1.0016	1.0015	1.0015	1.0015	1.0014	1.0014	1.0014	1.0013	1.0013	1.0013	1.0012	1.0012	1.0011	1.0011
0.4	1.0027	1.0027	1.0026	1.0026	1.0025	1.0024	1.0024	1.0023	1.0023	1.0022	1.0022	1.0021	1.0021	1.0021	1.002	1.002	1.0019	1.0019	1.0018	1.0018	1.0017	1.0017	1.0016	1.0016	1.0015	1.0014
0.5	1.0033	1.0033	1.0033	1.0032	1.0031	1.0030	1.0030	1.0029	1.0029	1.0028	1.0027	1.0027	1.0026	1.0026	1.0025	1.0025	1.0024	1.0024	1.0023	1.0022	1.0022	1.0021	1.002	1.002	1.0019	1.0018
0.6	1.0040	1.0040	1.0040	1.0039	1.0038	1.0036	1.0036	1.0035	1.0034	1.0034	1.0033	1.0032	1.0032	1.0031	1.003	1.003	1.0029	1.0028	1.0028	1.0027	1.0026	1.0025	1.0025	1.0024	1.0023	1.0022
0.7	1.0047	1.0047	1.0046	1.0045	1.0044	1.0043	1.0042	1.0041	1.0040	1.0039	1.0038	1.0038	1.0037	1.0036	1.0036	1.0035	1.0034	1.0033	1.0032	1.0032	1.0031	1.003	1.0029	1.0028	1.0027	1.0026
0.8	1.0054	1.0053	1.0053	1.0052	1.0050	1.0049	1.0048	1.0047	1.0046	1.0045	1.0044	1.0043	1.0042	1.0042	1.0041	1.004	1.0039	1.0038	1.0037	1.0036	1.0035	1.0034	1.0033	1.0032	1.003	1.0029
0.9	1.0060	1.0060	1.0060	1.0058	1.0057	1.0055	1.0054	1.0053	1.0052	1.0050	1.0049	1.0048	1.0048	1.0047	1.0046	1.0045	1.0044	1.0043	1.0042	1.0041	1.004	1.0038	1.0037	1.0036	1.0034	1.0033
1.0	1.0067	1.0067	1.0066	1.0065	1.0063	1.0061	1.0060	1.0059	1.0058	1.0056	1.0055	1.0054	1.0053	1.0052	1.0051	1.005	1.0049	1.0048	1.0047	1.0045	1.0044	1.0043	1.0041	1.004	1.0038	1.0037
1.1	1.0074	1.0074	1.0073	1.0071	1.0069	1.0067	1.0066	1.0065	1.0063	1.0062	1.0060	1.0059	1.0058	1.0057	1.0056	1.0055	1.0054	1.0053	1.0051	1.005	1.0049	1.0047	1.0046	1.0044	1.0042	1.0041
1.2	1.0080	1.0080	1.0080	1.0078	1.0076	1.0073	1.0072	1.0071	1.0069	1.0068	1.0066	1.0065	1.0064	1.0062	1.0061	1.006	1.0059	1.0058	1.0056	1.0055	1.0053	1.0052	1.005	1.0048	1.0046	1.0044
1.3	1.0087	1.0087	1.0086	1.0084	1.0082	1.0080	1.0078	1.0077	1.0075	1.0073	1.0071	1.0070	1.0069	1.0068	1.0066	1.0065	1.0064	1.0062	1.0061	1.0059	1.0058	1.0056	1.0054	1.0052	1.005	1.0048
1.4	1.0094	1.0094	1.0093	1.0091	1.0089	1.0086	1.0084	1.0083	1.0081	1.0079	1.0077	1.0076	1.0074	1.0073	1.0072	1.007	1.0069	1.0067	1.0066	1.0064	1.0062	1.006	1.0058	1.0056	1.0054	1.0052
1.5	1.0101	1.0101	1.0100	1.0097	1.0095	1.0092	1.0090	1.0089	1.0087	1.0085	1.0082	1.0081	1.0080	1.0078	1.0077	1.0076	1.0074	1.0072	1.007	1.0069	1.0067	1.0065	1.0063	1.006	1.0058	1.0056
1.6	1.0108	1.0107	1.0106	1.0104	1.0101	1.0098	1.0096	1.0095	1.0093	1.0090	1.0088	1.0087	1.0085	1.0084	1.0082	1.0081	1.0079	1.0077	1.0075	1.0073	1.0071	1.0069	1.0067	1.0065	1.0062	1.006
1.7	1.0114	1.0114	1.0113	1.0110	1.0108	1.0104	1.0103	1.0101	1.0099	1.0096	1.0094	1.0092	1.0091	1.0089	1.0088	1.0086	1.0084	1.0082	1.008	1.0078	1.0076	1.0074	1.0071	1.0069	1.0066	1.0064
1.8	1.0121	1.0121	1.0120	1.0117	1.0114	1.0111	1.0109	1.0107	1.0104	1.0102	1.0099	1.0098	1.0096	1.0094	1.0093	1.0091	1.0089	1.0087	1.0085	1.0083	1.008	1.0078	1.0076	1.0073	1.007	1.0068
1.9	1.0128	1.0128	1.0126	1.0123	1.0121	1.0117	1.0115	1.0113	1.0110	1.0108	1.0105	1.0103	1.0102	1.0100	1.0098	1.0096	1.0094	1.0092	1.009	1.0088	1.0085	1.0083	1.008	1.0077	1.0074	1.0071

$\frac{h_w}{P_{f2}}$	0.1	0.2	0.3	0.40	0.45	0.50	0.52	0.54	0.56	0.58	0.60	0.61	0.62	0.63	0.64	0.65	0.66	0.67	0.68	0.69	0.7	0.71	0.72	0.73	0.74	0.75
2.0	1.0135	1.0134	1.0133	1.0130	1.0127	1.0123	1.0121	1.0119	1.0116	1.0114	1.0110	1.0109	1.0107	1.0105	1.0103	1.0101	1.0099	1.0097	1.0095	1.0092	1.009	1.0087	1.0084	1.0081	1.0078	1.0075
2.1	1.0142	1.0141	1.0140	1.0136	1.0134	1.0129	1.0127	1.0125	1.0122	1.0119	1.0116	1.0114	1.0112	1.0111	1.0109	1.0106	1.0104	1.0102	1.01	1.0097	1.0094	1.0092	1.0089	1.0086	1.0083	1.0079
2.2	1.0148	1.0148	1.0147	1.0143	1.0140	1.0136	1.0133	1.0131	1.0128	1.0125	1.0122	1.0120	1.0118	1.0116	1.0114	1.0112	1.0109	1.0107	1.0104	1.0102	1.0099	1.0096	1.0093	1.009	1.0087	1.0083
2.3	1.0155	1.0155	1.0154	1.0150	1.0146	1.0142	1.0140	1.0137	1.0134	1.0131	1.0127	1.0126	1.0124	1.0121	1.0119	1.0117	1.0114	1.0112	1.0109	1.0106	1.0104	1.0101	1.0098	1.0094	1.0091	1.0087
2.4	1.0162	1.0162	1.0160	1.0156	1.0153	1.0148	1.0146	1.0143	1.0140	1.0137	1.0133	1.0131	1.0129	1.0127	1.0124	1.0122	1.012	1.0117	1.0114	1.0111	1.0108	1.0105	1.0102	1.0098	1.0095	1.0091
2.5	1.0169	1.0168	1.0167	1.0163	1.0159	1.0154	1.0152	1.0149	1.0146	1.0142	1.0139	1.0137	1.0134	1.0132	1.013	1.0127	1.0125	1.0122	1.0119	1.0116	1.0113	1.011	1.0106	1.0103	1.0099	1.0095
2.6	1.0176	1.0175	1.0174	1.0170	1.0166	1.0161	1.0158	1.0155	1.0152	1.0148	1.0144	1.0142	1.0140	1.0138	1.0135	1.0133	1.013	1.0127	1.0124	1.0121	1.0118	1.0114	1.0111	1.0107	1.0103	1.0099
2.7	1.0182	1.0182	1.0180	1.0176	1.0172	1.0167	1.0164	1.0161	1.0158	1.0154	1.0150	1.0148	1.0146	1.0143	1.014	1.0138	1.0135	1.0132	1.0129	1.0126	1.0122	1.0119	1.0115	1.0111	1.0107	1.0103
2.8	1.0189	1.0189	1.0187	1.0183	1.0179	1.0173	1.0170	1.0167	1.0164	1.0160	1.0156	1.0154	1.0151	1.0148	1.0146	1.0143	1.014	1.0137	1.0134	1.0131	1.0127	1.0124	1.012	1.0116	1.0112	1.0107
2.9	1.0196	1.0196	1.0194	1.0189	1.0185	1.0180	1.0177	1.0173	1.0170	1.0166	1.0162	1.0159	1.0157	1.0154	1.0151	1.0148	1.0145	1.0142	1.0139	1.0136	1.0132	1.0128	1.0124	1.012	1.0116	1.0111
3.0	1.0203	1.0203	1.0201	1.0196	1.0192	1.0186	1.0183	1.0180	1.0176	1.0172	1.0167	1.0165	1.0162	1.0160	1.0157	1.0154	1.015	1.0147	1.0144	1.014	1.0137	1.0133	1.0129	1.0124	1.012	1.0116
3.1	1.0210	1.0210	1.0208	1.0203	1.0198	1.0192	1.0189	1.0186	1.0182	1.0178	1.0173	1.0170	1.0168	1.0165	1.0162	1.0159	1.0156	1.0152	1.0149	1.0145	1.0141	1.0137	1.0133	1.0129	1.0124	1.012
3.2	1.0217	1.0216	1.0214	1.0209	1.0205	1.0198	1.0195	1.0192	1.0188	1.0184	1.0179	1.0176	1.0173	1.0170	1.0167	1.0164	1.0161	1.0158	1.0154	1.015	1.0146	1.0142	1.0138	1.0133	1.0128	1.0124
3.3	1.0224	1.0223	1.0221	1.0216	1.0211	1.0205	1.0202	1.0198	1.0194	1.0189	1.0184	1.0182	1.0179	1.0176	1.0173	1.0175	1.0171	1.0168	1.0164	1.016	1.0156	1.0151	1.0147	1.0142	1.0137	1.0132
3.4	1.0230	1.0230	1.0228	1.0223	1.0218	1.0211	1.0208	1.0204	1.0200	1.0195	1.0190	1.0187	1.0184	1.0181	1.0178	1.0175	1.0171	1.0168	1.0164	1.016	1.0156	1.0151	1.0147	1.0142	1.0137	1.0132
3.5	1.0237	1.0237	1.0235	1.0229	1.0224	1.0217	1.0214	1.0210	1.0216	1.0201	1.0196	1.0193	1.0910	1.0187	1.0184	1.018	1.0177	1.0173	1.0169	1.0165	1.016	1.0156	1.0151	1.0146	1.0141	1.0136
3.6	1.0244	1.0244	1.0242	1.0236	1.0231	1.0224	1.0220	1.0216	1.0212	1.0207	1.0202	1.0199	1.0196	1.0192	1.0189	1.0186	1.0182	1.0178	1.0174	1.017	1.0165	1.0161	1.0156	1.0151	1.0146	1.014
3.7	1.0251	1.0251	1.0248	1.0243	1.0237	1.0230	1.0226	1.0222	1.0218	1.0213	1.0207	1.0204	1.0201	1.0198	1.0195	1.0191	1.0187	1.0183	1.0179	1.0175	1.017	1.0165	1.016	1.0155	1.015	1.0144
3.8	1.0258	1.0258	1.0255	10249	1.0244	1.0236	1.0233	1.0229	1.0224	1.0219	1.0213	1.0210	1.0207	1.0204	1.02	1.0196	1.0192	1.0188	1.0184	1.018	1.0175	1.017	1.0165	1.016	1.0154	1.0148
3.9	1.0265	1.0264	1.0262	1.0256	1.0250	1.0243	1.0239	1.0235	1.0230	1.0225	1.0219	1.0216	1.0213	1.0209	1.0206	1.0202	1.0198	1.0194	1.0189	1.0185	1.018	1.0175	1.017	1.0164	1.0159	1.0153
4.0	1.0272	1.0271	1.0269	1.0263	1.0257	1.0249	1.0245	1.0241	1.0236	1.0231	1.0225	1.0222	1.0218	1.0215	1.0211	1.0207	1.0203	1.0199	1.0194	1.019	1.0185	1.018	1.0174	1.0169	1.0163	1.0157

Table C–5 Y_m Expansion Factors—Flange Taps (Static Pressure Mean of Upstream and Downstream)

$$\beta = \frac{d}{D}$$

$\frac{h_w}{P_{f2}}$	0.1	0.2	0.3	0.40	0.45	0.50	0.52	0.54	0.56	0.58	0.60	0.61	0.62	0.63	0.64	0.65	0.66	0.67	0.68	0.69	0.7	0.71	0.72	0.73	0.74	0.75
0	1	1	1	1	1	1	1	1	1	1	1	1	1	1	1	1	1	1	1	1	1	1	1	1	1	1
0.1	0.9998	0.9998	0.9998	0.9997	0.9997	0.9997	0.9997	0.9997	0.9997	0.9996	0.9996	0.9996	0.9996	0.9996	0.9996	0.9996	0.9996	0.9996	0.9996	0.9995	0.9995	0.9995	0.9995	0.9995	0.9995	0.9994
0.2	0.9995	0.9995	0.9995	0.9995	0.9994	0.9994	0.9994	0.9994	0.9993	0.9993	0.9993	0.9993	0.9992	0.9992	0.9992	0.9992	0.9992	0.9991	0.9991	0.9991	0.9991	0.999	0.999	0.999	0.9989	0.9989
0.3	0.9993	0.9993	0.9993	0.9992	0.9992	0.9991	0.9991	0.9990	0.9990	0.9990	0.9989	0.9989	0.9989	0.9988	0.9988	0.9988	0.9987	0.9987	0.9987	0.9986	0.9986	0.9986	0.9985	0.9985	0.9984	0.9984
0.4	0.9991	0.9990	0.9990	0.9990	0.9989	0.9988	0.9988	0.9987	0.9987	0.9986	0.9986	0.9985	0.9985	0.9984	0.9984	0.9984	0.9983	0.9983	0.9982	0.9982	0.9981	0.9981	0.998	0.998	0.9979	0.9978
0.5	0.9988	0.9988	0.9988	0.9987	0.9986	0.9985	0.9985	0.9984	0.9984	0.9983	0.9982	0.9982	0.9981	0.9981	0.9980	0.998	0.9979	0.9979	0.9978	0.9977	0.9977	0.9976	0.9975	0.9975	0.9974	0.9973
0.6	0.9986	0.9986	0.9986	0.9984	0.9984	0.9982	0.9982	0.9981	0.9980	0.9979	0.9978	0.9978	0.9977	0.9977	0.9976	0.9976	0.9975	0.9974	0.9974	0.9973	0.9972	0.9971	0.997	0.997	0.9969	0.9968
0.7	0.9984	0.9984	0.9983	0.9982	0.9981	0.9980	0.9979	0.9978	0.9977	0.9976	0.9975	0.9974	0.9974	0.9973	0.9972	0.9972	0.9971	0.997	0.9969	0.9968	0.9968	0.9966	0.9966	0.9964	0.9964	0.9962
0.8	0.9982	0.9981	0.9981	0.9980	0.9978	0.9977	0.9976	0.9975	0.9974	0.9973	0.9971	0.9971	0.9970	0.9969	0.9968	0.9968	0.9967	0.9966	0.9965	0.9964	0.9963	0.9962	0.9961	0.996	0.9958	0.9957
0.9	0.9979	0.9979	0.9978	0.9977	0.9976	0.9974	0.9973	0.9972	0.9971	0.9969	0.9968	0.9967	0.9966	0.9966	0.9965	0.9964	0.9963	0.9962	0.9961	0.996	0.9958	0.9957	0.9956	0.9955	0.9953	0.9952
1.0	0.9977	0.9977	0.9976	0.9974	0.9973	0.9971	0.9970	0.9969	0.9968	0.9966	0.9964	0.9964	0.9963	0.9962	0.9961	0.996	0.9959	0.9958	0.9956	0.9955	0.9954	0.9952	0.9951	0.995	0.9948	0.9946
1.1	0.9975	0.9975	0.9974	0.9972	0.9970	0.9968	0.9967	0.9966	0.9964	0.9963	0.9961	0.9960	0.9959	0.9958	0.9957	0.9956	0.9955	0.9953	0.9952	0.9951	0.9949	0.9948	0.9946	0.9945	0.9943	0.9941
1.2	0.9972	0.9972	0.9972	0.9970	0.9968	0.9965	0.9964	0.9963	0.9961	0.9959	0.9958	0.9956	0.9955	0.9954	0.9953	0.9952	0.9951	0.9949	0.9948	0.9946	0.9945	0.9943	0.9942	0.994	0.9938	0.9936
1.3	0.9970	0.9970	0.9969	0.9967	0.9965	0.9962	0.9961	0.9960	0.9958	0.9956	0.9954	0.9953	0.9952	0.9951	0.9949	0.9948	0.9947	0.9945	0.9944	0.9942	0.994	0.9939	0.9937	0.9935	0.9933	0.9931
1.4	0.9968	0.9968	0.9967	0.9965	0.9963	0.9960	0.9958	0.9957	0.9955	0.9952	0.9951	0.9950	0.9948	0.9947	0.9946	0.9944	0.9943	0.9941	0.9939	0.9938	0.9936	0.9934	0.9932	0.993	0.9928	0.9926
1.5	0.9966	0.9966	0.9965	0.9962	0.9960	0.9957	0.9955	0.9954	0.9952	0.9950	0.9946	0.9946	0.9945	0.9943	0.9942	0.994	0.9939	0.9937	0.9935	0.9933	0.9931	0.9929	0.9927	0.9925	0.9923	0.992
1.6	0.9964	0.9964	0.9962	0.9960	0.9957	0.9954	0.9952	0.9951	0.9949	0.9946	0.9944	0.9942	0.9941	0.9940	0.9938	0.9936	0.9935	0.9933	0.9931	0.9929	0.9927	0.9925	0.9922	0.992	0.9918	0.9915
1.7	0.9962	0.9961	0.9960	0.9957	0.9955	0.9951	0.9950	0.9948	0.9946	0.9943	0.9940	0.9939	0.9938	0.9936	0.9934	0.9932	0.9931	0.9929	0.9927	0.9925	0.9922	0.992	0.9918	0.9915	0.9913	0.991
1.8	0.9959	0.9959	0.9958	0.9955	0.9952	0.9949	0.9947	0.9945	0.9942	0.9940	0.9937	0.9936	0.9934	0.9932	0.993	0.9929	0.9927	0.9925	0.9923	0.992	0.9918	0.9916	0.9913	0.991	0.9908	0.9905

$\frac{h_w}{P_{f2}}$	0.1	0.2	0.3	0.40	0.45	0.50	0.52	0.54	0.56	0.58	0.60	0.61	0.62	0.63	0.64	0.65	0.66	0.67	0.68	0.69	0.7	0.71	0.72	0.73	0.74	0.75
1.9	0.9957	0.9957	0.9956	0.9953	0.9950	0.9946	0.9944	0.9942	0.9939	0.9937	0.9934	0.9932	0.9930	0.9929	0.9927	0.9925	0.9923	0.9921	0.9918	0.9916	0.9914	0.9911	0.9908	0.9906	0.9903	0.99
2.0	0.9955	0.9955	0.9954	0.9950	0.9947	0.9943	0.9941	0.9939	0.9936	0.9934	0.9930	0.9929	0.9927	0.9925	0.9923	0.9921	0.9919	0.9917	0.9914	0.9912	0.9909	0.9907	0.9904	0.9901	0.9898	0.9895
2.1	0.9953	0.9953	0.9951	0.9948	0.9945	0.9940	0.9938	0.9936	0.9933	0.9930	0.9927	0.9925	0.9923	0.9922	0.9919	0.9917	0.9915	0.9913	0.991	0.9908	0.9905	0.9902	0.9899	0.9896	0.9893	0.989
2.2	0.9951	0.9951	0.9949	0.9946	0.9942	0.9938	0.9936	0.9933	0.9930	0.9927	0.9924	0.9922	0.9920	0.9918	0.9916	0.9914	0.9911	0.9909	0.9906	0.9903	0.9901	0.9898	0.9895	0.9891	0.9888	0.9885
2.3	0.9949	0.9948	0.9947	0.9943	0.9940	0.9935	0.9933	0.9930	0.9927	0.9924	0.9920	0.9918	0.9916	0.9914	0.9912	0.991	0.9907	0.9905	0.9902	0.9899	0.9896	0.9893	0.989	0.9887	0.9883	0.9988
2.4	0.9947	0.9946	0.9945	0.9941	0.9937	0.9932	0.9930	0.9927	0.9924	0.9921	0.9917	0.9915	0.9913	0.9911	0.9908	0.9906	0.9903	0.9901	0.9898	0.9895	0.9892	0.9889	0.9885	0.9882	0.9878	0.9874
2.5	0.9945	0.9944	0.9943	0.9939	0.9935	0.9930	0.9927	0.9924	0.9921	0.9918	0.9914	0.9912	0.9910	0.9907	0.9905	0.9902	0.99	0.9897	0.9894	0.9891	0.9888	0.9884	0.9881	0.9877	0.9873	0.987
2.6	0.9943	0.9942	0.9941	0.9936	0.9932	0.9927	0.9924	0.9922	0.9918	0.9915	0.9911	0.9908	0.9906	0.9904	0.9901	0.9898	0.9896	0.9893	0.989	0.9887	0.9883	0.988	0.9876	0.9872	0.9868	0.9864
2.7	0.9940	0.9940	0.9938	0.9934	0.9930	0.9924	0.9922	0.9919	0.9915	0.9912	0.9907	0.9905	0.9903	0.9900	0.9898	0.9895	0.9892	0.9889	0.9886	0.9882	0.9879	0.9875	0.9872	0.9868	0.9864	0.986
2.8	0.9938	0.9938	0.9936	0.9932	0.9928	0.9922	0.9919	0.9916	0.9912	0.9908	0.9904	0.9902	0.9899	0.9897	0.9894	0.9891	0.9888	0.9885	0.9882	0.9878	0.9875	0.9871	0.9867	0.9863	0.9859	0.9854
2.9	0.9936	0.9936	0.9934	0.9929	0.9925	0.9919	0.9916	0.9913	0.9910	0.9905	0.9901	0.9898	0.9896	0.9893	0.989	0.9887	0.9884	0.9881	0.9878	0.9874	0.987	0.9867	0.9863	0.9858	0.9854	0.985
3.0	0.9934	0.9934	0.9932	0.9927	0.9923	0.9917	0.9914	0.9910	0.9906	0.9902	0.9898	0.9895	0.9892	0.9890	0.9887	0.9884	0.9881	0.9877	0.9874	0.987	0.9866	0.9862	0.9858	0.9854	0.9849	0.9845
3.1	0.9932	0.9932	0.9930	0.9925	0.9920	0.9914	0.9911	0.9908	0.9904	0.9899	0.9894	0.9892	0.9889	0.9886	0.9883	0.988	0.9877	0.9873	0.987	0.9866	0.9862	0.9858	0.9854	0.9849	0.9844	0.984
3.2	0.9930	0.9930	0.9928	0.9923	0.9918	0.9912	0.9908	0.9905	0.9901	0.9896	0.9891	0.9889	0.9886	0.9883	0.988	0.9876	0.9873	0.987	0.9866	0.9862	0.9858	0.9854	0.9849	0.9845	0.984	0.9835
3.3	0.9928	0.9928	0.9926	0.9920	0.9916	0.9909	0.9906	0.9902	0.9898	0.9893	0.9888	0.9885	0.9882	0.9879	0.9876	0.9873	0.9869	0.9866	0.9862	0.9858	0.9854	0.9849	0.9845	0.984	0.9835	0.983
3.4	0.9926	0.9926	0.9924	0.9918	0.9913	0.9906	0.9903	0.9899	0.9895	0.9890	0.9885	0.9882	0.9879	0.9876	0.9873	0.9869	0.9866	0.9862	0.9858	0.9854	0.985	0.9845	0.984	0.9835	0.983	0.9825
3.5	0.9924	0.9924	0.9922	0.9916	0.9911	0.9904	0.9900	0.9896	0.9892	0.9887	0.9882	0.9879	0.9876	0.9872	0.9869	0.9866	0.9862	0.9858	0.9854	0.985	0.9845	0.9841	0.9836	0.9831	0.9826	0.982
3.6	0.9922	0.9922	0.992	0.9914	0.9909	0.9901	0.9898	0.9894	0.9889	0.9884	0.9879	0.9876	0.9872	0.9869	0.9866	0.9862	0.9858	0.9854	0.985	0.9846	0.9841	0.9836	0.9831	0.9826	0.9821	0.9815
3.7	0.9921	0.992	0.9918	0.9912	0.9906	0.9899	0.9895	0.9891	0.9886	0.9881	0.9876	0.9872	0.9869	0.9866	0.9862	0.9858	0.9855	0.985	0.9846	0.9842	0.9837	0.9832	0.9827	0.9822	0.9816	0.981
3.8	0.9919	0.9918	0.9916	0.991	0.9904	0.9896	0.9893	0.9888	0.9884	0.9878	0.9872	0.9869	0.9866	0.9862	0.9859	0.9855	0.9851	0.9847	0.9842	0.9838	0.9833	0.9828	0.9823	0.9817	0.9812	0.9806
3.9	0.9917	0.9916	0.9914	0.9907	0.9902	0.9894	0.989	0.9886	0.9881	0.9875	0.9869	0.9866	0.9863	0.9859	0.9855	0.9851	0.9847	0.9843	0.9838	0.9834	0.9829	0.9824	0.9818	0.9813	0.9807	0.9801
4	0.9915	0.9914	0.9912	0.9905	0.9899	0.9891	0.9887	0.9883	0.9878	0.9872	0.9866	0.9863	0.9859	0.9856	0.9852	0.9848	0.9844	0.9839	0.9835	0.983	0.9825	0.9819	0.9814	0.9808	0.9802	0.9796

Table C-6 F_b Basic Orifice Factors—Pipe Taps

Basic temperature = 60 °F
Flowing temperature = 60 °F
Base pressure = 14.73 psia
Specific gravity = 1.0

Pipe Sizes—Nominal and Published Inside Diameters, in

Orifice Diameter		2			3			4	
(in)	1.689	1.939	2.067	2.300	2.626	2.900	3,068	3.152	3.430
0.250	12.850	12.813	12.800	12.782	12.765	12.753	12.748	12.745	12.737
0.375	29.359	29.097	29.005	28.882	28.771	28.710	28.682	28.669	28.634
0.500	53.703	52.816	52.401	52.019	51.591	51.353	51.243	51.196	51.064
0.625	87.212	84.919	84.083	82.922	81.795	81.142	80.835	80.703	80.332
0.750	132.23	126.86	124.99	122.45	120.06	118.67	118.00	117.70	116.86
0.875	192.74	181.02	177.08	171.92	167.23	164.58	163.31	162.76	161.17
1.000	275.45	251.10	243.27	233.30	224.56	219.76	217.52	216.55	213.79
1.125	391.93	342.98	327.98	309.43	293.79	285.48	281.66	280.02	275.42
1.250		465.99	437.99	404.52	377.36	363.41	357.12	354.45	347.03
1.375			583.96	524.68	478.68	455.82	445.74	441.48	429.83
1.500				679.10	602.45	565.79	549.94	543.31	525.40
1.625					755.34	697.43	672.95	662.81	635.76
1.750					946.99	856.37	819.05	803.77	763.51
1.875						1,050.4	993.98	971.19	911.98
2.000						1,290.7	1,205.6	1,171.8	1,085.5
2.125							1,465.1	1,415.0	1,289.7
2.250									1,532.0
2.375									1,822.8

Orifice Diameter (in)	4		6				8		
	3.826	4.026	4.897	5.189	5.761	6.065	7.625	7.981	8.071
0.250	12.727	12.722							
0.375	26.598	28.584							
0.500	50.936	50.886	50.739	50.705	50.652	50.628			
0.625	79.974	79.835	79.436	79.349	79.217	79.162			
0.750	116.05	115.73	114.81	114.61	114.32	114.20			
0.875	159.57	158,94	157.11	156.71	156.13	155.89	155.10	154.99	154.96
1.000	211.03	209.91	206.62	205.91	204.84	204.41	203.00	202.80	202.75
1.125	270.90	269.10	263.71	262.51	260.71	259.98	257.62	257.28	257.20
1.250	339.87	337.05	328.73	326.85	324.02	322.86	319.10	318.56	318.44
1.375	418.79	414.51	402.06	399.30	395.08	393.33	387.62	386.81	386.62
1.500	508.76	502.38	484.20	480.23	474.20	471.69	463.39	462.19	461.92
1.625	611.11	601.80	575.73	570.14	561.73	558.24	546.61	544.92	544.53
1.750	727.54	714.16	677.38	669.63	658.08	653.33	637.51	635.19	634.65
1.875	860.17	841.19	789.99	779.40	763.77	757.39	736.34	733.23	732.52
2.000	1,011.7	985.04	914.57	900.28	879.38	870.93	843.34	839.29	838.35
2.125	1,185.3	1,148.4	1,052.3	1,033.2	1,005.6	994.52	958.78	953.58	952.38
2.250	1,385.4	1,334.4	1,204.7	1,179.4	1,143.2	1,128.8	1,083.0	1,076.4	1,074.9
2.375	1,617.2	1,547.3	1,373.4	1,340.2	1,293.1	1,274.6	1,216.3	1,208.0	1,206.1
2.500	1,887.6	1,792.3	1,560.5	1,517.2	1,456.4	1,432.7	1,359.2	1,348.8	1,346.5
2.625	2,206.0	2,075,9	1,768.3	1,712.3	1,634.3	1,604.3	1,512.0	1,499.2	1,496.3
2.750		2,407.0	1,999.8	1,927.6	1,828.3	1,790.3	1,675.4	1,659.7	1,656.1
2.875			2,258.5	2,165,9	2,039.9	1,992.2	1,849.9	1,830.6	1,826.3
3.000			2,548.6	2,430.2	2,271.2	2,211.6	2,036.0	2,012.7	2,007.3
3.125			2,875.2	2,724.4	2,524.3	2,450.1	2,234.7	2,206.4	2,199.9
3.250			3,244.8	3,052.8	2,801.8	2,709.9	2,446.5	2,412.4	2,404.7
3.375			3,665.6	3,420.9	3,106.9	2,993.3	2,672.5	2,631.6	2,622.3
3.500				3,835.7	3,443.0	3,303.0	2,913.7	2,864.7	2,853.7
3.625				4,305.7	3,914.4	3,642.3	3,171.1	3,112.7	3,099.6
3.750					4,226.3	4,014.8	3,446.0	3,376.6	3,361.0
3.875					4,684.9	4,425.1	3,739.9	3,657.6	3,639.2
4.000					5,197.7	4,878.4	4,054.2	3,957.0	3,935.2
4.250							4,751.4	4,616.6	4,586.6
4.500							5,554.7	5,369.0	5,327.9
4.750							6,485.3	6,231.1	6,175.2
5.000							7,571.4	7,224.3	7,148.7
5.250							8,850.3	8,376.3	8,274.0
5.500								9,723.8	9,585.1

Orifice Diameter (in)	10			12			18		
	9.564	10.020	10.136	11.376	11.938	12.090	14.688	15.000	15.250
1.000	202.16								
1.125	256.22	256.01	255.96						
1.250	316.90	316.56	316.49	315.84	315.57	315.51			
1.375	384.29	383.79	383.68	382.66	382.30	382.22			
1.500	458.52	457.79	457.63	456.16	455.64	455.52	453.92	453.78	
1.625	539.72	538.69	538.45	536.38	535.66	535.48	533.27	533.07	532.93
1.750	628.03	626.61	626.29	623.44	622.45	622.20	619.18	618.92	618.73
1.875	723.61	721.70	721.27	717.43	716.10	715.78	711.73	711.39	711.13
2.000	826.63	824.12	823.54	818.48	816.73	816.30	810.99	810.53	810.19
2.125	937.28	934.02	933.27	926.72	924.44	923.88	917.01	916.43	915.99
2.250	1,055.7	1,051.6	1,050.6	1,042.3	1,039.4	1,038.7	1,029.9	1,092.6	1,028.6
2.375	1,182.2	1,177.0	1,175.8	1,165.3	1,161.6	1,160.7	1,149.7	1,148.8	1,148.1
2.500	1,316.9	1,310.5	1,309.0	1,295.9	1,291.4	1,290.2	1.276.5	1.275.4	1,274.5
2.625	1,460.0	1,452.1	1,450.3	1,434.3	1,428.7	1,427.4	1,410.5	1,409.1	1,408.0
2.750	1,611.8	1,602.3	1,600.1	1,580.7	1,573.9	1,572.2	1,551.7	1,549.9	1,548.6
2.875	1,772.5	1,761.0	1,758.4	1,735.1	1,726.9	1,724.9	1,700.1	1,698.1	1,696.5
3.000	1,942.5	1,928.8	1,925.6	1,897.8	1,888.1	1,885.7	1,856.1	1,853.6	1,851.7
3.125	2,122.1	2,1057	2,102.0	2,069.0	2,057.5	2,054.7	2,019.5	2,016.6	2,014.3
3.250	2,311.6	2,292.2	2,287.8	2,248.9	2,235.4	2,232.1	2,190.7	2,187.2	2,184.5
3.375	2,511.5	2,488.6	2,483.4	2,437.7	2,421.8	2,418.0	2,369.6	2,365.5	2,362.4
3.500	2,722.3	2,695.3	2,689.1	2,635.6	2,617.2	2,612.6	2,556.5	2,551.7	2,548.1
3.625	2,944.3	2,912.7	2,905.5	2,843.0	2,821.6	2,816.3	2,751.4	2,745.9	2,741.7
3.750	3,178.1	3,141.2	3,132.7	3,060.2	3,035.3	3,029.3	2,954.5	2,948.1	2,943.3
3.875	3,424.3	3,381.3	3,371.5	3,287.4	3,258.7	3,251.7	3,165.9	3,158.6	3,153.1
4.000	3,683.5	3,633.5	3,622.1	3,524.9	3,492.0	3,483.9	3,385.8	3,377.5	3,371.2
4.250	4,243.8	4,176.8	4,161.6	4,032.8	3,989.5	3,979.0	3,851.6	3,840.9	3,832.8
4.500	4,865.1	4,776.2	4,756.1	4,587.1	4,530.8	4,517.2	4,353.4	4,339.8	4,329.6
4.750	5,554.9	5,437.9	5,411.5	5,191.5	5,119.0	5,119.0	4,892.9	4,875.8	4,862.9
5.000	6,322.2	6,169.2	6,134.9	5,850.6	5,757.8	5,757,8	5,471.9	5,450.5	5,434.3
5.250	7,177.7	6,978.9	6,934.4	6,569.4	6,451.5	6,451.5	6,092.5	6,065.9	6,045.9
5.500	8,134.1	7,877.2	7,820.0	7,354.1	7,205.1	7,205.1	6,757.0	6,724.1	6,699.4
5.750	9,207.0	8,876.3	8,803.1	8,211.4	8,024.2	8,024.2	7,468.0	7,427.6	7,397.4
6.000	10,415	9,991.2	9,897.8	9,149.5	8,915.4	8,915.4	8,228.5	8,179.2	8,142.3
6.250	11,783	11,240	11,121	10,178	9,886.1	9,886.1	9,041.6	8,981.7	8,937.0
6.500	13,340	12,644	12,492	11,307	10,945	10,945	9,911.2	9,838.7	9,764.7

Orifice Diameter (in)	10			12			18		
	9.564	10.020	10.136	11.376	11.938	12.090	14.688	15.000	15.250
6.750		14,230	14,038	12,550	12,103	12,103	10,841	10,754	10,689
7.000		16,035	15,790	13,923	13,923	13,371	11,837	11,732	11,654
7.250				15,442	14,762	14,604	12,902	12,777	12,684
7.500				17,131	16,294	16,101	14,044	13,894	13,783
7.750				19,017	17,986	17,750	15,268	15,090	14,959
8.000					19,861	19,572	16,583	16,371	16,216
8.250					21,947	21,593	17,996	17,746	17,561
8.500							19,517	19,221	19,003
8.750							21,156	20,807	20,551
9.000							22,926	22,515	22,214
9.250							24,841	24,356	24,003
9.500							26,917	26,346	25,932
9.750							29,172	28,501	28,014
10.000							31,629	30,839	30,268
10.250							34,315	33,383	32,713
10.500								36,160	35,372

Orifice Diameter (in)	20			24			30		
	18.814	19.000	19.250	22.626	23.000	23.250	28.628	29.000	29.250
2.000	806.71	806.57	806.40						
2.125	911.51	911.35	911.13						
2.250	1,022.9	1,022.7	1,022.4						
2.375	1,141.0	1,140.7	1,140.4	1,136.8	1,136.5	1,136.3			
2.500	1,265.7	1,265.4	1,265.0	1,260.6	1,260.2	1,259.9			
2.625	1,397.2	1,396.8	1,396.3	1,390.9	1,390.5	1,390.2			
2.750	1,535.5	1,535.0	1,534.4	1,527.9	1,527.3	1,527.0			
2.875	1,680.7	1,680.1	1,679.3	1,671.5	1,670.9	1,670.4	1,663.8		
3.000	1,832.7	1,832.1	1,831.2	1,821.9	1,821.1	1,820.6	1,812.7	1,812.3	1,812.0
3.125	1,991.8	1,991.0	1,990.0	1,978.9	1,978.0	1,977.4	1,968.1	1,967.7	1,967.4
3.250	2,158.0	2,157.0	2,155.8	2,142.8	2,141.7	2,141.0	2,130.2	2,129.6	2,129.3
3.375	2,331.3	2,330.2	2,328.7	2,313.5	2,312.3	2,311.5	2,298.8	2,298.2	2,297.7
3.500	2,511.9	2,510.6	2,508.8	2,491.2	2,489.7	2,488.8	2,474.1	2,473.3	2,472.9

Orifice Diameter (in)	20			24			30		
	18.814	19.000	19.250	22.626	23.000	23.250	28.628	29.000	29.250
3.625	2,699.7	2,698.2	2,696.2	2,675.8	2,674.0	2,673.0	2,656.0	2,655.2	2.654.6
3.750	2,895.0	2,893.2	2,890.9	2,867.4	2,865.4	2,864.1	2,844.6	2,843.7	2,843.0
3,875	3,097.7	3,095.7	3,093.0	3,066.0	3,063.8	3,062.3	3,040.0	3,038.9	3,038.2
4.000	3,308.0	3,305.7	3,302.7	3,271.8	3,269.2	3,267.6	3,242.2	3,240.9	3,240.1
4.250	3,751.6	3,748.7	3,744.8	3,705.0	3,701.7	3,699.6	3,666.9	3.665.3	3,664.3
4.500	4,226.8	4,223.0	4,218.1	4,167.6	4,163.4	4,160.7	4,119.3	4,117.3	4,116.0
4.750	4,734.1	4,729.4	4,723.3	4,660.0	4,654.8	4,651.4	4,599.6	4,597.1	4,595.4
5.000	5,274.6	5,268.7	5,261.1	5,183.0	5,176.4	5,172.3	5,108.2	5,105.0	5,103.0
5.250	5,849.0	5,841.9	5,832.6	5,737.1	5,729.1	5,723.9	5,645.4	5,641.4	5,639.1
5.500	6,458.6	6,449.9	6,438.7	6,322.9	6,313.2	6,307.0	6,211.8	6,207.2	6,204.2
5.750	7,104.4	7,094.0	7,080.4	6,941.3	6,929.7	6,922.2	6,807.7	6,802.1	6,798.5
6.000	7,787.9	7,775.4	7,759.1	7,592.8	7,579.0	7,570.1	7,433.6	7,426.9	7,422.6
6.250	8,510.4	8,495.4	8,476.0	8,278.3	8,262.0	8,251.5	8,089.9	8,082.0	8,076.9
6.500	9,273.4	9,255.6	9,232.5	8,998.8	8,979.5	8,967.1	8,777.2	8,768.0	8,761.9
6.750	10,079	10,058	10,030	9,755.0	9,732.4	9,717.9	9,496.0	9,485.2	9,478.1
7.000	10,928	10,903	10,871	10,548	10,522	10,505	10,247	10,234	10,226
7.250	11,823	11,794	11,756	11,379	11,348	11,329	11,030	11,016	11,006
7.500	12,767	12,733	12,689	12,249	12,214	12,191	11,847	11,830	11,819
7.750	13,762	13,722	13,670	13,160	13,119	13,093	12,697	12,678	12,665
8.000	14,810	14,763	14,703	14,113	14,065	14,035	13,582	13,560	13,546
8.250	15,914	15,860	15,791	15,109	15,054	15,020	14,501	14,477	14,461
8.500	17,078	17,015	16,935	16,150	16,087	16,048	15,457	15,429	15,411
8.750	18,305	18,232	18,129	17,237	17,166	17,121	16,450	16,418	16,397
9.000	19,598	19,515	19,408	18,373	18,292	18,241	17,480	17,444	17,421
9.250	20,963	20,866	20,743	19,560	19,468	19,409	18,548	18,508	18,482
9.500	22,402	22,292	22,151	20,800	20,695	20,628	19,656	19,611	19,582
9.750	23,923	23,796	23,634	22,094	21,976	21,900	20,805	20,754	20,721
10.000	25,529	25,384	25,198	23,447	23,312	23,227	21,995	21,938	21,901
10.250	27,227	27,061	26,849	24,859	24,708	24,612	23,228	23,165	23,124
10.500	29,023	28,834	28,592	26,335	26,164	26,056	24,505	24,434	24,358
10.750	30,925	30,709	30,434	27,878	27,685	27,563	25,827	25,749	25,698

Orifice Diameter (in)	20			24			30		
	18.814	**19.000**	**19.250**	**22.626**	**23.000**	**23.250**	**28.628**	**29.000**	**29.250**
11.000	32,940	32,694	32,381	29,490	29,273	29,136	27,196	27,109	27,052
11.250	35,078	34,798	34,443	31,175	30,932	30,779	28,613	28,156	28,453
11.500	37,348	37,030	36,626	32,938	32,666	32,494	30,080	29,972	29,903
11.750	39,761	39,400	38,941	34,783	34,478	34,285	31,598	31,479	31,402
12.000	42,330	41,920	41,399	36,714	36,373	36,158	33,169	33,038	32,953
12.500	47,991	47,461	46,790	40,855	40,429	40,161	36,478	36,318	36,215
13.000	54,463	53,778	52,914	45,406	44,877	44,544	40,024	39,829	39,704
13.500				50,420	49,763	49,352	43,823	43,589	43,437
14.000				55,959	55,147	54,638	47,898	47,615	47,433
14.500				62,099	61,094	60,468	52,271	51,932	51,714
15.000				68,929	67,687	66,915	56,967	56,562	56,301
15.500				76,562	75,025	74,074	62,017	61,533	61,223
16.000					83,231	82,055	67,453	66,878	66,509
16.500							73,314	72,630	72,193
17.000							79,641	78,831	78,313
17.500							86,485	85,525	84,913
18.000							93,900	92,765	92,042
18.500							101,950	100,610	99,758
19.000							110,720	109,130	108,130
19.500							120,300	118,420	117,230
20.000							130,780	128,560	127,150

Table C–7 "*b*" Values for Reynolds Number Factor F_r Determination—Pipe Taps

Pipe Sizes—Nominal and Published Inside Diameters, in

Orifice Diameter (in)	2			3			4		
	1.689	1.939	2.067	2.300	2.626	2.900	3.068	3.152	3.438
0.250	0.1105	0.1091	0.1087	0.1081	0.1078	0.1078	0.1080	0.1081	0.1084
0.375	0.0890	0.0878	0.0877	0.0879	0.0888	0.0898	0.0905	0.0908	0.0918
0.500	0.0758	0.0734	0.0729	0.0728	0.0737	0.0750	0.0758	0.0763	0.0778
0.625	0.0693	0.0647	0.0635	0.0624	0.0624	0.0634	0.0642	0.0646	0.0662
0.750	0.0675	0.0608	0.0586	0.0559	0.0546	0.0548	0.0552	0.0555	0.0568
0.875	0.0684	0.0602	0.0570	0.0528	0.0497	0.0488	0.0488	0.0489	0.0496
1.000	0.0702	0.0614	0.0576	0.0522	0.0473	0.0452	0.0445	0.0443	0.0443
1.125	0.0708	0.0635	0.0595	0.0532	0.0469	0.0435	0.0422	0.0417	0.0407
1.250		0.0650	0.0616	0.0552	0.0478	0.0434	0.0414	0.0406	0.0387
1.375			0.0629	0.0574	0.0496	0.0443	0.0418	0.0408	0.0379
1.500				0.0590	0.0518	0.0460	0.0431	0.0418	0.0382
1.625					0.0539	0.0482	0.0450	0.0435	0.0392
1.750					0.0553	0.0504	0.0471	0.0456	0.0408
1.875						0.0521	0.0492	0.0477	0.0427
2.000						0.0532	0.0508	0.0495	0.0448
2.125							0.0519	0.0509	0.0467
2.250									0.0483
2.375									0.0494

Orifice Diameter (in)	4		6				8		
	3.826	**4.026**	**4.897**	**5.189**	**5.761**	**6.065**	**7.625**	**7.981**	**8.071**
0.250	0.1087	0.1091							
0.375	0.0932	0.0939							
0.500	0.0799	0.0810	0.0850	0.0862	0.0883	0.0895			
0.625	0.0685	0.0697	0.0747	0.0762	0.0789	0.0802			
0.750	0.0590	0.0602	0.0655	0.0672	0.0703	0.0718			
0.875	0.0513	0.0524	0.0575	0.0592	0.0625	0.0642	0.0716	0.0730	0.0733
1.000	0.0453	0.0461	0.0506	0.0523	0.0556	0.0573	0.0652	0.0668	0.0662
1.125	0.0408	0.0412	0.0448	0.0464	0.0495	0.0512	0.0592	0.0609	0.0613
1.250	0.0376	0.0377	0.0401	0.0413	0.0442	0.0458	0.0538	0.0555	0.0560
1.375	0.0358	0.0353	0.0363	0.0373	0.0397	0.0412	0.0489	0.0506	0.0510
1.500	0.0350	0.0340	0.0334	0.0340	0.0360	0.0372	0.0445	0.0462	0.0466
1.625	0.0351	0.0336	0.0313	0.0315	0.0329	0.0339	0.0404	0.0421	0.0425
1.75	0.0358	0.034	0.03	0.0298	0.0304	0.0311	0.0369	0.0384	0.0388
1.875	0.0371	0.0349	0.0293	0.0287	0.0285	0.029	0.0338	0.0352	0.0355
2	0.0388	0.0363	0.0292	0.0281	0.0273	0.0273	0.0311	0.0323	0.0327
2.125	0.0407	0.036	0.0297	0.0281	0.0265	0.0262	0.0288	0.0298	0.0301
2.25	0.0427	0.0398	0.0305	0.0285	0.0261	0.0258	0.0268	0.0277	0.028
2.375	0.0445	0.0417	0.0316	0.0293	0.0262	0.0253	0.0252	0.0259	0.0261
2.5	0.046	0.0435	0.033	0.0304	0.0267	0.0254	0.0239	0.0244	0.0246
2.625	0.0472	0.045	0.0345	0.0317	0.0274	0.0258	0.023	0.0232	0.0233
2.75		0.0462	0.0362	0.0331	0.0264	0.0265	0.0224	0.0224	0.0224
2.875			0.0379	0.0347	0.0295	0.0274	0.022	0.0218	0.0218
3			0.0395	0.0364	0.0308	0.0285	0.0219	0.0214	0.0213
3.125			0.041	0.038	0.0323	0.0297	0.022	0.0213	0.0211
3.25			0.0422	0.0394	0.0338	0.0311	0.0223	0.0214	0.0212
3.375			0.0432	0.0408	0.0353	0.0325	0.0228	0.0216	0.0214
3.5				0.0419	0.0367	0.0339	0.0235	0.0221	0.0218
3.625				0.0428	0.0381	0.0354	0.0243	0.0227	0.0224
3.75					0.0393	0.0367	0.0252	0.0234	0.023
3.875					0.0404	0.038	0.0262	0.0243	0.0238
4					0.0413	0.0391	0.0273	0.0252	0.0246
4.25							0.0296	0.0273	0.0268
4.5							0.0321	0.0296	0.029
4.75							0.0344	0.032	0.0314
5							0.0364	0.0342	0.0336
5.25							0.0381	0.0361	0.0356
5.5								0.0377	0.0372

Orifice Diameter (in)		10			12			16	
	9.564	10.02	10.136	11.376	11.938	12.09	14.688	15	15.25
1	0.0728								
1.125	0.0674	0.069	0.0694						
1.25	0.0624	0.0641	0.0646	0.0687	0.0704	0.0708			
1.375	0.0576	0.0594	0.0599	0.0643	0.0661	0.0666			
1.5	0.0532	0.055	0.0555	0.0601	0.062	0.0625	0.0697	0.0705	
1.625	0.049	0.0509	0.0514	0.0561	0.058	0.0585	0.0662	0.067	0.0676
1.75	0.0452	0.0471	0.0476	0.0523	0.0543	0.0548	0.0628	0.0636	0.0642
1.875	0.0417	0.0436	0.044	0.0488	0.0508	0.0513	0.0594	0.0603	0.061
2	0.0385	0.0403	0.0407	0.0454	0.0475	0.048	0.0563	0.0572	0.0578
2.125	0.0355	0.0372	0.0377	0.0423	0.0443	0.0449	0.0532	0.0541	0.0548
2.25	0.0329	0.0345	0.0349	0.0394	0.0414	0.0419	0.0503	0.0512	0.0519
2.375	0.0305	0.032	0.0324	0.0367	0.0387	0.0392	0.0475	0.0484	0.0492
2.5	0.0283	0.0298	0.0301	0.0342	0.0361	0.0366	0.0449	0.0458	0.0466
2.625	0.0265	0.0277	0.0281	0.0319	0.0337	0.0342	0.0424	0.0433	0.044
2.75	0.0248	0.026	0.0262	0.0298	0.0316	0.032	0.04	0.0409	0.0417
2.875	0.0234	0.0244	0.0246	0.0279	0.0295	0.03	0.0378	0.0387	0.0394
3	0.0222	0.023	0.0232	0.0262	0.0277	0.0281	0.0356	0.0365	0.0372
3.125	0.0212	0.0218	0.022	0.0244	0.026	0.0264	0.0336	0.0345	0.0352
3.25	0.0204	0.0209	0.0221	0.0232	0.0245	0.0249	0.0317	0.0326	0.0332
3.375	0.0199	0.0201	0.0202	0.022	0.0232	0.0235	0.03	0.0308	0.0314
3.5	0.0195	0.0195	0.0196	0.021	0.022	0.0222	0.0263	0.0291	0.0297
3.625	0.0193	0.0191	0.0191	0.02	0.0209	0.0212	0.0368	0.0275	0.0281
3.75	0.0192	0.0188	0.0188	0.0193	0.02	0.0202	0.0254	0.0261	0.0267
3.875	0.0193	0.0187	0.0186	0.0187	0.0192	0.0194	0.024	0.0247	0.0253
4	0.0195	0.0187	0.0186	0.0182	0.0185	0.0187	0.0228	0.0235	0.024
4.25	0.0203	0.0192	0.0189	0.0176	0.0196	0.0177	0.0207	0.0213	0.0217
4.5	0.0215	0.02	0.0197	0.0175	0.0172	0.0171	0.019	0.0194	0.0198
4.75	0.023	0.0212	0.0208	0.0178	0.0171	0.017	0.0176	0.018	0.0182
5	0.0248	0.0228	0.0223	0.0185	0.0174	0.0173	0.0166	0.0168	0.017
5.25	0.0267	0.0244	0.0239	0.0194	0.0181	0.0178	0.016	0.0161	0.0162
5.5	0.0287	0.0263	0.0257	0.0207	0.019	0.0186	0.0156	0.0156	0.0156
5.75	0.0307	0.0282	0.0276	0.0221	0.0202	0.0197	0.0155	0.0154	0.0153
6	0.0326	0.0302	0.0295	0.0231	0.0215	0.021	0.0157	0.0154	0.0153
6.25	0.0343	0.032	0.0316	0.0253	0.023	0.0224	0.0161	0.0157	0.0154
6.5	0.0358	0.0336	0.0331	0.027	0.0246	0.0239	0.0167	0.0162	0.0159

Orifice Diameter (in)	10			12			16		
	9.564	**10.02**	**10.136**	**11.376**	**11.938**	**12.09**	**14.688**	**15**	**15.25**
6.75		0.0351	0.0346	0.0288	0.0262	0.0256	0.0174	0.0169	0.0164
7		0.0363	0.0359	0.0304	0.0279	0.0272	0.0184	0.0177	0.0172
7.25				0.032	0.0295	0.0288	0.0195	0.0187	0.0181
7.5				0.0334	0.031	0.0304	0.0206	0.0198	0.0191
7.75				0.0347	0.0325	0.0318	0.0219	0.0209	0.0202
8					0.0338	0.0332	0.0232	0.0222	0.0214
8.25					0.0349	0.0344	0.0246	0.0235	0.0227
8.5							0.0259	0.0248	0.024
8.75							0.0273	0.0262	0.0253
9							0.0286	0.0276	0.0267
9.25							0.0299	0.0288	0.028
9.5							0.0311	0.03	0.0292
9.75							0.0322	0.0312	0.0304
10							0.0332	0.0323	0.0315
10.25							0.0341	0.0333	0.0326
10.5								0.0341	0.0335

Orifice Diameter (in)	20			24			30		
	18.814	**19**	**19.25**	**22.626**	**23**	**23.25**	**28.628**	**29**	**29.25**
2	0.0663	0.0667	0.0672						
2.125	0.0635	0.0639	0.0644						
2.25	0.0309	0.0613	0.0618						
2.375	0.0583	0.0588	0.0593	0.0658	0.0665	0.0669			
2.5	0.0558	0.0562	0.0568	0.0635	0.0642	0.0646			
2.625	0.0534	0.0539	0.0544	0.0613	0.062	0.0624			
2.75	0.051	0.0515	0.052	0.0591	0.0598	0.0603			
2.875	0.0488	0.0492	0.0498	0.057	0.0577	0.0582	0.0667		
3	0.0466	0.047	0.0476	0.0549	0.0556	0.0561	0.0649	0.0654	0.0657
3.125	0.0445	0.0449	0.0455	0.0529	0.0536	0.0541	0.063	0.0636	0.0639
3.25	0.0425	0.0429	0.0435	0.0509	0.0516	0.0521	0.0613	0.0616	0.0622
3.375	0.0406	0.041	0.0416	0.049	0.0497	0.0502	0.0595	0.0601	0.0604
3.5	0.0387	0.0391	0.0397	0.0471	0.0479	0.0484	0.0578	0.0584	0.0587
3.625	0.0369	0.0373	0.0379	0.0454	0.0461	0.0466	0.0561	0.0567	0.0571
3.75	0.0352	0.0356	0.0362	0.0436	0.0444	0.0449	0.0545	0.055	0.0554
3.875	0.0336	0.034	0.0346	0.0419	0.0427	0.0432	0.0528	0.0534	0.0538

Orifice Diameter (in)		20			24			30	
	18.814	19	19.25	22.626	23	23.25	28.628	29	29.25
4	0.032	0.0324	0.033	0.0403	0.0411	0.0416	0.0513	0.0518	0.0522
4.25	0.0291	0.0295	0.0301	0.0372	0.038	0.0385	0.0482	0.0488	0.0492
4.5	0.0265	0.0269	0.0274	0.0343	0.0351	0.0356	0.0453	0.0459	0.0463
4.75	0.0242	0.0246	0.025	0.0316	0.0324	0.0328	0.0425	0.0431	0.0435
5	0.0221	0.0225	0.0229	0.0292	0.0299	0.0303	0.0399	0.0405	0.0409
5.25	0.0203	0.0206	0.021	0.0269	0.0276	0.028	0.0374	0.038	0.0384
5.5	0.0188	0.019	0.0194	0.248	0.0255	0.0259	0.035	0.0356	0.036
5.750	0.0175	0.0177	0.0180	0.0230	0.0236	0.0240	0.0328	0.0334	0.0338
6.000	0.0164	0.0165	0.0168	0.0212	0.0218	0.0222	0.0307	0.0313	0.0317
6.250	0.0155	0.0156	0.0158	0.0197	0.0202	0.0206	0.0287	0.0293	0.0297
6.500	0.0148	0.0149	0.0150	0.0184	0.0189	0.0192	0.0269	0.0274	0.0278
6.750	0.0143	0.0144	0.0145	0.0172	0.0176	0.0179	0.0252	0.0257	0.026
7.000	0.0141	0.0141	0.0141	0.0162	0.0166	0.0168	0.0236	0.0241	0.0244
7.250	0.0141	0.0140	0.0139	0.0153	0.0156	0.0158	0.0221	0.0226	0.0229
7.500	0.0140	0.0140	0.0139	0.0146	0.0148	0.0150	0.0207	0.0212	0.0215
7.750	0.0142	0.0141	0.0140	0.0140	0.0142	0.0144	0.0195	0.0199	0.0202
8.000	0.0146	0.0144	0.0142	0.0136	0.0138	0.0138	0.0183	0.0187	0.019
8.250	0.0151	0.0148	0.0146	0.0133	0.0134	0.0132	0.0173	0.0177	0.0179
8.500	0.0156	0.0154	0.0151	0.0132	0.0132	0.0130	0.0164	0.0167	0.0169
8.750	0.0163	0.0160	0.0157	0.0131	0.0130	0.0130	0.0155	0.0158	0.0161
9.000	0.0171	0.0168	0.0163	0.0131	0.0130	0.0130	0.0148	0.0151	0.0153
9.250	0.0180	0.0176	0.0171	0.0133	0.0130	0.0133	0.0142	0.0144	0.0146
9.500	0.0189	0.0185	0.0180	0.0136	0.0133	0.0132	0.0136	0.0138	0.014
9.750	0.0198	0.0194	0.0189	0.0139	0.0136	0.0134	0.0132	0.0133	0.0134
10.000	0.0209	0.0204	0.0198	0.0143	0.0140	0.0135	0.0128	0.0129	0.013
10.250	0.0219	0.0214	0.0208	0.0148	0.0144	0.0142	0.0125	0.0126	0.0127
10.500	0.0230	0.0225	0.0219	0.0154	0.0150	0.0147	0.0123	0.0124	0.0124
10.750	0.0241	0.0236	0.0229	0.0160	0.0155	0.0152	0.0122	0.0122	0.0122
11.000	0.0252	0.0247	0.0240	0.0168	0.0162	0.0158	0.0121	0.0121	0.0121
11.250	0.0263	0.0261	0.0251	0.0175	0.0169	0.0165	0.0122	0.0121	0.0121
11.500	0.0273	0.0268	0.0262	0.0183	0.0176	0.0172	0.0122	0.0121	0.0122
11.750	0.0284	0.0278	0.0272	0.0191	0.0184	0.0180	0.0124	0.0123	0.0122
12.000	0.0293	0.0288	0.0282	0.0200	0.0192	0.0190	0.0126	0.0124	0.0123
12.500	0.0312	0.0307	0.0301	0.0218	0.0210	0.0204	0.0132	0.0130	0.0128
13.000	0.0327	0.0323	0.0318	0.0236	0.0228	0.0222	0.0140	0.0137	0.0135

Orifice Diameter (in)	20			24			30		
	18.814	**19**	**19.25**	**22.626**	**23**	**23.25**	**28.628**	**29**	**29.25**
13.500				0.0254	0.0246	0.0240	0.0150	0.0146	0.0143
14.000				0.0272	0.0264	0.0258	0.0161	0.0156	0.0153
14.500				0.0289	0.0280	0.0275	0.0173	0.0168	0.0165
15.000				0.0304	0.0296	0.0291	0.0166	0.0181	0.0177
15.500				0.0310	0.0311	0.0306	0.0200	0.0194	0.0190
16.000					0.0323	0.0318	0.0215	0.0209	0.0204
16.500							0.0230	0.0223	0.0219
17.000							0.0244	0.0238	0.0233
17.500							0.0259	0.0252	0.0248
18.000							0.0272	0.0266	0.0261
18.500							0.0286	0.0279	0.0275
19.000							0.0298	0.0292	0.0288
19.500							0.0309	0.0303	0.0299
20.000							0.0318	0.0313	0.0310

Table C–8 Y_1 Expansion Factors—Pipe Taps (Static Pressure Taken from Upstream Taps)

$$\beta = \frac{d}{D}$$

$\frac{h_w}{P_{f1}}$	0.10	0.20	0.30	0.40	0.45	0.50	0.52	0.54	0.56	0.58	0.60	0.61	0.62	0.63	0.64	0.65	0.66	0.67	0.68	0.69	0.70
0.1	0.999	0.9989	0.9988	0.9985	0.9984	0.9982	0.9981	0.998	0.9979	0.9978	0.9977	0.9976	0.9976	0.9975	0.9974	0.9973	0.9972	0.9971	0.997	0.9969	0.9968
0.2	0.9981	0.9979	0.9976	0.9971	0.9968	0.9964	0.9962	0.9961	0.9959	0.9957	0.9954	0.9953	0.9951	0.995	0.9948	0.9947	0.9945	0.9943	0.9941	0.9938	0.9935
0.3	0.9971	0.9968	0.9964	0.9956	0.9952	0.9946	0.9944	0.9941	0.9938	0.9935	0.9931	0.9929	0.9927	0.9925	0.9923	0.992	0.9917	0.9914	0.9911	0.9907	0.9903
0.4	0.9962	0.9958	0.9951	0.9942	0.9936	0.9928	0.9925	0.9921	0.9917	0.9913	0.9908	0.9906	0.9903	0.99	0.9897	0.9893	0.989	0.9886	0.9881	0.9876	0.9871
0.5	0.9952	0.9947	0.9939	0.9927	0.9919	0.991	0.9906	0.9902	0.9897	0.9891	0.9885	0.9882	0.9879	0.9875	0.9871	0.9867	0.9862	0.9857	0.9851	0.9845	0.9839
0.6	0.9943	0.9937	0.9927	0.9913	0.9903	0.9892	0.9887	0.9882	0.9876	0.987	0.9862	0.9859	0.9854	0.985	0.9845	0.984	0.9834	0.9828	0.9822	0.9814	0.9806
0.7	0.9933	0.9926	0.9915	0.9898	0.9887	0.9874	0.9869	0.9862	0.9856	0.9848	0.984	0.9835	0.983	0.9825	0.9819	0.9813	0.9807	0.98	0.9792	0.9784	0.9774
0.8	0.9923	0.9916	0.9903	0.9883	0.9871	0.9857	0.985	0.9843	0.9835	0.9826	0.9817	0.9811	0.9806	0.98	0.9794	0.9787	0.9779	0.9771	0.9762	0.9753	0.9742
0.9	0.9914	0.9905	0.9891	0.9869	0.9855	0.9839	0.9831	0.9823	0.9814	0.9805	0.9794	0.9788	0.9782	0.9775	0.9768	0.976	0.9752	0.9742	0.9733	0.9722	0.971
1	0.9904	0.9895	0.9878	0.9854	0.9839	0.9821	0.9812	0.9803	0.9794	0.9783	0.9771	0.9764	0.9757	0.975	0.9742	0.9733	0.9724	0.9714	0.9703	0.9691	0.9677
1.1	0.9895	0.9884	0.9866	0.984	0.9823	0.9803	0.9794	0.9784	0.9773	0.9761	0.9748	0.9741	0.9733	0.9725	0.9716	0.9707	0.9696	0.9685	0.9673	0.966	0.9645
1.2	0.9885	0.9874	0.9854	0.9825	0.9807	0.9785	0.9775	0.9764	0.9752	0.9739	0.9725	0.9717	0.9709	0.97	0.969	0.968	0.9669	0.9757	0.9643	0.9629	0.9613
1.3	0.9876	0.9863	0.9842	0.9811	0.9791	0.9767	0.9756	0.9744	0.9732	0.9718	0.9702	0.9694	0.9685	0.9675	0.9664	0.9653	0.9641	0.9628	0.9614	0.9598	0.9581
1.4	0.9866	0.9853	0.983	0.9796	0.9775	0.9749	0.9737	0.9725	0.9711	0.9696	0.9679	0.967	0.966	0.965	0.9639	0.9627	0.9614	0.9599	0.9584	0.9567	0.9548
1.5	0.9857	0.9842	0.9818	0.9782	0.9758	0.9731	0.9719	0.9705	0.969	0.9674	0.9656	0.9646	0.9636	0.9625	0.9613	0.96	0.9586	0.9571	0.9554	0.9536	0.9516
1.6	0.9847	0.9832	0.9805	0.9767	0.9742	0.9713	0.97	0.9685	0.967	0.9652	0.9633	0.9623	0.9612	0.96	0.9587	0.9573	0.9558	0.9542	0.9525	0.9505	0.9484
1.7	0.9837	0.9821	0.9793	0.9752	0.9726	0.9695	0.9681	0.9666	0.9649	0.9631	0.961	0.9599	0.9587	0.9575	0.9561	0.9547	0.9531	0.9514	0.9495	0.9474	0.9452
1.8	0.9828	0.9811	0.9781	0.9738	0.971	0.9677	0.9662	0.9646	0.9628	0.9609	0.9587	0.9576	0.9563	0.955	0.9535	0.952	0.9503	0.9485	0.9465	0.9443	0.9419
1.9	0.9818	0.98	0.9769	0.9723	0.9694	0.9659	0.9643	0.9626	0.9608	0.9587	0.9565	0.9552	0.9539	0.9525	0.951	0.9493	0.9476	0.9456	0.9435	0.9412	0.9387
2	0.9809	0.979	0.9757	0.9709	0.9678	0.9641	0.9625	0.9607	0.9587	0.9566	0.9542	0.9529	0.9515	0.95	0.9484	0.9467	0.9448	0.9428	0.9406	0.9381	0.9355
2.1	0.9799	0.9778	0.9745	0.9694	0.9662	0.9623	0.9606	0.9587	0.9566	0.9544	0.9519	0.9505	0.949	0.9475	0.9458	0.944	0.942	0.9399	0.9376	0.9351	0.9323

$\frac{h_w}{P_{f1}}$	0.10	0.20	0.30	0.40	0.45	0.50	0.52	0.54	0.56	0.58	0.60	0.61	0.62	0.63	0.64	0.65	0.66	0.67	0.68	0.69	0.70
2.2	0.979	0.9768	0.9732	0.968	0.9646	0.9605	0.9587	0.9567	0.9546	0.9522	0.9496	0.9481	0.9466	0.945	0.9432	0.9413	0.9393	0.9371	0.9346	0.932	0.929
2.3	0.978	0.9758	0.972	0.9665	0.963	0.9587	0.9568	0.9548	0.95	0.95	0.9473	0.9458	0.9442	0.9425	0.9406	0.9387	0.9365	0.9342	0.9317	0.9289	0.9258
2.4	0.977	0.9747	0.9708	0.965	0.9613	0.957	0.955	0.9528	0.9505	0.9479	0.945	0.9434	0.9418	0.94	0.9381	0.936	0.9338	0.9313	0.9287	0.9258	0.9226
2.5	0.9761	0.9737	0.9696	0.9636	0.9597	0.9552	0.9531	0.9508	0.9484	0.9457	0.9427	0.9411	0.9393	0.9375	0.9355	0.9333	0.931	0.9285	0.9257	0.9227	0.9194
2.6	0.9751	0.9726	0.9684	0.9621	0.9581	0.9534	0.9512	0.9489	0.9463	0.9435	0.9404	0.9387	0.9369	0.9350	0.9329	0.9307	0.9282	0.9256	0.9227	0.9196	0.9161
2.7	0.9742	0.9716	0.9672	0.9607	0.9565	0.9516	0.9493	0.9469	0.9443	0.9414	0.9381	0.9364	0.9345	0.9325	0.9303	0.9280	0.9255	0.9277	0.9198	0.9165	0.9129
2.8	0.9732	0.9705	0.9659	0.9592	0.9549	0.9498	0.9475	0.9449	0.9422	0.9392	0.9358	0.9340	0.9321	0.9300	0.9277	0.9253	0.9227	0.9199	0.9168	0.9134	0.9097
2.9	0.9723	0.9695	0.9647	0.9578	0.9533	0.9480	0.9456	0.9430	0.9401	0.9370	0.9335	0.9316	0.9296	0.9275	0.9252	0.9227	0.9200	0.9170	0.9138	0.9103	0.9064
3.0	0.9713	0.9684	0.9635	0.9563	0.9517	0.9462	0.9437	0.9410	0.9381	0.9348	0.9312	0.9293	0.9272	0.9250	0.9226	0.9200	0.9172	0.9142	0.9108	0.9072	0.9032
3.1	0.9704	0.9674	0.9623	0.9549	0.9501	0.9444	0.9418	0.9390	0.9360	0.9327	0.9290	0.9269	0.9248	0.9225	0.9200	0.9173	0.9144	0.9113	0.9079	0.9041	0.9000
3.2	0.9694	0.9663	0.9611	0.9534	0.9485	0.9426	0.9400	0.9371	0.9339	0.9305	0.9267	0.9246	0.9223	0.9200	0.9174	0.9147	0.9117	0.9084	0.9049	0.9010	0.8968
3.3	0.9684	0.9653	0.9599	0.9519	0.9469	0.9408	0.9381	0.9351	0.9319	0.9283	0.9244	0.9222	0.9199	0.9175	0.9148	0.9120	0.9089	0.9056	0.9019	0.8979	0.8935
3.4	0.9675	0.9642	0.9587	0.9505	0.9452	0.9390	0.9362	0.9331	0.9298	0.9261	0.9221	0.9199	0.9175	0.9150	0.9122	0.9093	0.9062	0.9027	0.8990	0.8948	0.8903
3.5	0.9665	0.9632	0.9574	0.9490	0.9436	0.9372	0.9343	0.9312	0.9277	0.9240	0.9198	0.9175	0.9151	0.9125	0.9097	0.9067	0.9034	0.8999	0.8960	0.8918	0.8871
3.6	0.9656	0.9621	0.9562	0.9476	0.9420	0.9354	0.9324	0.9292	0.9257	0.9218	0.9175	0.9151	0.9126	0.9100	0.9071	0.9040	0.9006	0.8970	0.8930	0.8887	0.8839
3.7	0.9646	0.9611	0.9550	0.9461	0.9404	0.9336	0.9306	0.9272	0.9236	0.9196	0.9152	0.9128	0.9102	0.9075	0.9045	0.9013	0.8979	0.8941	0.8900	0.8856	0.8806
3.8	0.9637	0.9600	0.9538	0.9447	0.9388	0.9318	0.9287	0.9253	0.9216	0.9175	0.9129	0.9104	0.9078	0.9050	0.9019	0.8987	0.8951	0.8913	0.8871	0.8825	0.8774
3.9	0.9627	0.9590	0.9526	0.9432	0.9372	0.9301	0.9268	0.9233	0.9195	0.9153	0.9106	0.9081	0.9054	0.9025	0.8993	0.8960	0.8924	0.8884	0.8841	0.8794	0.8742
4.0	0.9617	0.9579	0.9514	0.9417	0.9356	0.9283	0.9249	0.9213	0.9174	0.9131	0.9083	0.9057	0.9029	0.9000	0.8968	0.8933	0.8896	0.8826	0.8811	0.9763	0.8710

Table C–9 Y_2 Expansion Factors—Pipe Taps (Static Pressure Taken from Downstream Taps)

$$\beta = \frac{d}{D}$$

$\frac{h_w}{P_{f2}}$	0.10	0.20	0.30	0.40	0.45	0.50	0.52	0.54	0.56	0.58	0.60	0.61	0.62	0.63	0.64	0.65	0.66	0.67	0.68	0.69	0.70
0.1	1.0008	1.0008	1.0006	1.0003	1.0002	1.0000	0.9999	0.9998	0.9997	0.9996	0.9995	0.9994	0.9994	0.9993	0.9992	0.9991	0.999	0.9989	0.9988	0.9987	0.9986
0.2	1.0017	1.0015	1.0012	1.0007	1.0004	1.0000	0.9999	0.9997	0.9995	0.9993	0.999	0.9989	0.9988	0.9986	0.9985	0.9983	0.9981	0.9979	0.9977	0.9974	0.9972
0.3	1.0025	1.0023	1.0018	1.001	1.0006	1.0000	0.9998	0.9995	0.9992	0.9989	0.9986	0.9984	0.9982	0.9979	0.9977	0.9974	0.9972	0.9969	0.9965	0.9962	0.9958
0.4	1.0034	1.003	1.0024	1.0014	1.0008	1.0001	0.9997	0.9994	0.999	0.9986	0.9981	0.9978	0.9976	0.9972	0.9969	0.9966	0.9962	0.9958	0.9954	0.9949	0.9944
0.5	1.0042	1.0038	1.003	1.0018	1.001	1.0001	0.9997	0.9992	0.9988	0.9982	0.9976	0.9973	0.997	0.9966	0.9962	0.9958	0.9953	0.9948	0.9942	0.9936	0.993
0.6	1.0051	1.0045	1.0036	1.0021	1.0012	1.0001	0.9996	0.9991	0.9985	0.9979	0.9972	0.9968	0.9964	0.9959	0.9954	0.9949	0.9944	0.9938	0.9931	0.9924	0.9916
0.7	1.0059	1.0053	1.0041	1.0025	1.0014	1.0002	0.9996	0.999	0.9983	0.9975	0.9967	0.9962	0.9958	0.9953	0.9947	0.9941	0.9935	0.9928	0.992	0.9912	0.9902
0.8	1.0068	1.006	1.0047	1.0028	1.0016	1.0002	0.9995	0.9988	0.998	0.9972	0.9962	0.9957	0.9952	0.9946	0.994	0.9933	0.9926	0.9918	0.9909	0.9899	0.9889
0.9	1.0076	1.0068	1.0053	1.0032	1.0018	1.0002	0.9995	0.9987	0.9978	0.9969	0.9958	0.9952	0.9946	0.994	0.9932	0.9925	0.9917	0.9908	0.9898	0.9887	0.9875
1	1.0085	1.0075	1.0059	1.0036	1.0021	1.0003	0.9994	0.9986	0.9976	0.9965	0.9954	0.9947	0.994	0.9933	0.9925	0.9917	0.9908	0.9898	0.9887	0.9875	0.9862
1.1	1.0093	1.0083	1.0065	1.0039	1.0023	1.0003	0.9994	0.9984	0.9974	0.9962	0.9949	0.9942	0.9935	0.9927	0.9918	0.9909	0.9899	0.9888	0.9876	0.9863	0.9848
1.2	1.0102	1.0091	1.0071	1.0043	1.0025	1.0004	0.9994	0.9983	0.9972	0.9959	0.9945	0.9937	0.9929	0.992	0.9911	0.9901	0.989	0.9878	0.9865	0.9851	0.9835
1.3	1.011	1.0098	1.0077	1.0047	1.0027	1.0004	0.9994	0.9982	0.997	0.9956	0.9941	0.9932	0.9924	0.9914	0.9904	0.9893	0.9881	0.9868	0.9854	0.9839	0.9822
1.4	1.0119	1.0106	1.0083	1.0051	1.003	1.0004	0.9993	0.9981	0.9968	0.9953	0.9936	0.9928	0.9918	0.9908	0.9897	0.9885	0.9872	0.9859	0.9844	0.9827	0.9809
1.5	1.0127	1.0113	1.0089	1.0054	1.0032	1.0005	0.9993	0.998	0.9966	0.995	0.9932	0.9923	0.9912	0.9902	0.989	0.9877	0.9864	0.9849	0.9833	0.9815	0.9796
1.6	1.0136	1.0121	1.0096	1.0058	1.0034	1.0006	0.9993	0.9979	0.9964	0.9947	0.9928	0.9918	0.9907	0.9896	0.9883	0.987	0.9855	0.984	0.9822	0.9804	0.9783
1.7	1.0144	1.0128	1.0102	1.0062	1.0036	1.0006	0.9992	0.9978	0.962	0.9944	0.9924	0.9913	0.9902	0.9889	0.9876	0.9862	0.9847	0.983	0.9812	0.9792	0.977
1.8	1.0153	1.0136	1.0108	1.0066	1.0039	1.0007	0.9992	0.9977	0.996	0.9941	0.992	0.9908	0.9896	0.9883	0.987	0.9854	0.9838	0.9821	0.9801	0.978	0.9757
1.9	1.0161	1.0144	1.0114	1.007	1.0041	1.0008	0.9992	0.9976	0.9958	0.9938	0.9916	0.9904	0.9891	0.9877	0.9863	0.9847	0.983	0.9811	0.9791	0.9769	0.9744

$\frac{h_w}{P_{f2}}$	0.10	0.20	0.30	0.40	0.45	0.50	0.52	0.54	0.56	0.58	0.60	0.61	0.62	0.63	0.64	0.65	0.66	0.67	0.68	0.69	0.70
2	1.017	1.0151	1.012	1.0073	1.0044	1.0008	0.9992	0.9975	0.9956	0.9935	0.9912	0.9899	0.9886	0.9872	0.9856	0.984	0.9822	0.9802	0.9781	0.9757	0.9732
2.1	1.0178	1.0159	1.0126	1.0077	1.0046	1.0009	0.9992	0.9974	0.9954	0.9932	0.9908	0.9895	0.9881	0.9866	0.9849	0.9832	0.9813	0.9793	0.977	0.9746	0.9719
2.2	1.0187	1.0167	1.0132	1.0081	1.0048	1.001	0.9992	0.9973	0.9952	0.9929	0.9904	0.989	0.9876	0.986	0.9843	0.9825	0.9805	0.9784	0.976	0.9734	0.9706
2.3	1.0195	1.0174	1.0138	1.0085	1.0051	1.001	0.9992	0.9972	0.995	0.9927	0.99	0.9886	0.987	0.9854	0.9836	0.9817	0.9797	0.9774	0.975	0.9723	0.9694
2.4	1.0204	1.0182	1.0144	1.0089	1.0053	1.0011	0.9992	0.9971	0.9949	0.9924	0.9896	0.9881	0.9865	0.9848	0.983	0.981	0.9789	0.9765	0.974	0.9712	0.9681
2.5	1.0212	1.0189	1.015	1.0093	1.0056	1.0012	0.9992	0.9971	0.9947	0.9921	0.9893	0.9877	0.986	0.9842	0.9823	0.9803	0.978	0.9756	0.973	0.9701	0.9669
2.6	1.0221	1.0197	1.0156	1.0097	1.0058	1.0013	0.9992	0.9970	0.9945	0.9919	0.9889	0.9873	0.9855	0.9837	0.9817	0.9796	0.9772	0.9747	0.9720	0.9690	0.9657
2.7	1.0229	1.0205	1.0162	1.0101	1.0061	1.0014	0.9992	0.9969	0.9944	0.9916	0.9885	0.9868	0.9850	0.9831	0.9811	0.9788	0.9764	0.9738	0.9710	0.9679	0.9644
2.8	1.0238	1.0212	1.0169	1.0104	1.0063	1.0014	0.9992	0.9968	0.9942	0.9914	0.9882	0.9864	0.9846	0.9826	0.9804	0.9781	0.9757	0.9730	0.9700	0.9668	0.9632
2.9	1.0246	1.0220	1.0175	1.0108	1.0066	1.0015	0.9992	0.9968	0.9941	0.9911	0.9878	0.9860	0.9841	0.982	0.9798	0.9774	0.9749	0.9721	0.9690	0.9657	0.9620
3.0	1.0255	1.0228	1.0181	1.0112	1.0068	1.0016	0.9993	0.9967	0.9939	0.9908	0.9874	0.9856	0.9836	0.9815	0.9792	0.9767	0.9741	0.9712	0.9681	0.9646	0.9608
3.1	1.0246	1.0235	1.0187	1.0116	1.0071	1.0017	0.9993	0.9966	0.9938	0.9906	0.9871	0.9852	0.9831	0.9809	0.9786	0.9760	0.9733	0.9703	0.9671	0.9636	0.9596
3.2	1.0272	1.0243	1.0193	1.0120	1.0074	1.0018	0.9993	0.9966	0.9936	0.9904	0.9867	0.9848	0.9826	0.9804	0.9780	0.9754	0.9725	0.9695	0.9661	0.9625	0.9584
3.3	1.0280	1.0250	1.0199	1.0124	1.0076	1.0019	0.9993	0.9965	0.9935	0.9901	0.9864	0.9843	0.9822	0.9798	0.9774	0.9747	0.9718	0.9686	0.9652	0.9614	0.9572
3.4	1.0289	1.0258	1.0206	1.0128	1.0079	1.0020	0.9994	0.9965	0.9933	0.9899	0.9860	0.9839	0.9817	0.9793	0.9768	0.9740	0.9710	0.9678	0.9642	0.9603	0.9561
3.5	1.0298	1.0266	1.0212	1.0133	1.0082	1.0021	0.9994	0.9964	0.9932	0.9896	0.9857	0..9835	0.9812	0.9788	0.9762	0.9733	0.9702	0.9669	0.9633	0.9593	0.9549
3.6	1.0306	1.0273	1.0218	1.0137	1.0084	1.0022	0.9994	0.9964	0.9931	0.9894	0.9854	0.9832	0.9808	0.9783	0.9756	0.9727	0.9695	0.9661	0.9623	0.9582	0.9537
3.7	1.0314	1.0281	1.0224	1.0141	1.0087	1.0024	0.9994	0.9963	0.9929	0.9892	0.9850	0.9828	0.9803	0.9778	0.9750	0.9720	0.9688	0.9652	0.9614	0.9572	0.9526
3.8	1.0323	1.0289	1.0230	1.0145	1.0090	1.0025	0.9995	0.9963	0.9928	0.9890	0.9847	0.9824	0.9799	0.9772	0.9744	0.9713	0.9680	0.9644	0.9605	0.9562	0.9514
3.9	1.0332	1.0296	1.0237	1.0149	1.0093	1.0026	0.9995	0.9963	0.9927	0.9888	0.9844	0.9820	0.9794	0.9767	0.9738	0.9707	0.9673	0.9636	0.9596	0.9551	0.9503
4.0	1.0340	1.0304	1.0243	1.0153	1.0095	1.0027	0.9996	0.9962	0.9926	0.9885	0.9840	0.9816	0.9790	0.9762	0.9732	0.9700	0.9665	0.9628	0.9586	0.9541	0.9491

Appendix D

The Minimum Gas Production Rate for Water Removal in Gas Wells

Data presented in this section were generated with Guo's method on the basis of the following parameter values:

Gas-specific gravity:	0.60 to 0.90 air =1
Hole inclination:	0°
Wellhead pressure:	100 to 900 psi
Wellhead temperature:	60 °F
Geothermal gradient:	0.01 °F/ft
Condensate gravity:	70 API
Condensate production rate:	0
Water-specific gravity:	1.08 water = 1
Water production rate:	0 to 50 bbl/day
Solid make:	0
Tubing ID:	1.66 to 3.5 in
Conduit wall roughness:	0.000015 in
Maximum interfacial tension:	60 dyne/cm

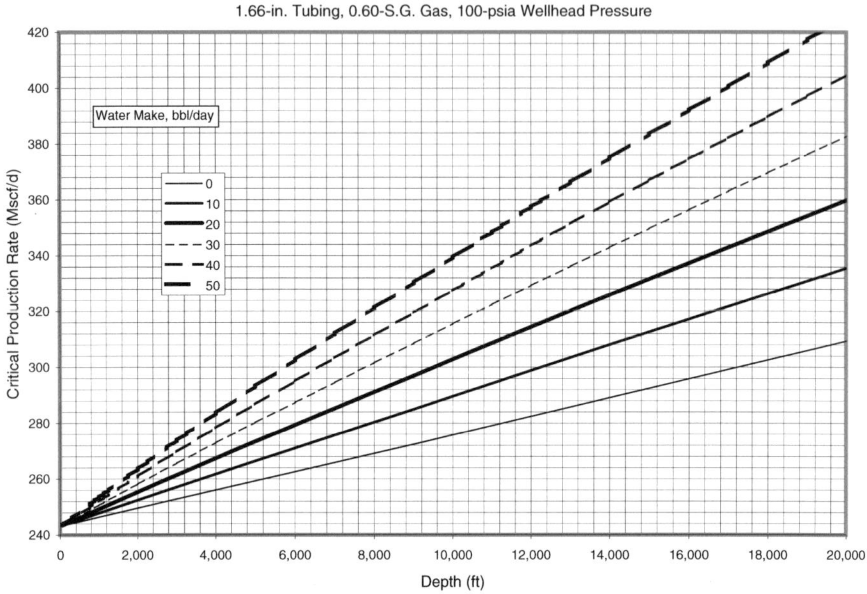

Figure D–1 **Critical gas production rate for water removal in 1.66-in tubing against 100 psia wellhead pressure, S.G. 0.60 gas.**

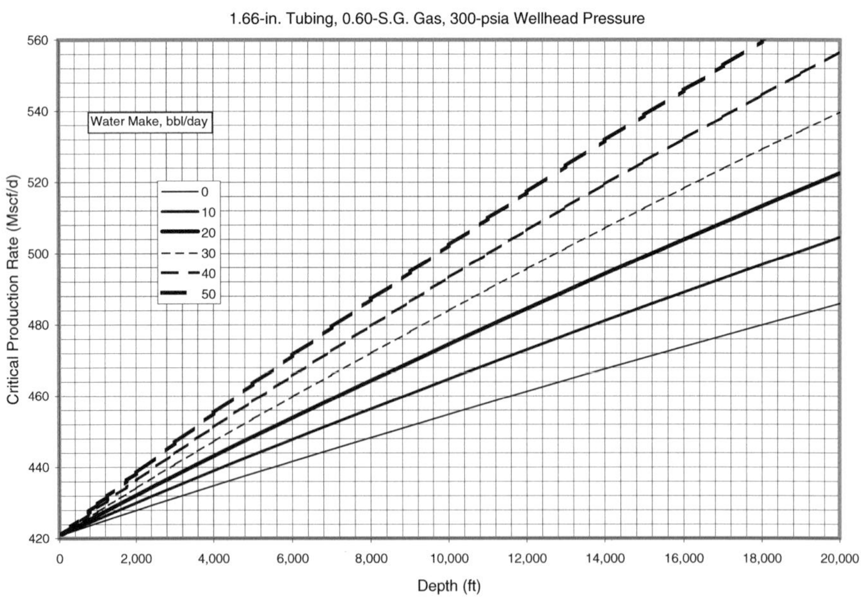

Figure D–2 **Critical gas production rate for water removal in 1.66-in tubing against 300 psia wellhead pressure, S.G. 0.60 gas.**

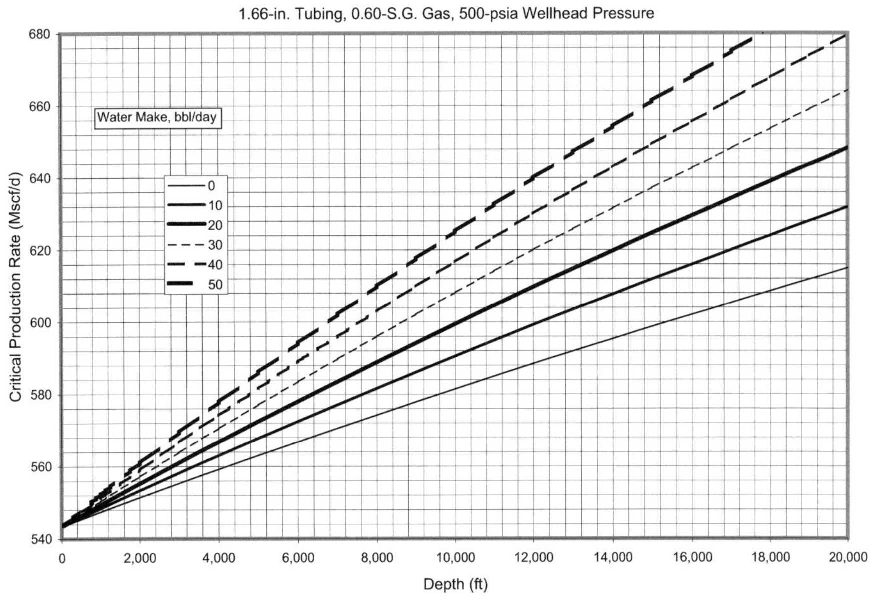

Figure D–3 **Critical gas production rate for water removal in 1.66-in tubing against 500 psia wellhead pressure, S.G. 0.60 gas.**

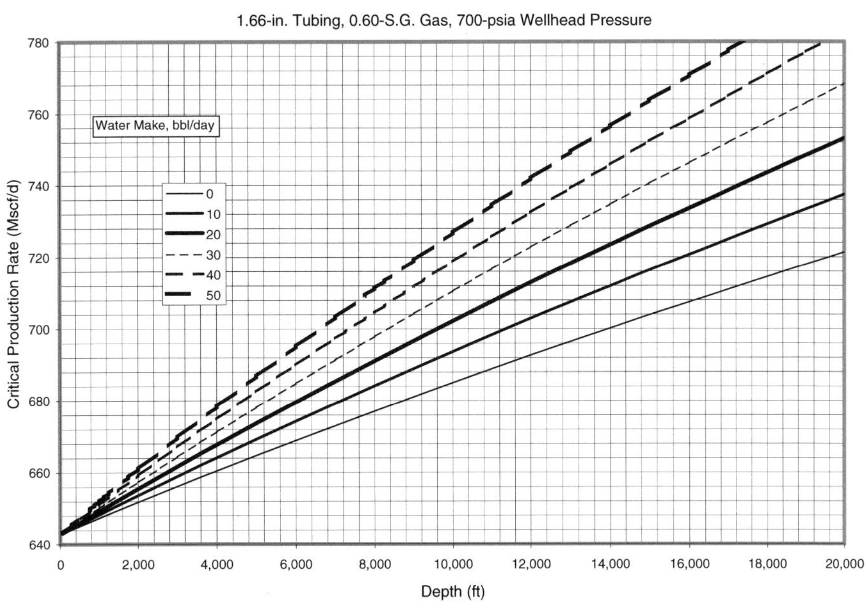

Figure D–4 **Critical gas production rate for water removal in 1.66-in tubing against 700 psia wellhead pressure, S.G. 0.60 gas.**

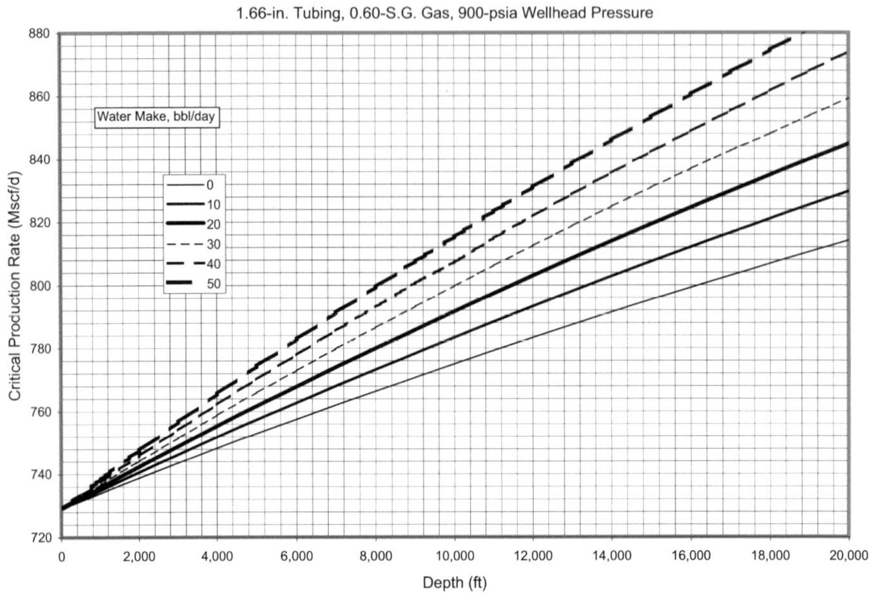

Figure D–5 **Critical gas production rate for water removal in 1.66-in tubing against 900 psia wellhead pressure, S.G. 0.60 gas.**

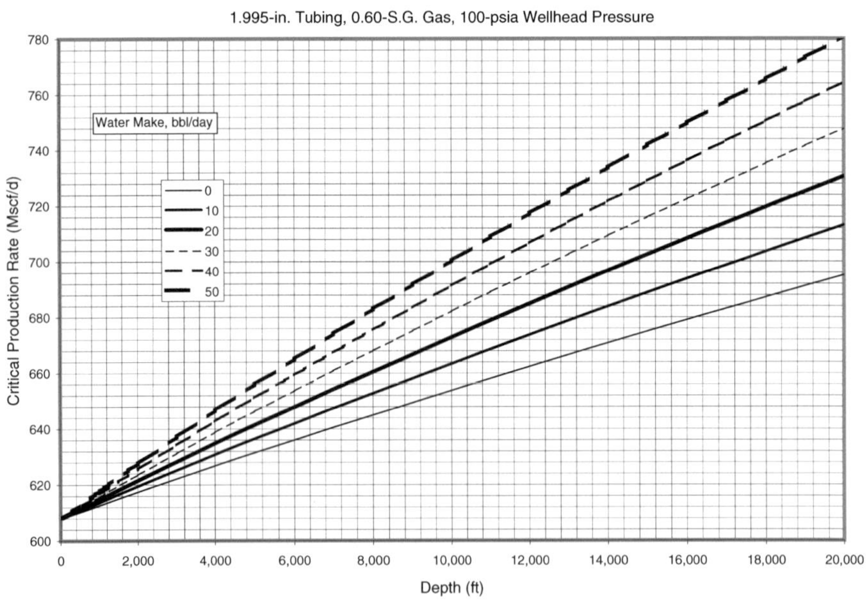

Figure D–6 **Critical gas production rate for water removal in 1.995-in tubing against 100 psia wellhead pressure, S.G. 0.60 gas.**

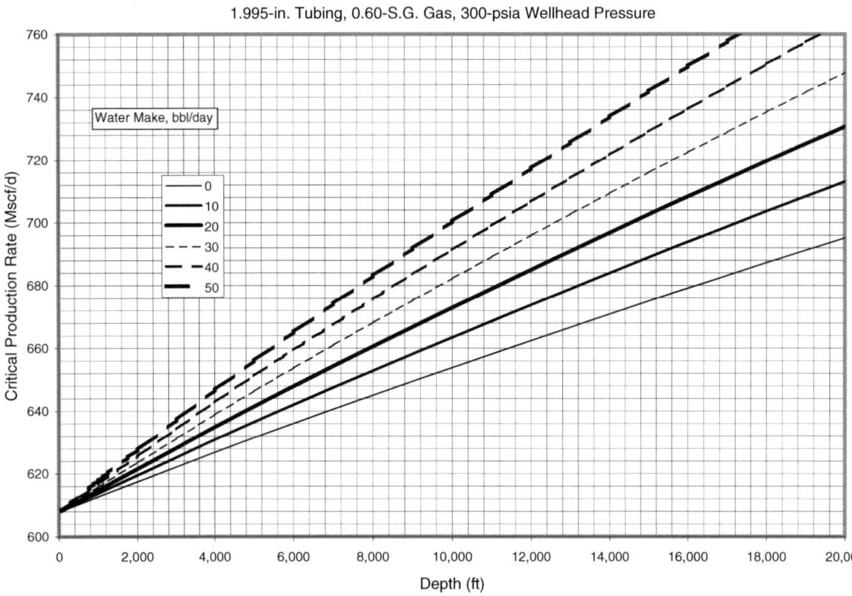

Figure D–7 **Critical gas production rate for water removal in 1.995-in tubing against 300 psia wellhead pressure, S.G. 0.60 gas.**

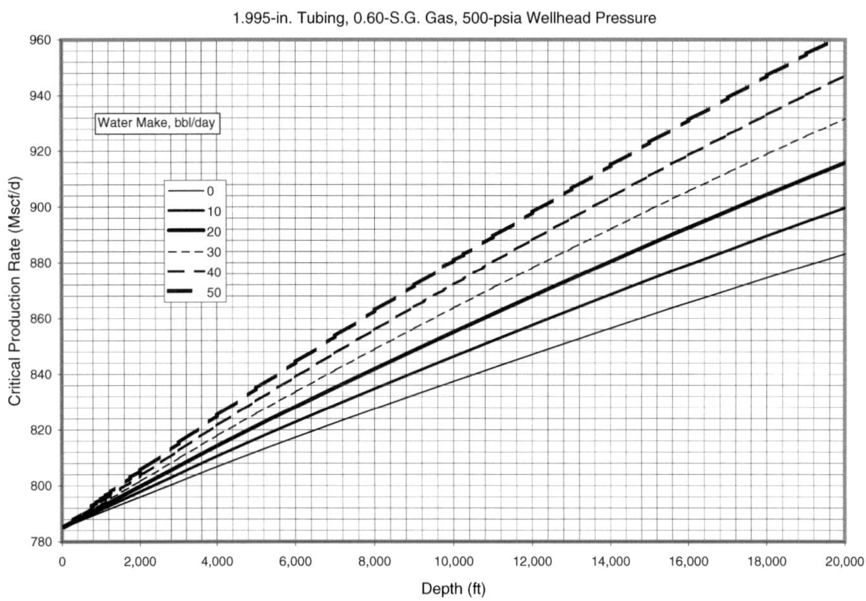

Figure D–8 **Critical gas production rate for water removal in 1.995-in tubing against 500 psia wellhead pressure, S.G. 0.60 gas.**

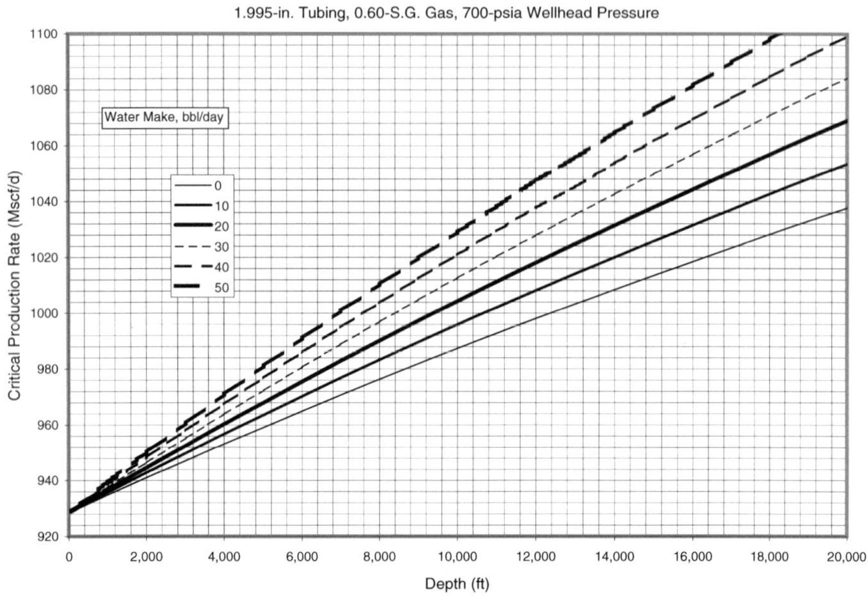

Figure D–9 **Critical gas production rate for water removal in 1.995-in tubing against 700 psia wellhead pressure, S.G. 0.60 gas.**

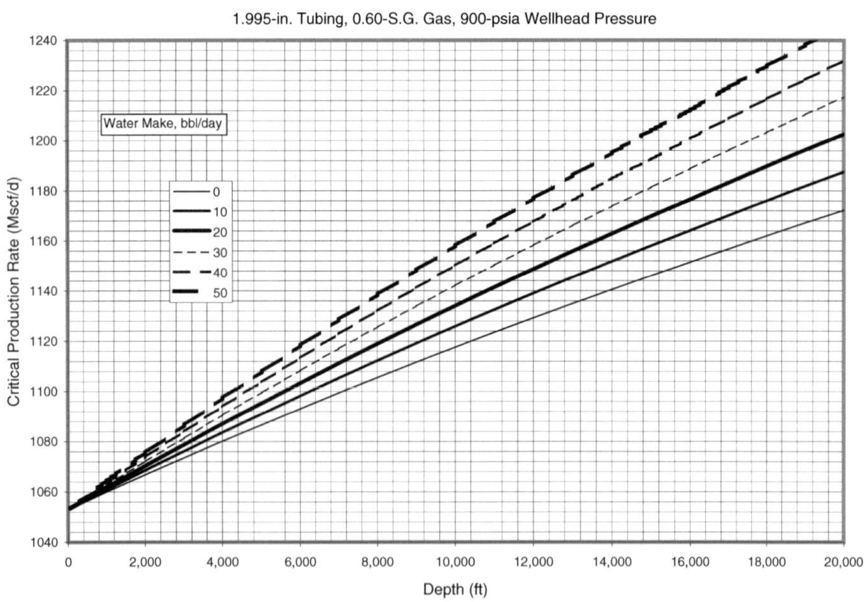

Figure D–10 **Critical gas production rate for water removal in 1.995-in tubing against 900 psia wellhead pressure, S.G. 0.60 gas.**

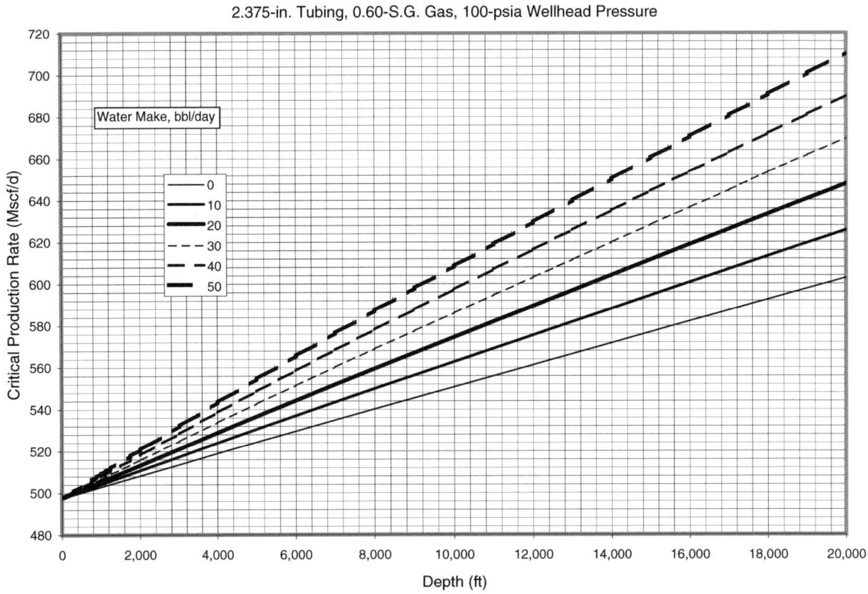

Figure D–11 **Critical gas production rate for water removal in 2.375-in tubing against 100 psia wellhead pressure, S.G. 0.60 gas.**

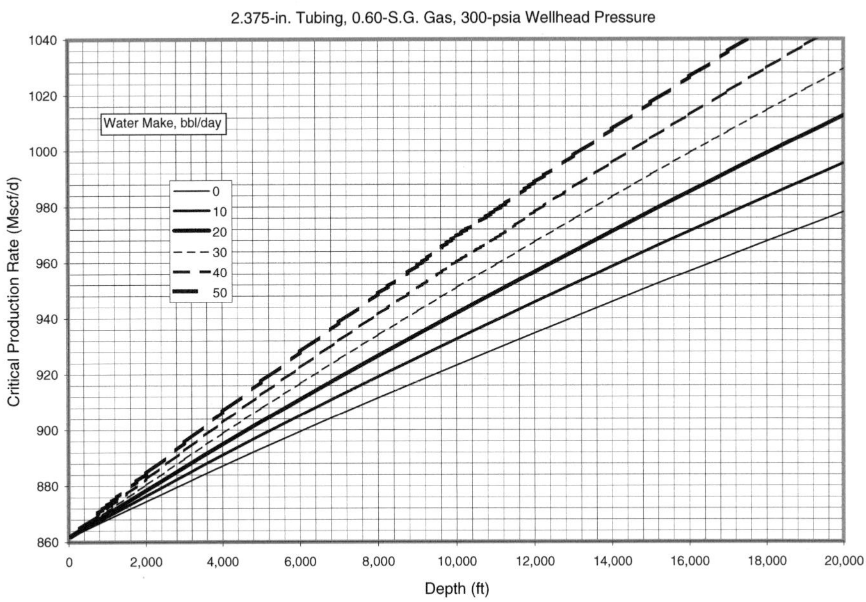

Figure D–12 **Critical gas production rate for water removal in 2.375-in tubing against 300 psia wellhead pressure, S.G. 0.60 gas.**

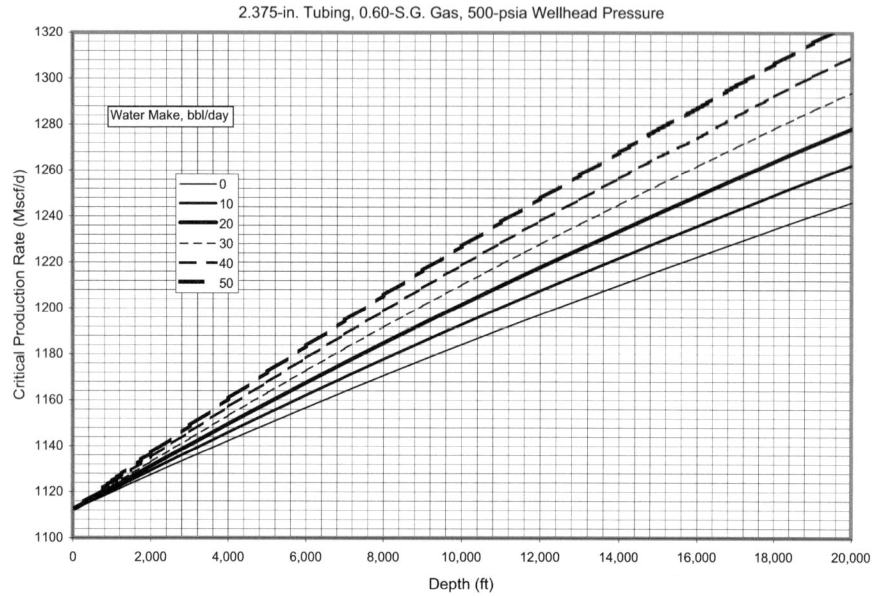

Figure D–13 **Critical gas production rate for water removal in 2.375-in tubing against 500 psia wellhead pressure, S.G. 0.60 gas.**

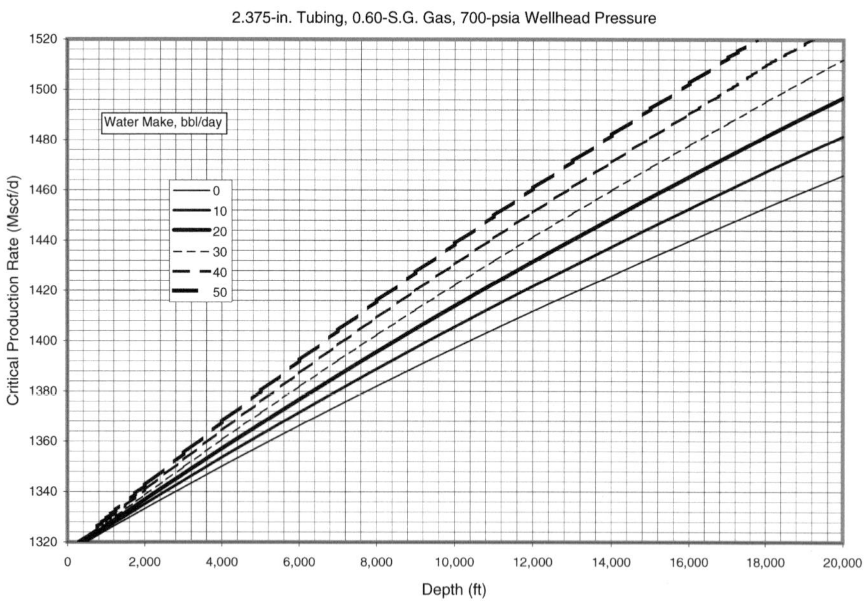

Figure D–14 **Critical gas production rate for water removal in 2.375-in tubing against 700 psia wellhead pressure, S.G. 0.60 gas.**

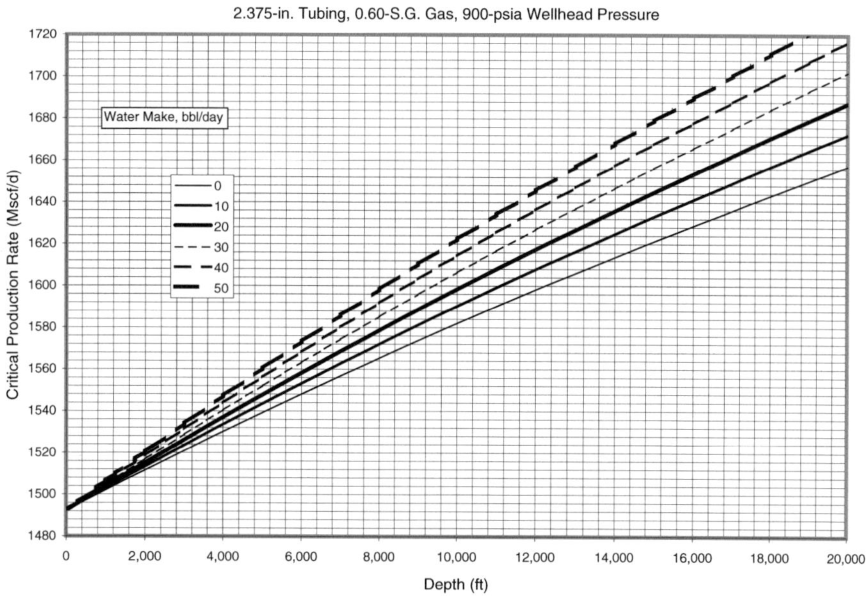

Figure D–15 **Critical gas production rate for water removal in 2.375-in tubing against 900 psia wellhead pressure, S.G. 0.60 gas.**

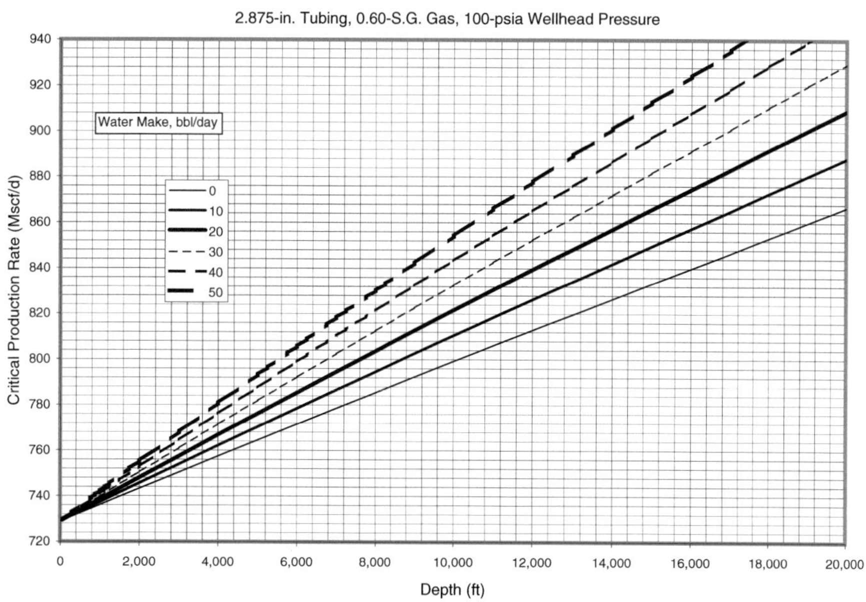

Figure D–16 **Critical gas production rate for water removal in 2.875-in tubing against 100 psia wellhead pressure, S.G. 0.60 gas.**

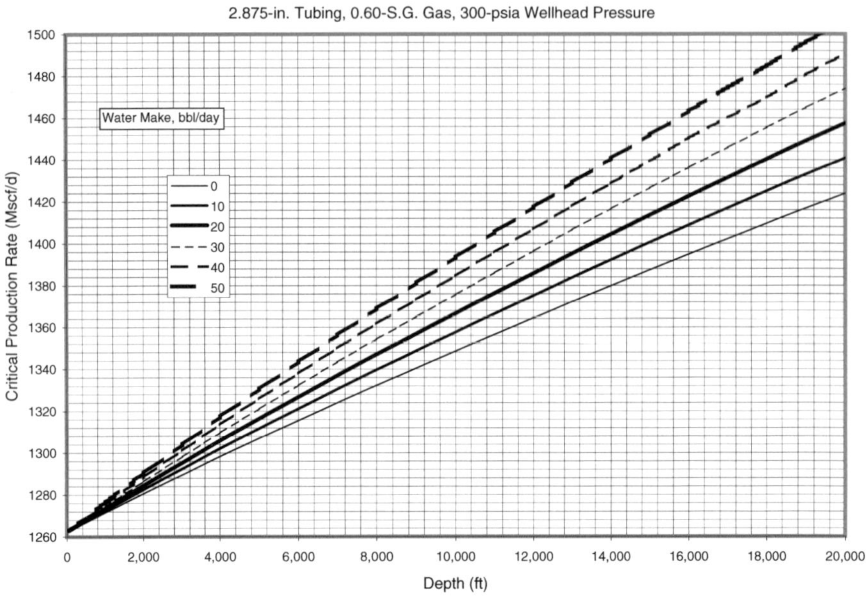

Figure D–17 **Critical gas production rate for water removal in 2.875-in tubing against 300 psia wellhead pressure, S.G. 0.60 gas.**

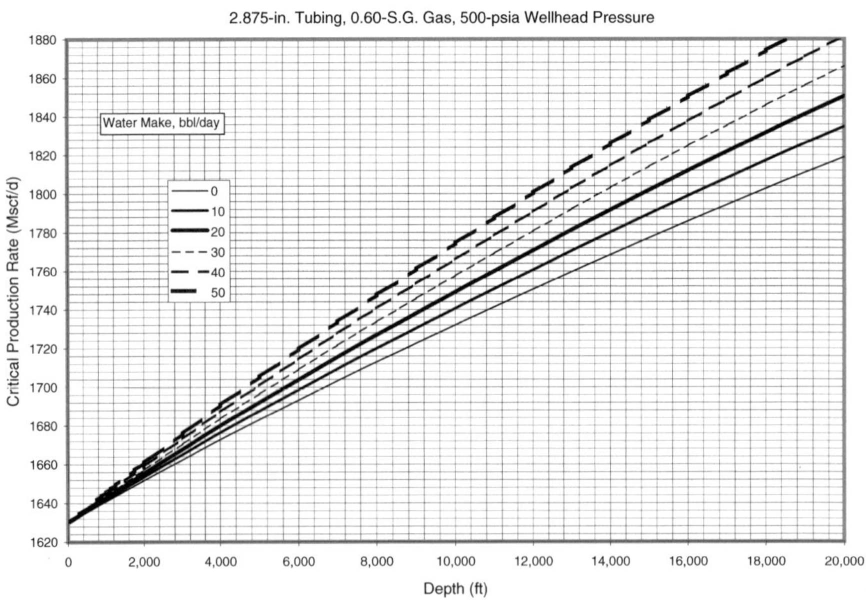

Figure D–18 **Critical gas production rate for water removal in 2.875-in tubing against 500 psia wellhead pressure, S.G. 0.60 gas.**

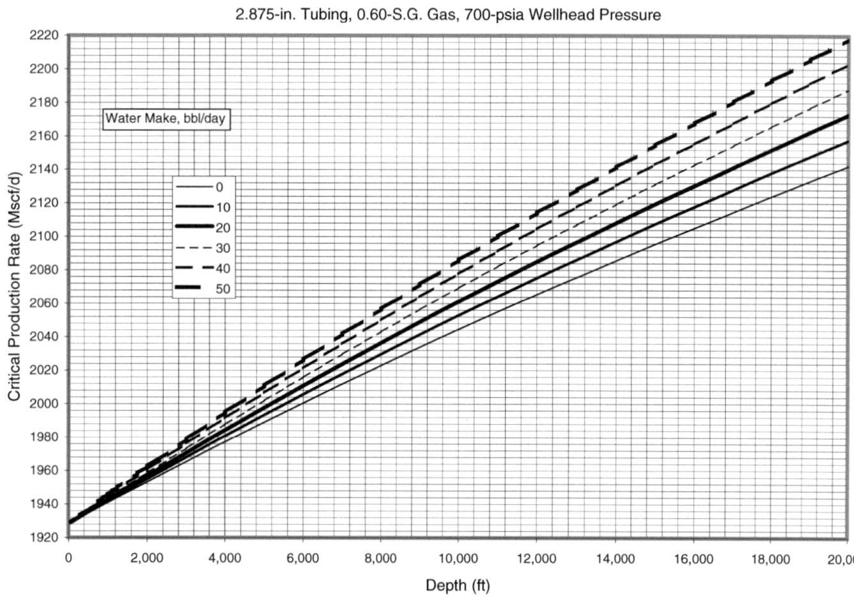

Figure D–19 **Critical gas production rate for water removal in 2.875-in tubing against 700 psia wellhead pressure, S.G. 0.60 gas.**

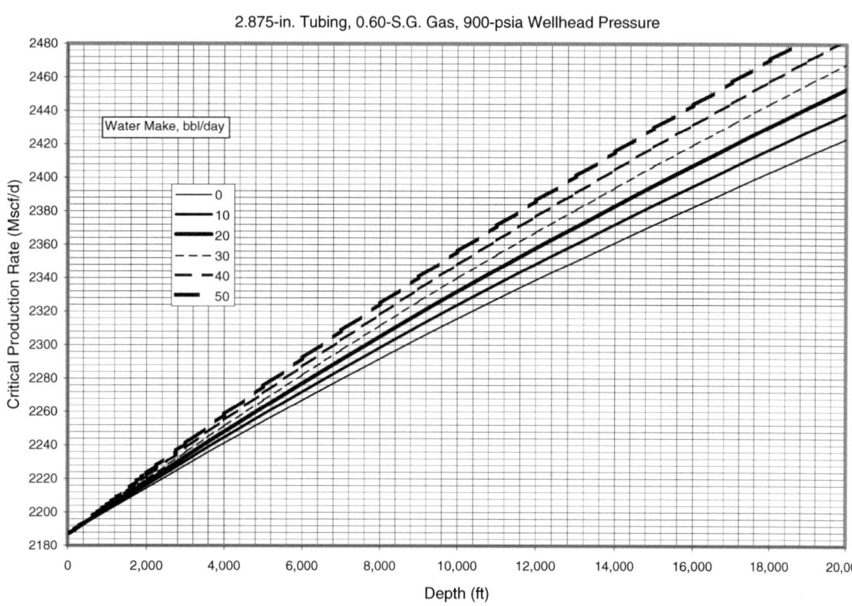

Figure D–20 **Critical gas production rate for water removal in 2.875-in tubing against 900 psia wellhead pressure, S.G. 0.60 gas.**

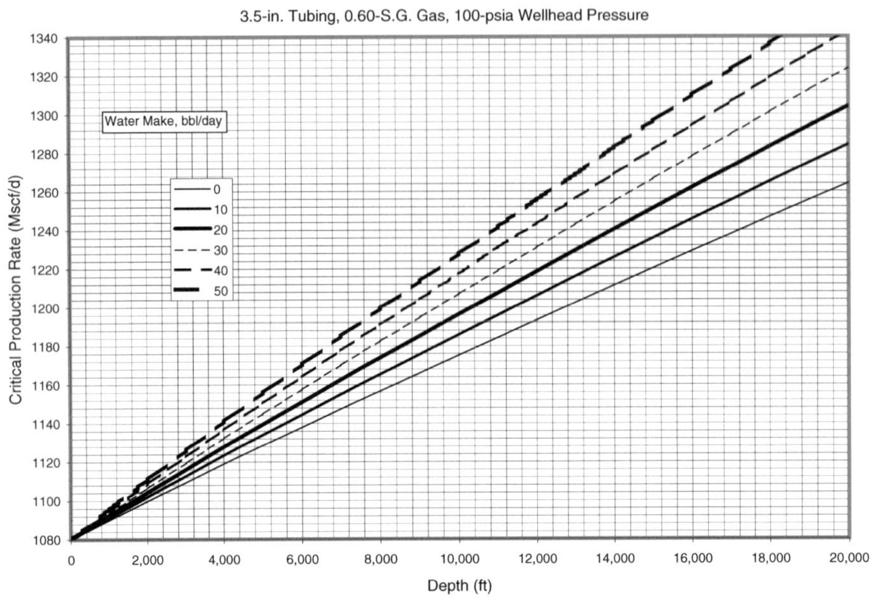

Figure D–21 **Critical gas production rate for water removal in 3.5-in tubing against 100 psia wellhead pressure, S.G. 0.60 gas.**

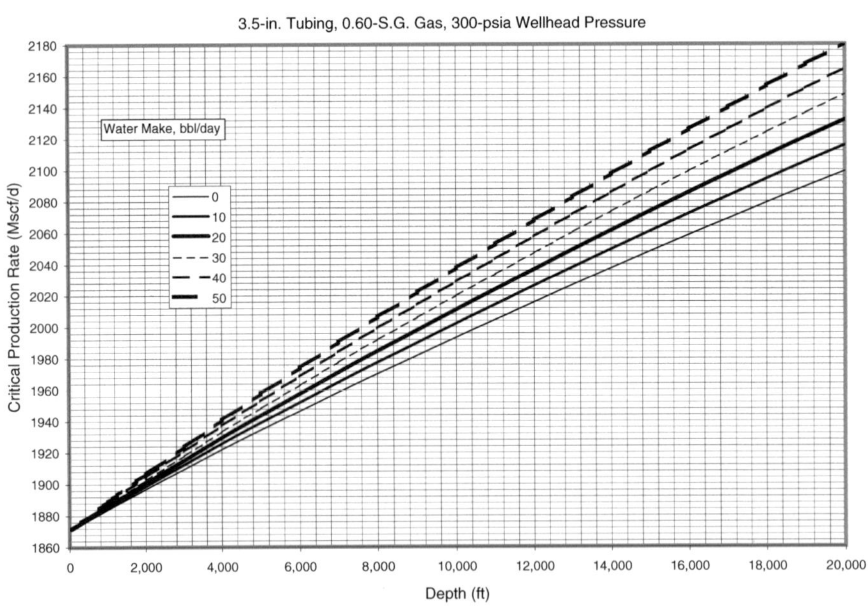

Figure D–22 **Critical gas production rate for water removal in 3.5-in tubing against 300 psia wellhead pressure, S.G. 0.60 gas.**

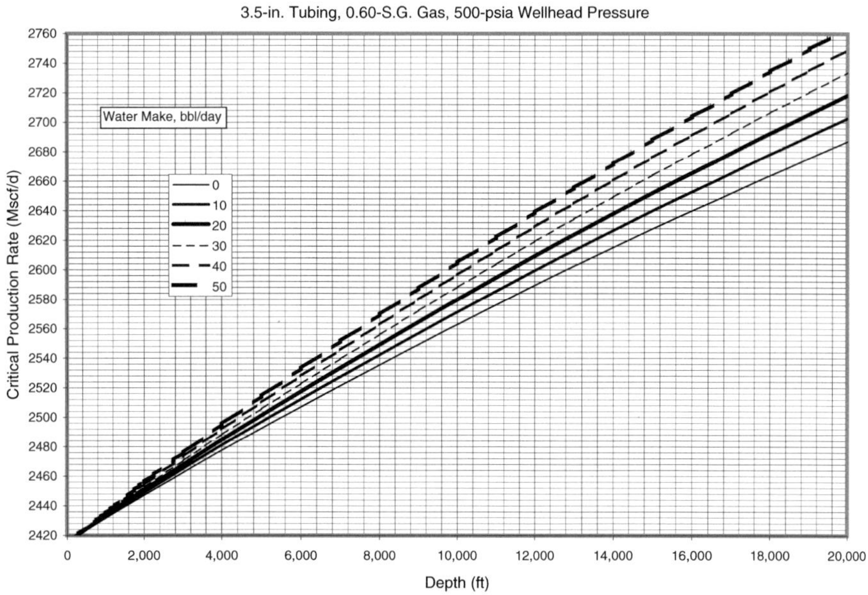

Figure D–23 **Critical gas production rate for water removal in 3.5-in tubing against 500 psia wellhead pressure, S.G. 0.60 gas.**

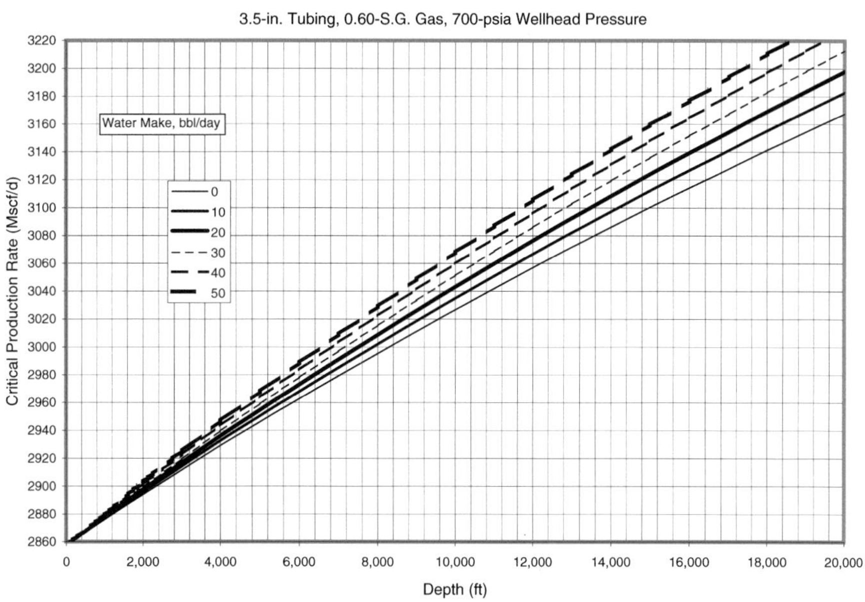

Figure D–24 **Critical gas production rate for water removal in 3.5-in tubing against 700 psia wellhead pressure, S.G. 0.60 gas.**

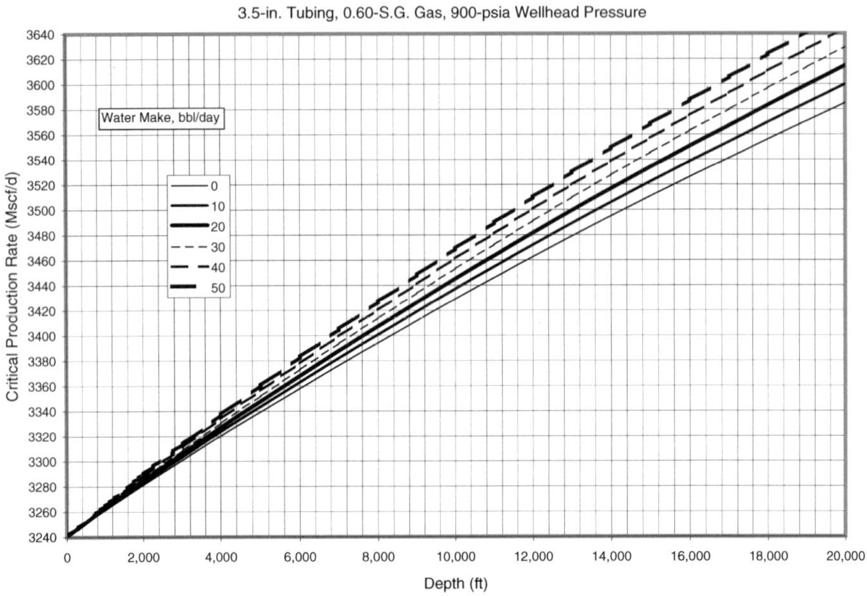

Figure D–25 **Critical gas production rate for water removal in 3.5-in tubing against 900 psia wellhead pressure, S.G. 0.60 gas.**

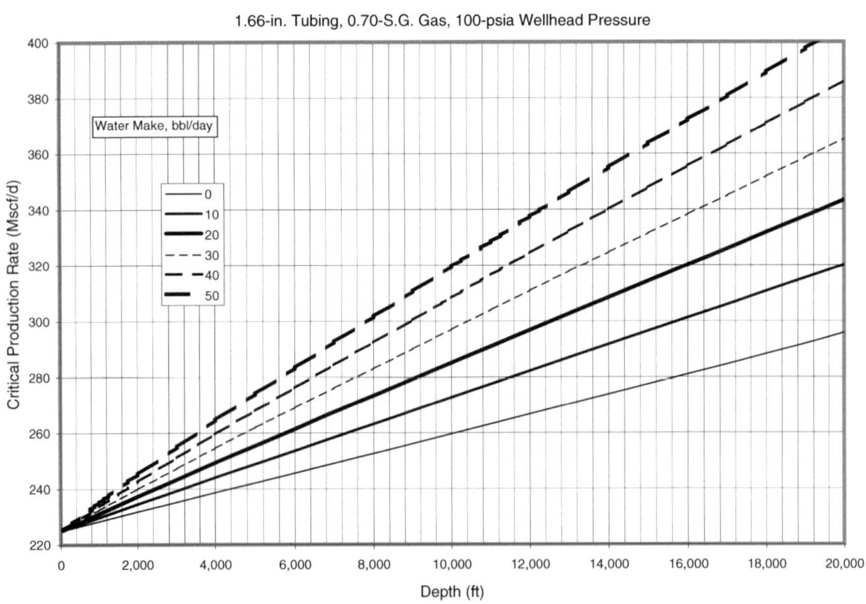

Figure D–26 **Critical gas production rate for water removal in 1.66-in tubing against 100 psia wellhead pressure, S.G. 0.70 gas.**

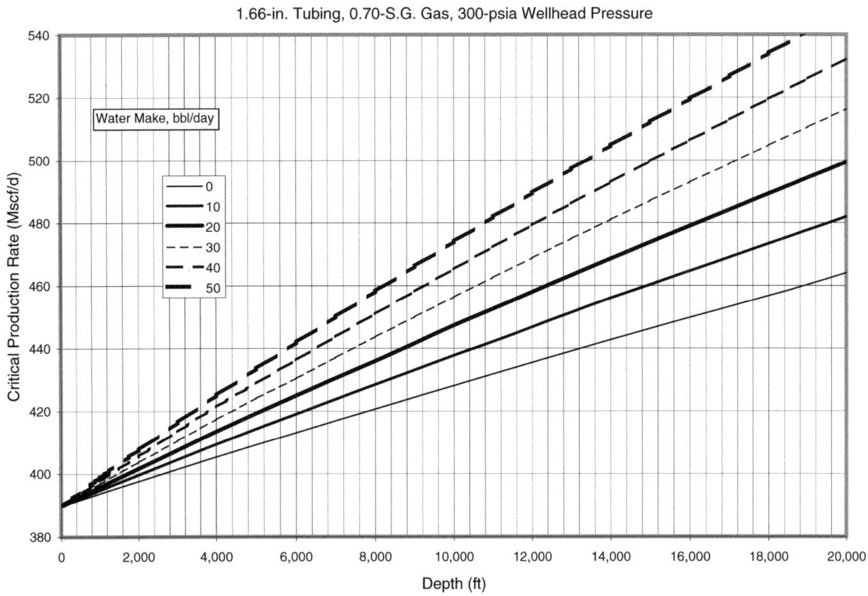

Figure D–27 **Critical gas production rate for water removal in 1.66-in tubing against 300 psia wellhead pressure, S.G. 0.70 gas.**

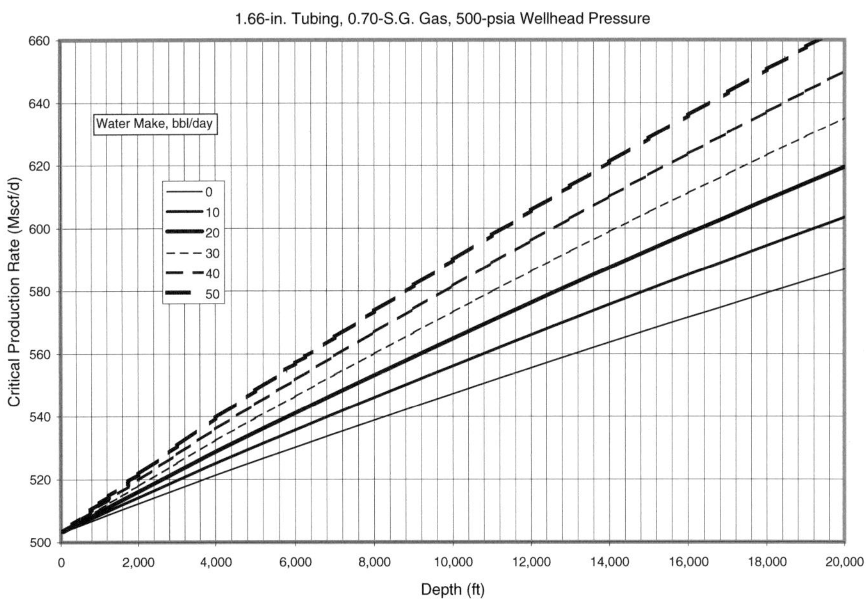

Figure D–28 **Critical gas production rate for water removal in 1.66-in tubing against 500 psia wellhead pressure, S.G. 0.70 gas.**

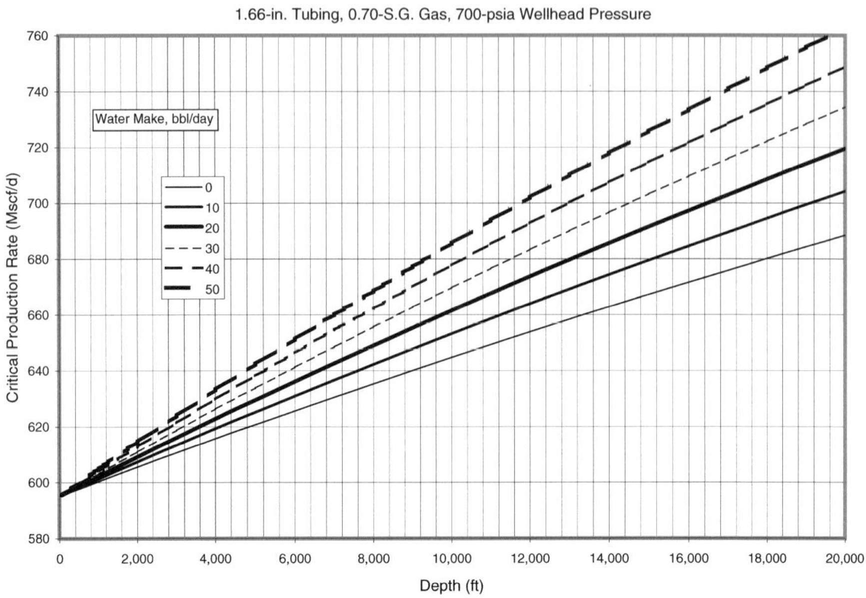

Figure D–29 Critical gas production rate for water removal in 1.66-in tubing against 700 psia wellhead pressure, S.G. 0.70 gas.

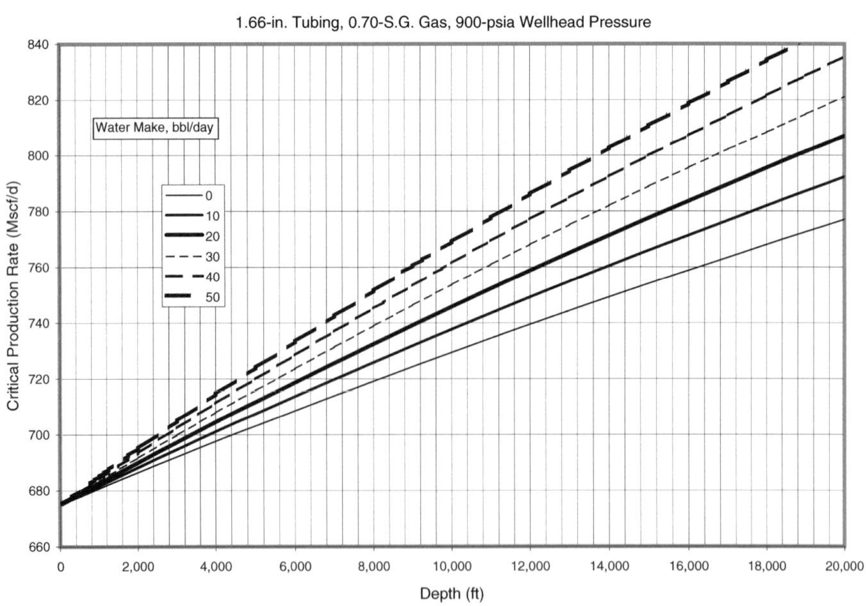

Figure D–30 Critical gas production rate for water removal in 1.66-in tubing against 900 psia wellhead pressure, S.G. 0.70 gas.

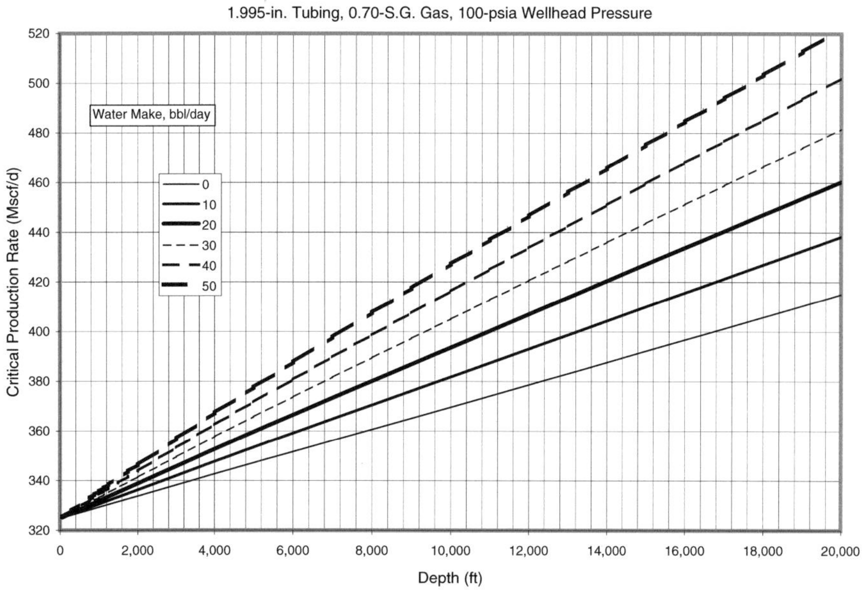

Figure D–31 **Critical gas production rate for water removal in 1.995-in tubing against 100 psia wellhead pressure, S.G. 0.70 gas.**

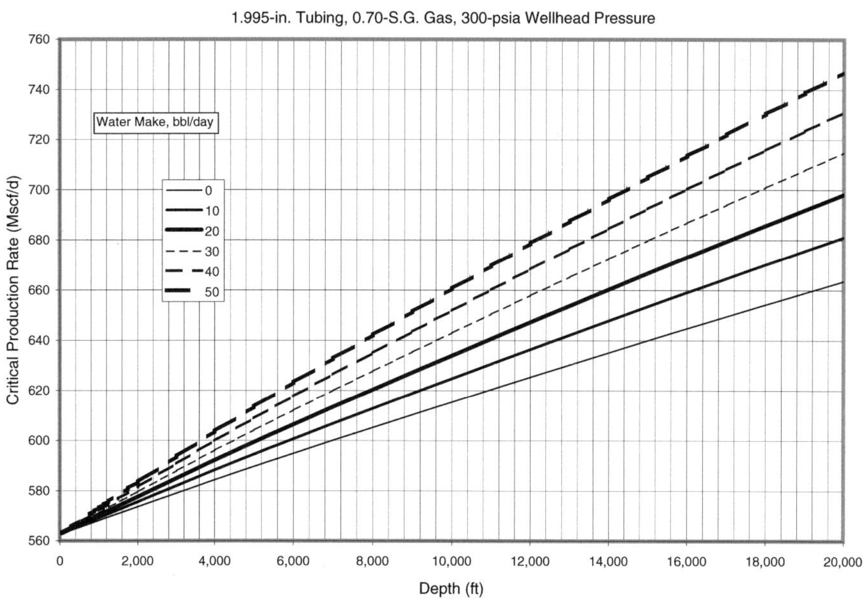

Figure D–32 **Critical gas production rate for water removal in 1.995-in tubing against 300 psia wellhead pressure, S.G. 0.70 gas.**

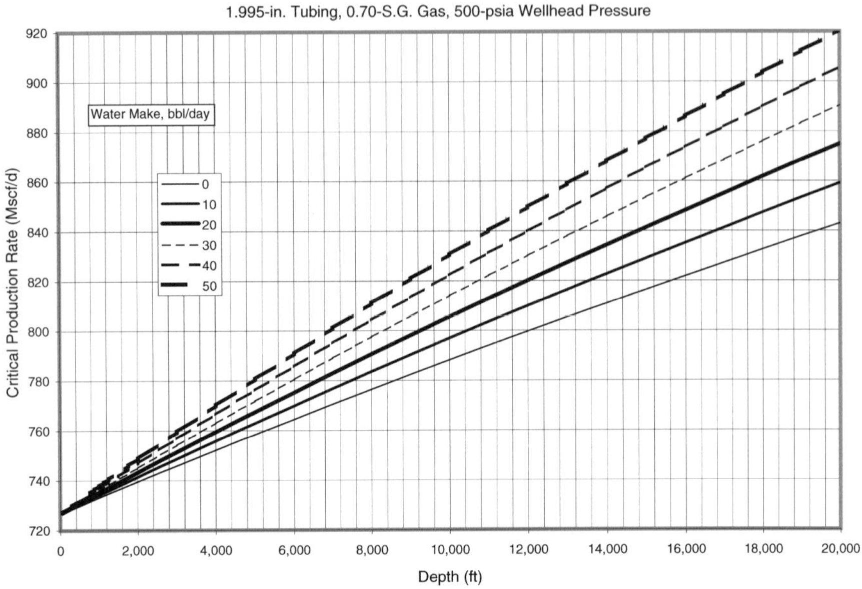

Figure D–33 **Critical gas production rate for water removal in 1.995-in tubing against 500 psia wellhead pressure, S.G. 0.70 gas.**

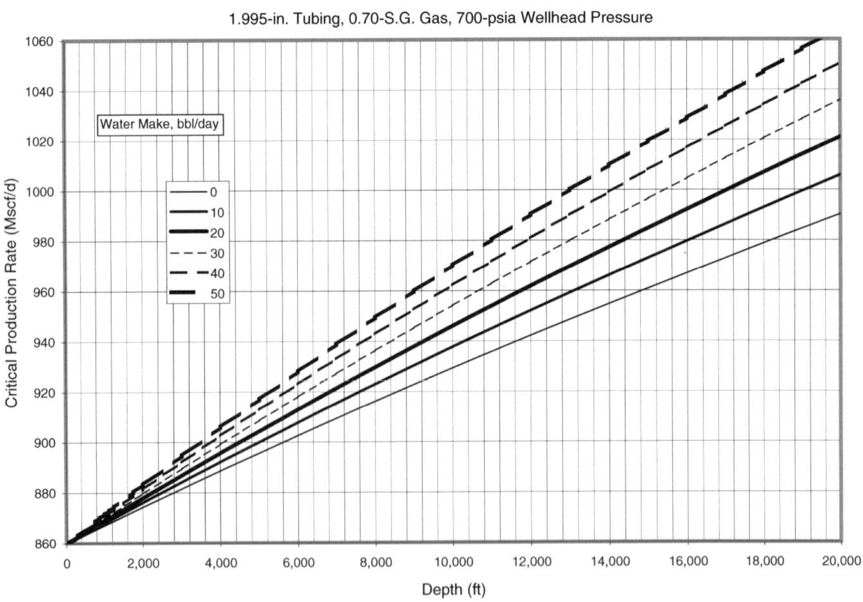

Figure D–34 **Critical gas production rate for water removal in 1.995-in tubing against 700 psia wellhead pressure, S.G. 0.70 gas.**

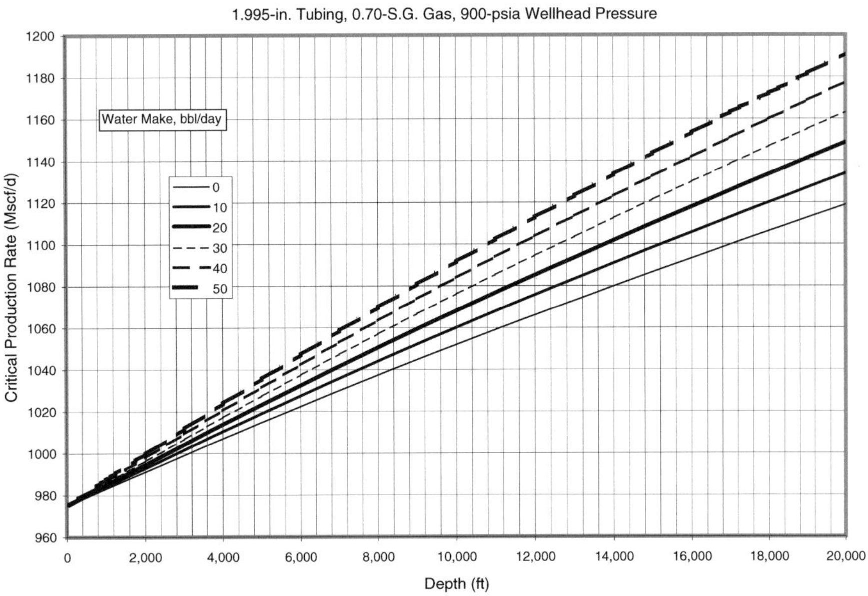

Figure D–35 **Critical gas production rate for water removal in 1.995-in tubing against 900 psia wellhead pressure, S.G. 0.70 gas.**

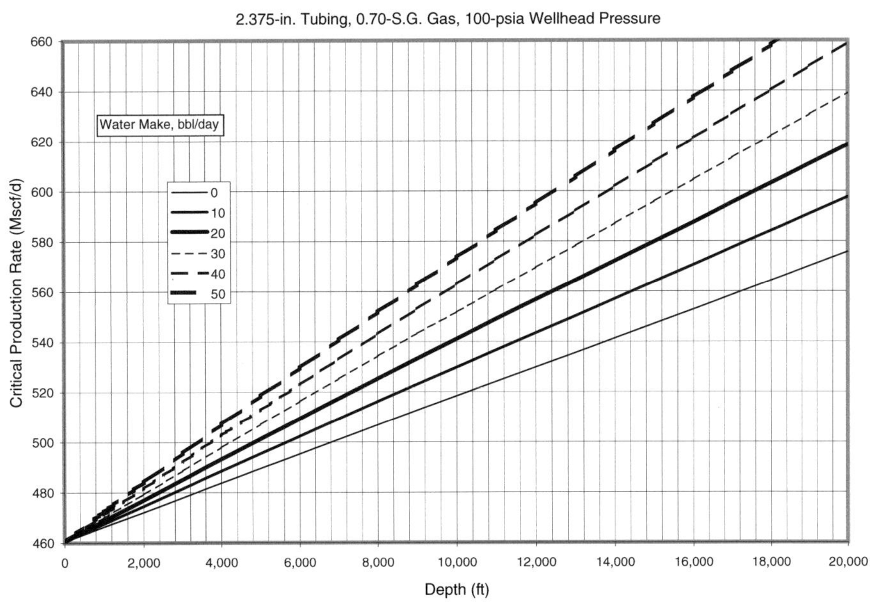

Figure D–36 **Critical gas production rate for water removal in 2.375-in tubing against 100 psia wellhead pressure, S.G. 0.70 gas.**

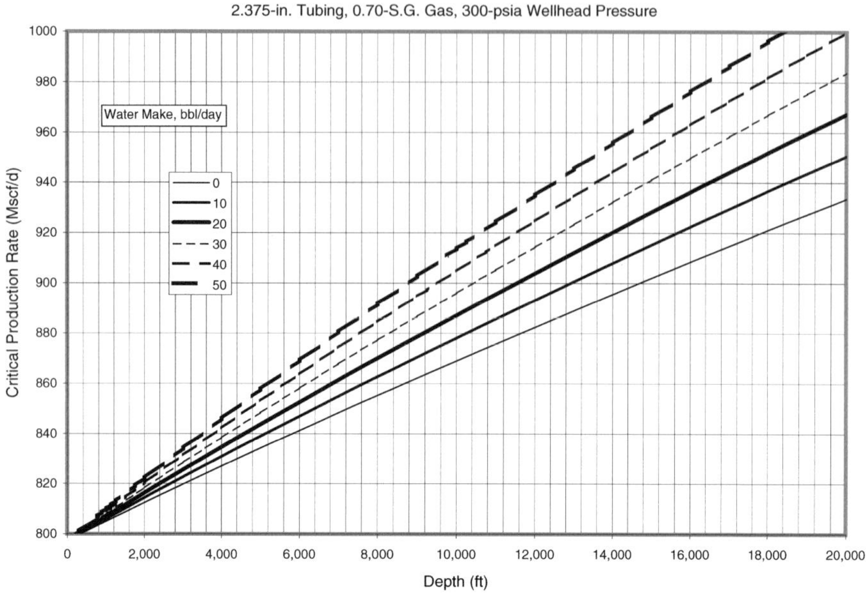

Figure D–37 **Critical gas production rate for water removal in 2.375-in tubing against 300 psia wellhead pressure, S.G. 0.70 gas.**

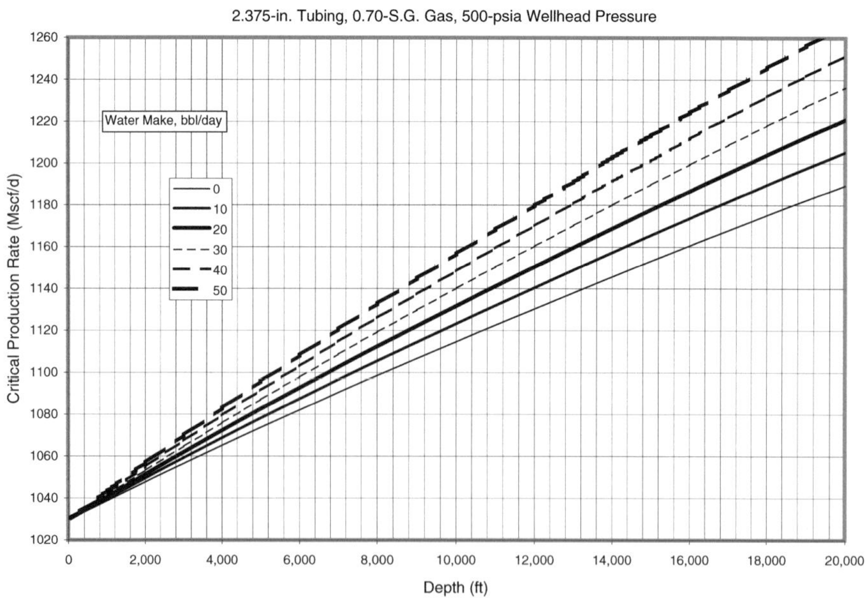

Figure D–38 **Critical gas production rate for water removal in 2.375-in tubing against 500 psia wellhead pressure, S.G. 0.70 gas.**

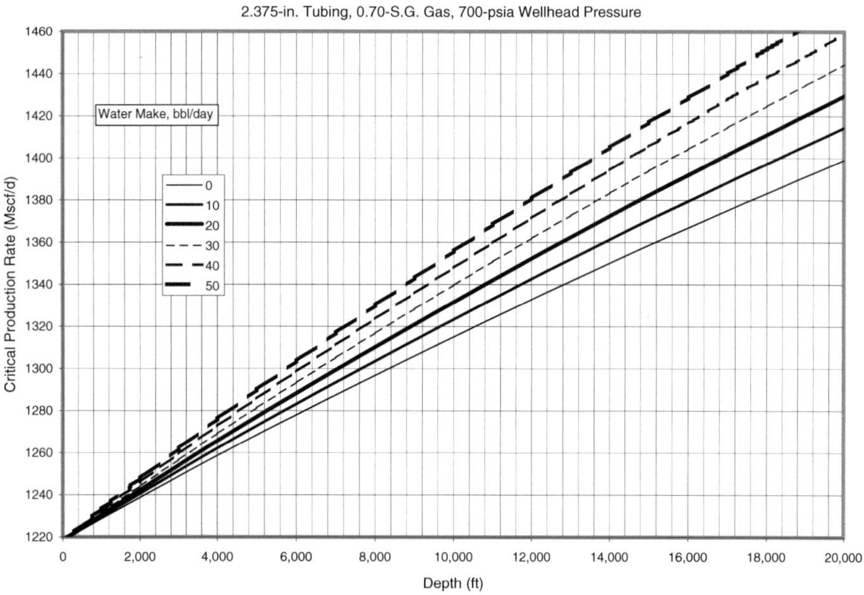

Figure D–39 **Critical gas production rate for water removal in 2.375-in tubing against 700 psia wellhead pressure, S.G. 0.70 gas.**

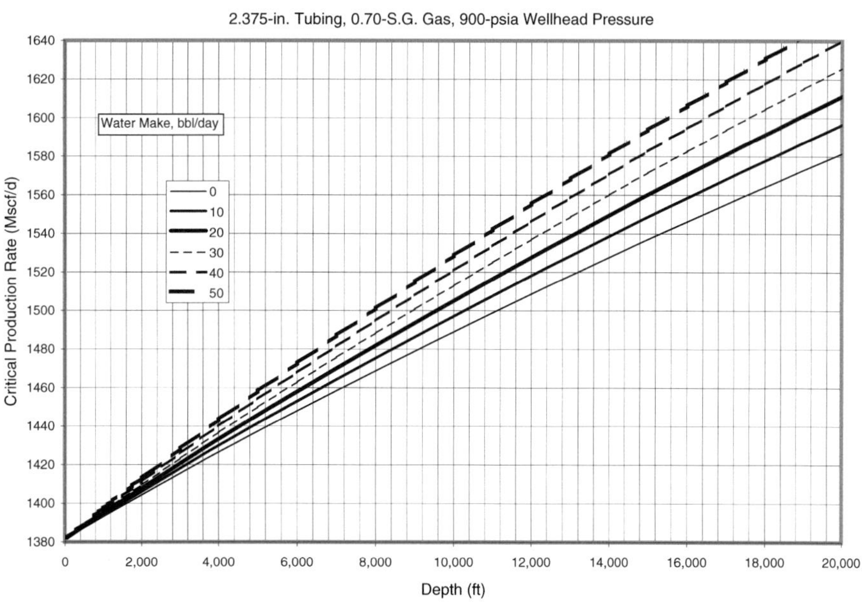

Figure D–40 **Critical gas production rate for water removal in 2.375-in tubing against 900 psia wellhead pressure, S.G. 0.70 gas.**

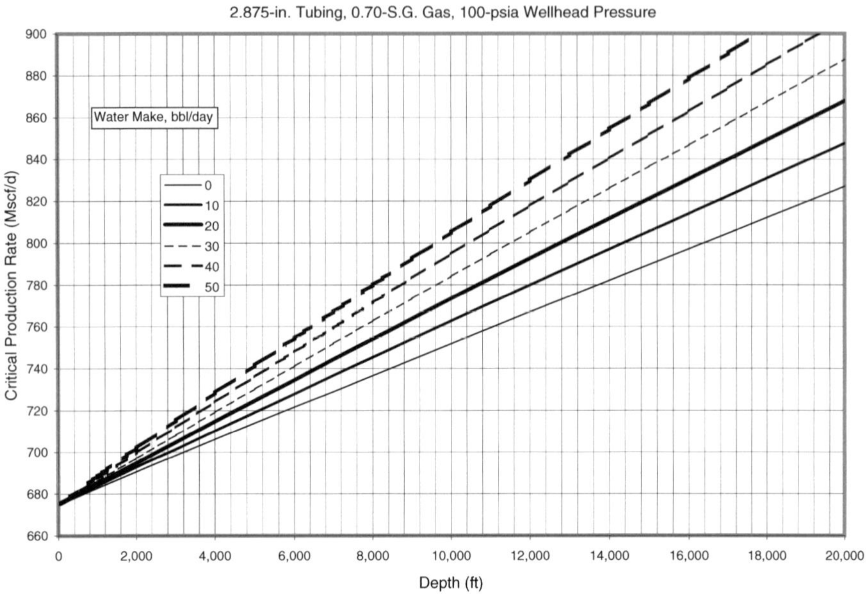

Figure D–41 **Critical gas production rate for water removal in 2.875-in tubing against 100 psia wellhead pressure, S.G. 0.70 gas.**

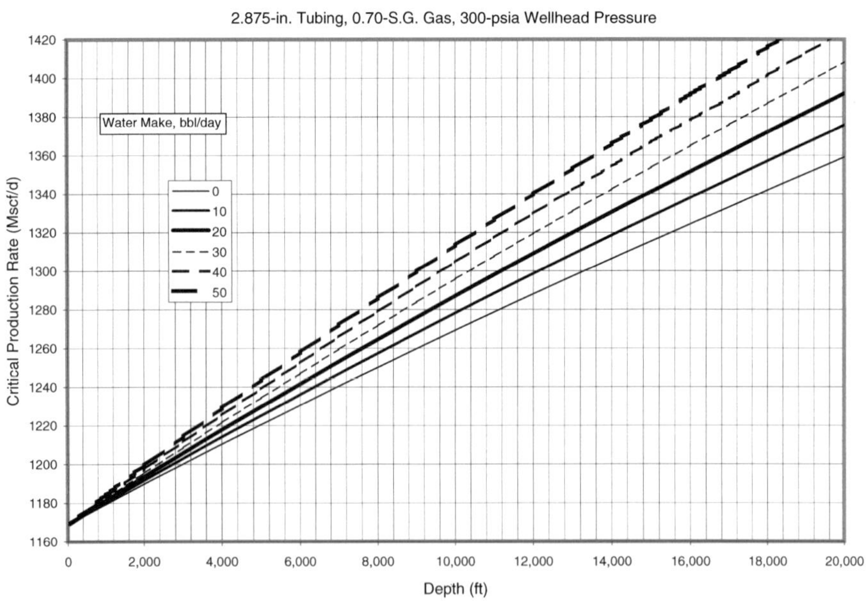

Figure D–42 **Critical gas production rate for water removal in 2.875-in tubing against 300 psia wellhead pressure, S.G. 0.70 gas.**

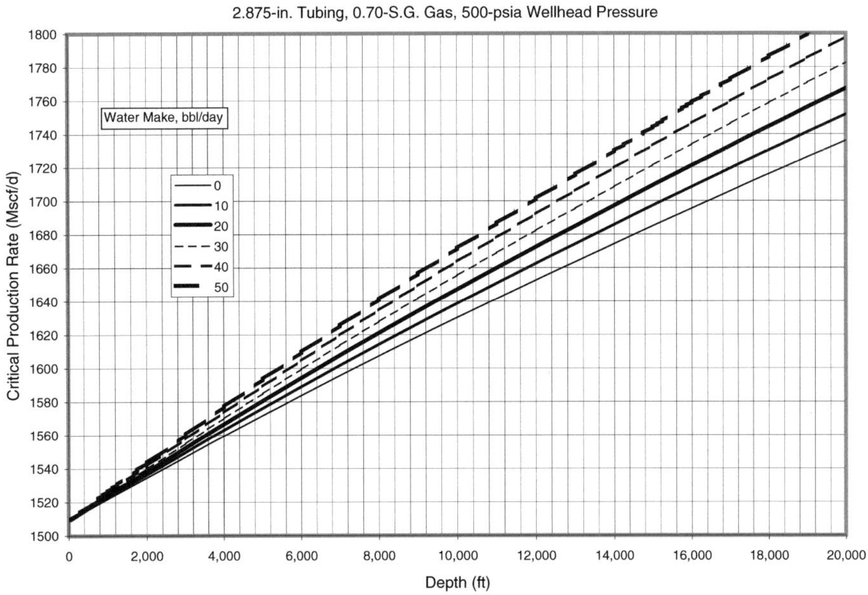

Figure D–43 **Critical gas production rate for water removal in 2.875-in tubing against 500 psia wellhead pressure, S.G. 0.70 gas.**

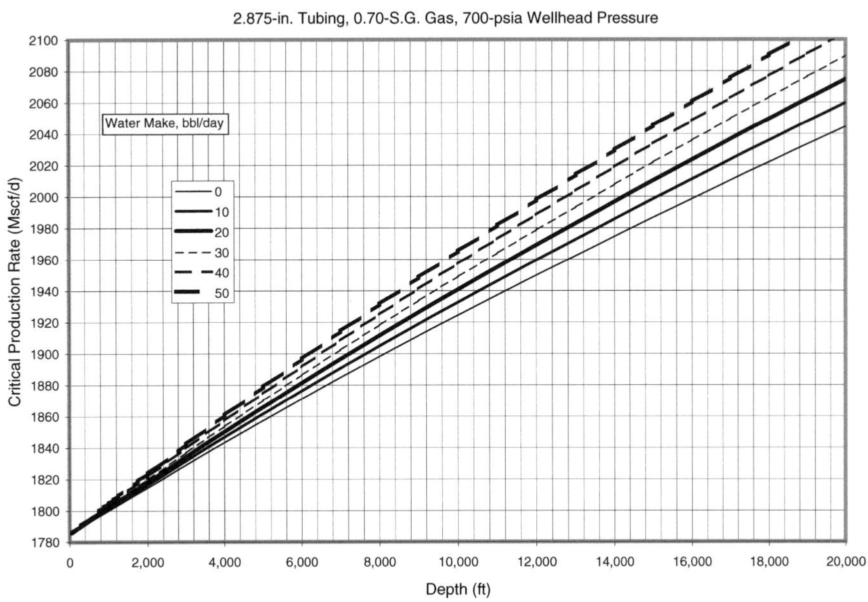

Figure D–44 **Critical gas production rate for water removal in 2.875-in tubing against 700 psia wellhead pressure, S.G. 0.70 gas.**

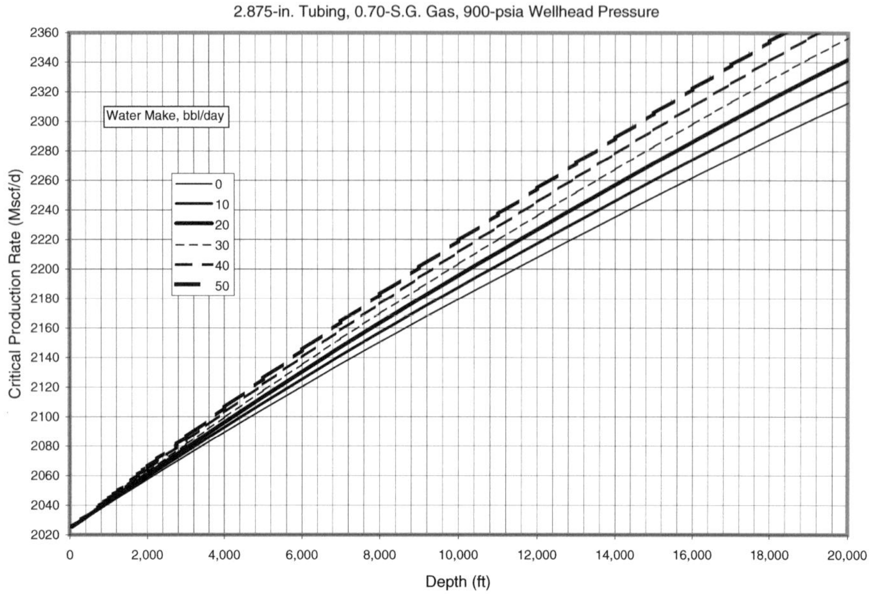

Figure D–45 **Critical gas production rate for water removal in 2.875-in tubing against 900 psia wellhead pressure, S.G. 0.70 gas.**

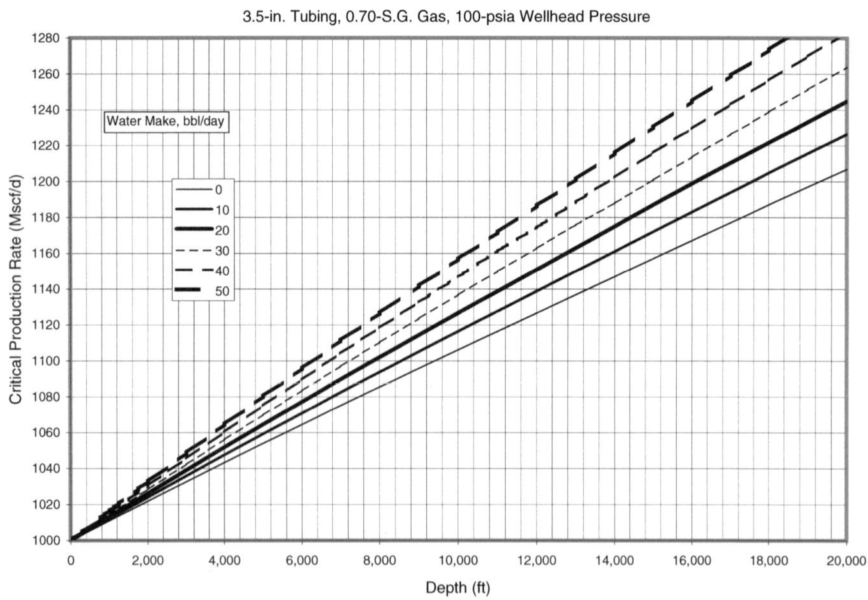

Figure D–46 **Critical gas production rate for water removal in 3.5-in tubing against 100 psia wellhead pressure, S.G. 0.70 gas.**

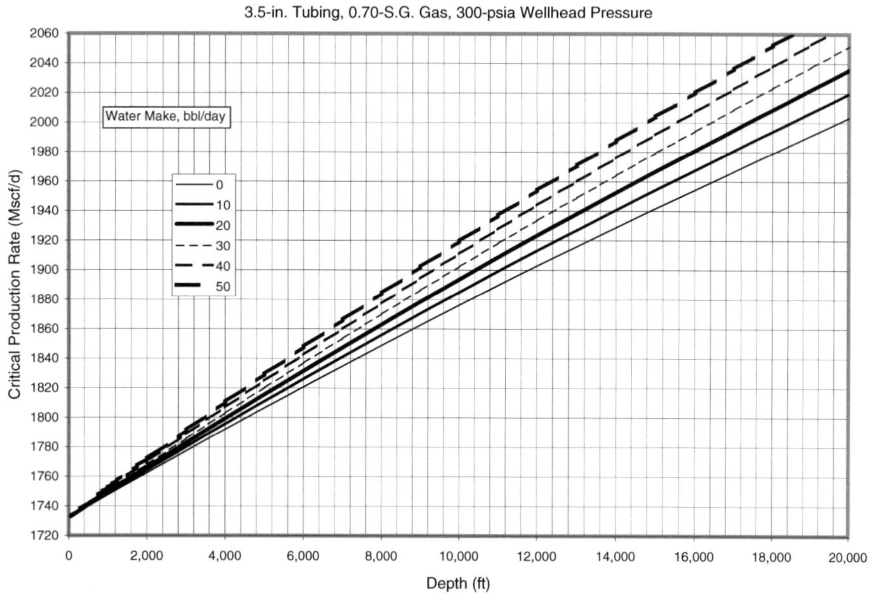

Figure D–47 **Critical gas production rate for water removal in 3.5-in tubing against 300 psia wellhead pressure, S.G. 0.70 gas.**

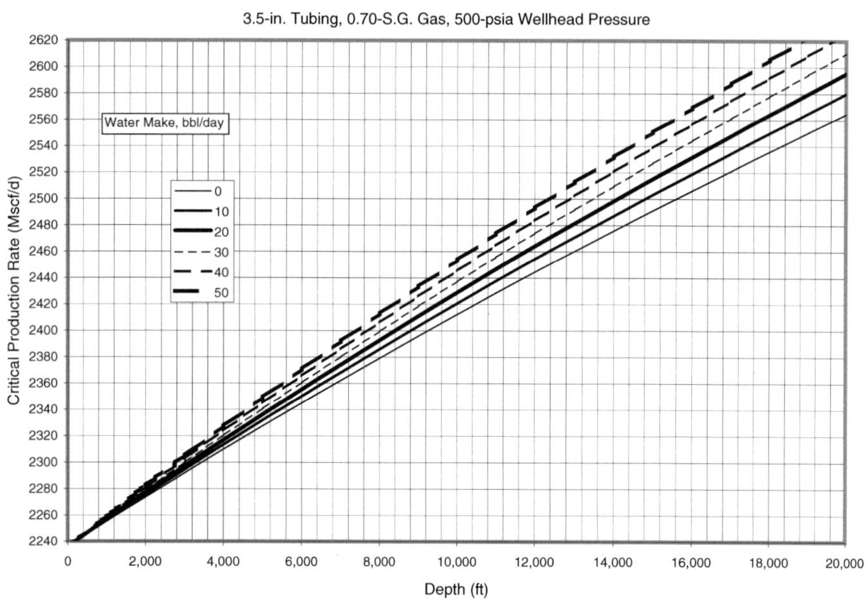

Figure D–48 **Critical gas production rate for water removal in 3.5-in tubing against 500 psia wellhead pressure, S.G. 0.70 gas.**

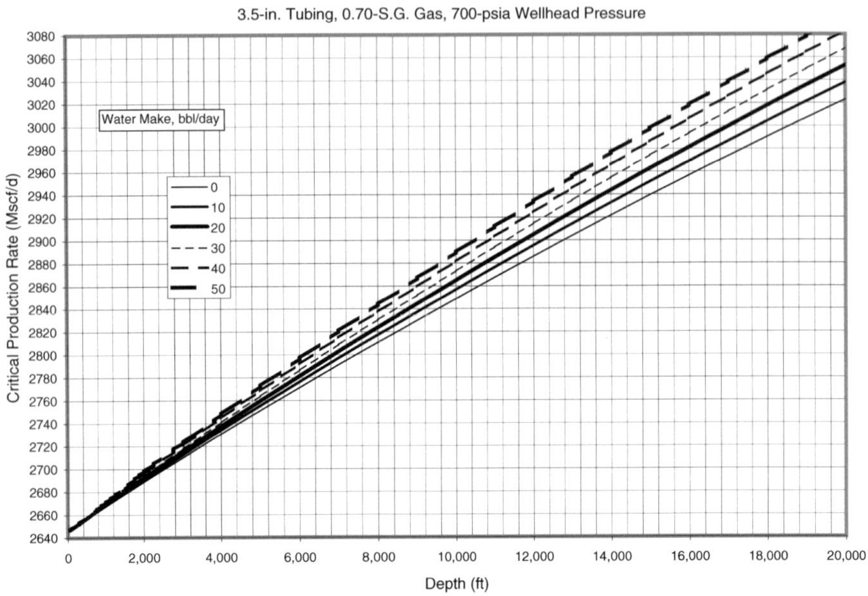

Figure D–49 **Critical gas production rate for water removal in 3.5-in tubing against 700 psia wellhead pressure, S.G. 0.70 gas.**

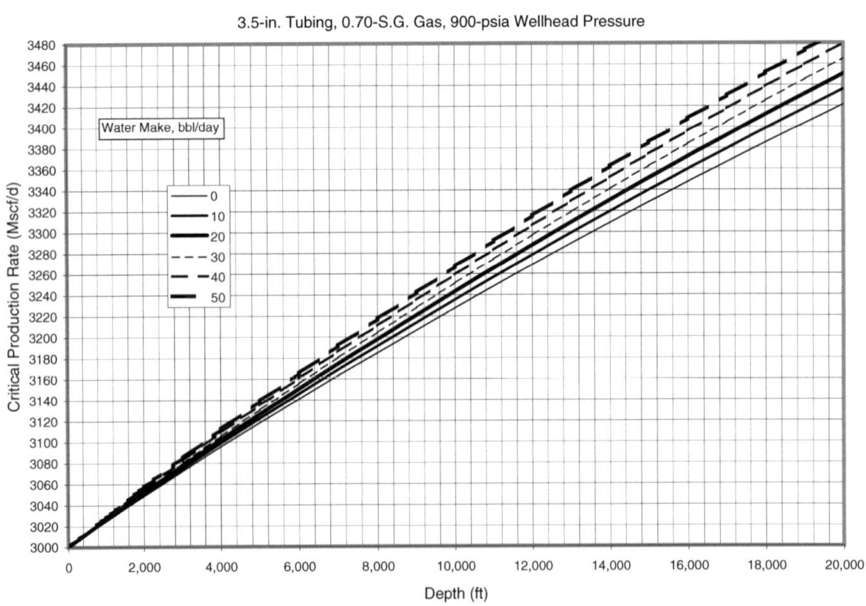

Figure D–50 **Critical gas production rate for water removal in 3.5-in tubing against 900 psia wellhead pressure, S.G. 0.70 gas.**

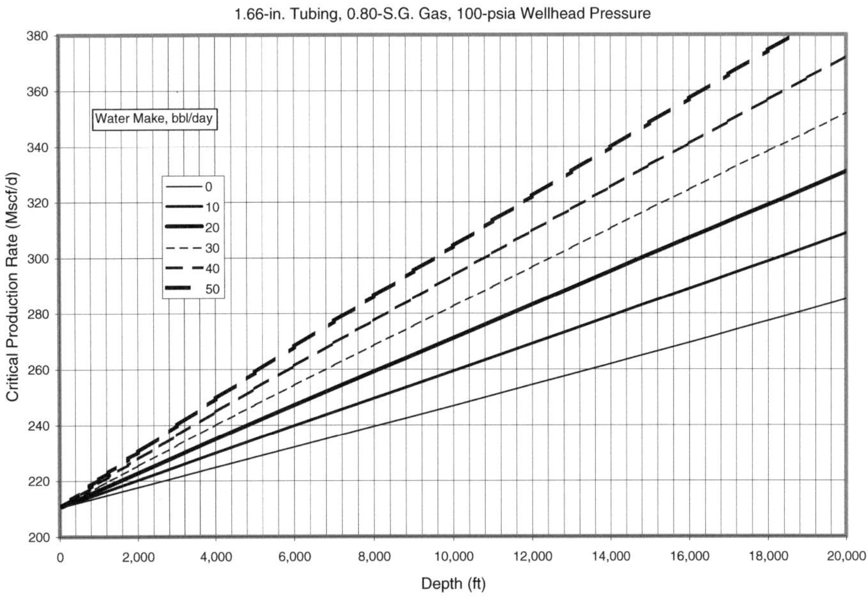

Figure D–51 Critical gas production rate for water removal in 1.66-in tubing against 100 psia wellhead pressure, S.G. 0.80 gas.

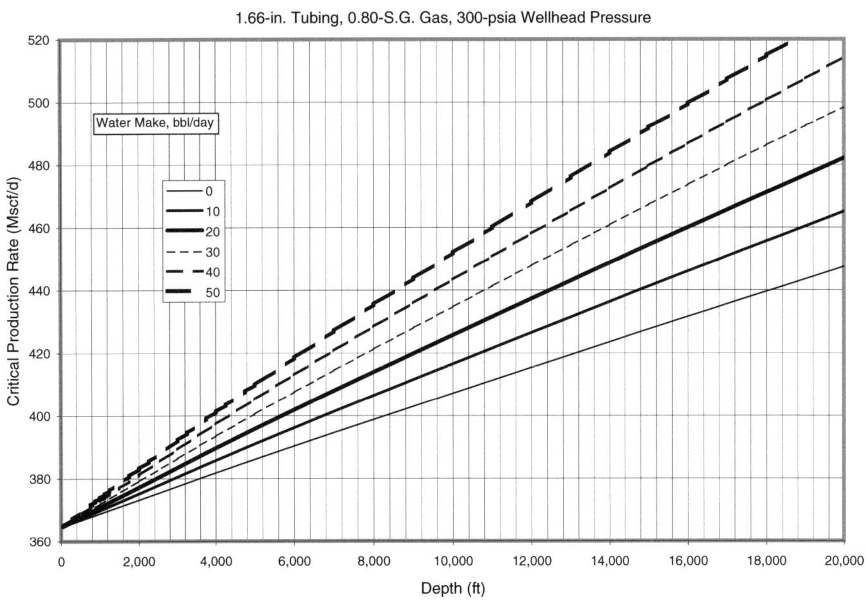

Figure D–52 Critical gas production rate for water removal in 1.66-in tubing against 300 psia wellhead pressure, S.G. 0.80 gas.

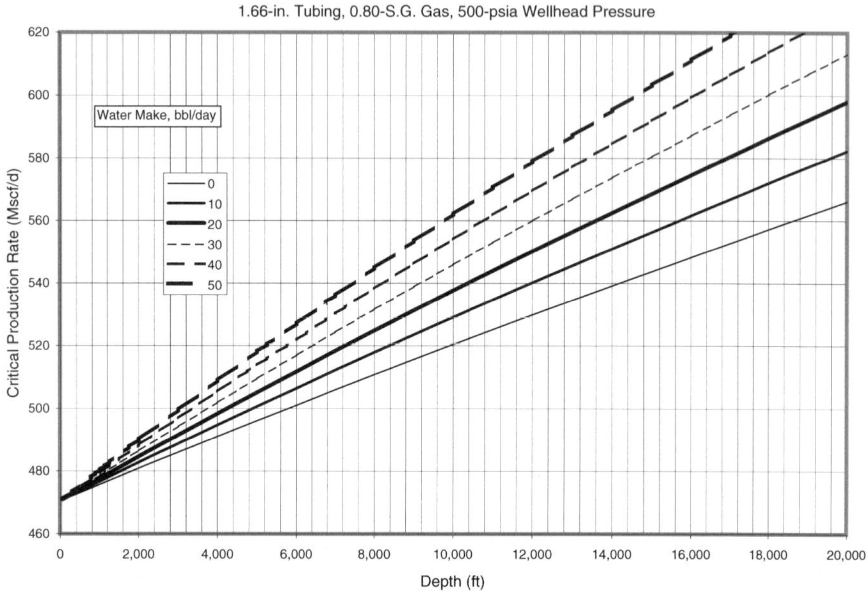

Figure D–53 **Critical gas production rate for water removal in 1.66-in tubing against 500 psia wellhead pressure, S.G. 0.80 gas.**

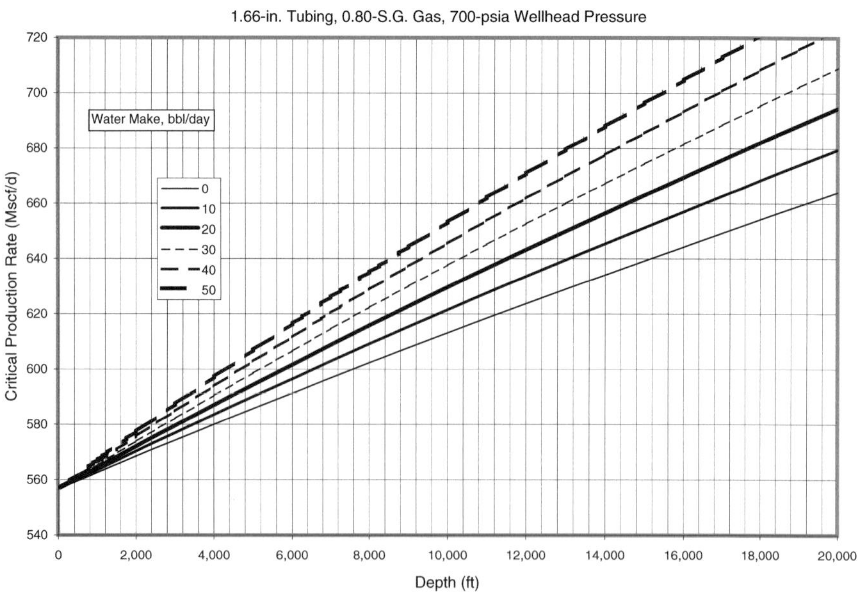

Figure D–54 **Critical gas production rate for water removal in 1.66-in tubing against 700 psia wellhead pressure, S.G. 0.80 gas.**

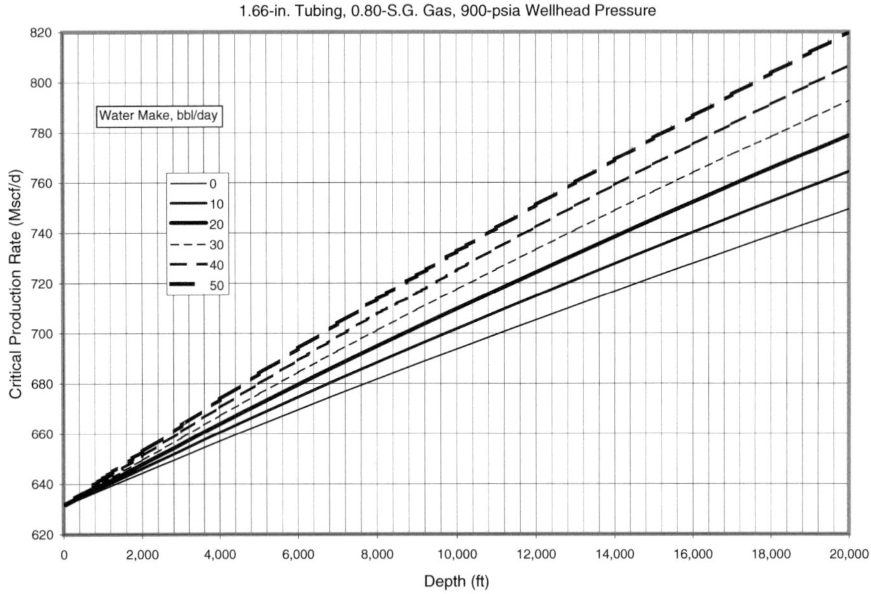

Figure D–55 **Critical gas production rate for water removal in 1.66-in tubing against 900 psia wellhead pressure, S.G. 0.80 gas.**

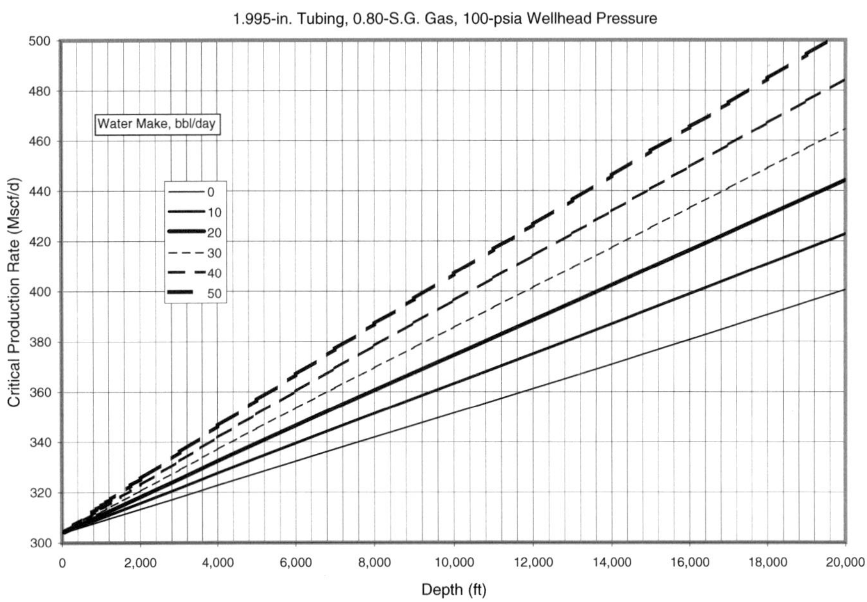

Figure D–56 **Critical gas production rate for water removal in 1.995-in tubing against 100 psia wellhead pressure, S.G. 0.80 gas.**

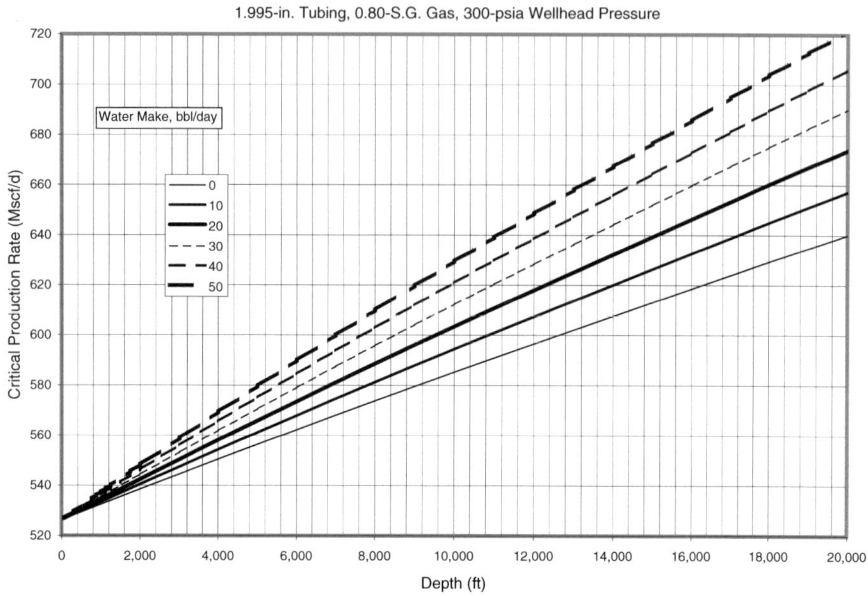

Figure D–57 **Critical gas production rate for water removal in 1.995-in tubing against 300 psia wellhead pressure, S.G. 0.80 gas.**

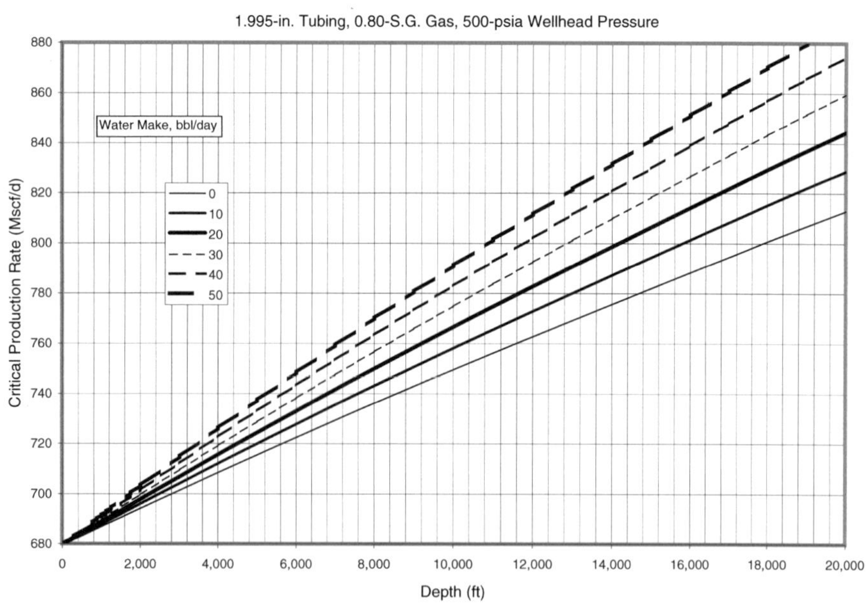

Figure D–58 **Critical gas production rate for water removal in 1.995-in tubing against 500 psia wellhead pressure, S.G. 0.80 gas.**

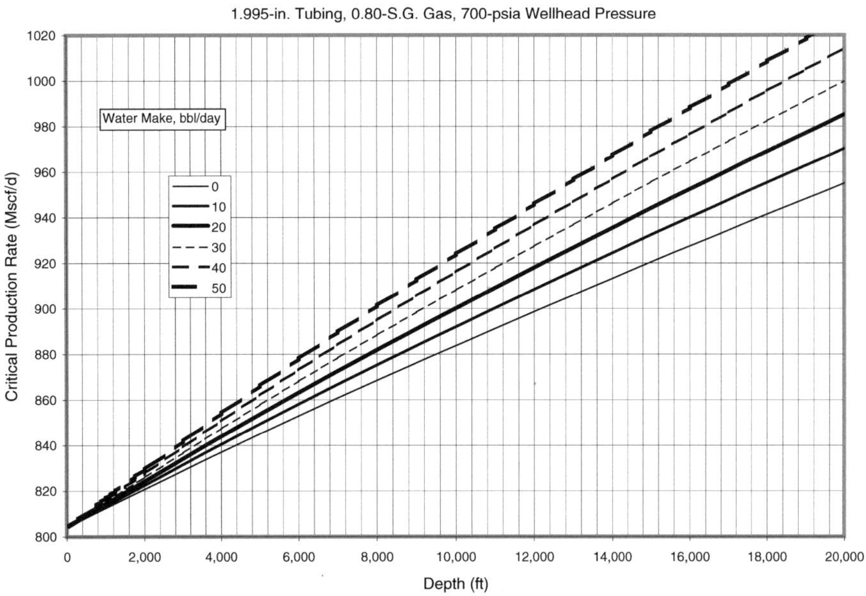

Figure D–59 **Critical gas production rate for water removal in 1.995-in tubing against 700 psia wellhead pressure, S.G. 0.80 gas.**

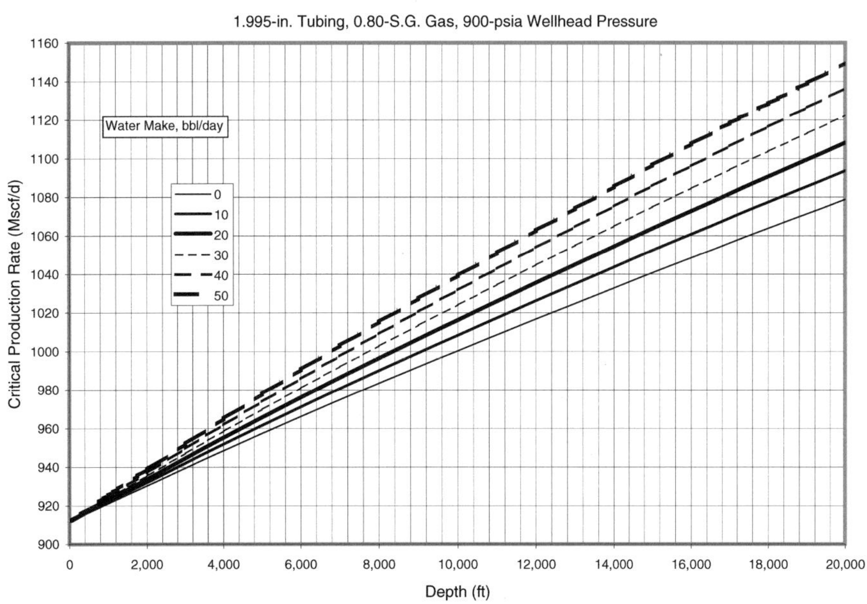

Figure D–60 **Critical gas production rate for water removal in 1.995-in tubing against 900 psia wellhead pressure, S.G. 0.80 gas.**

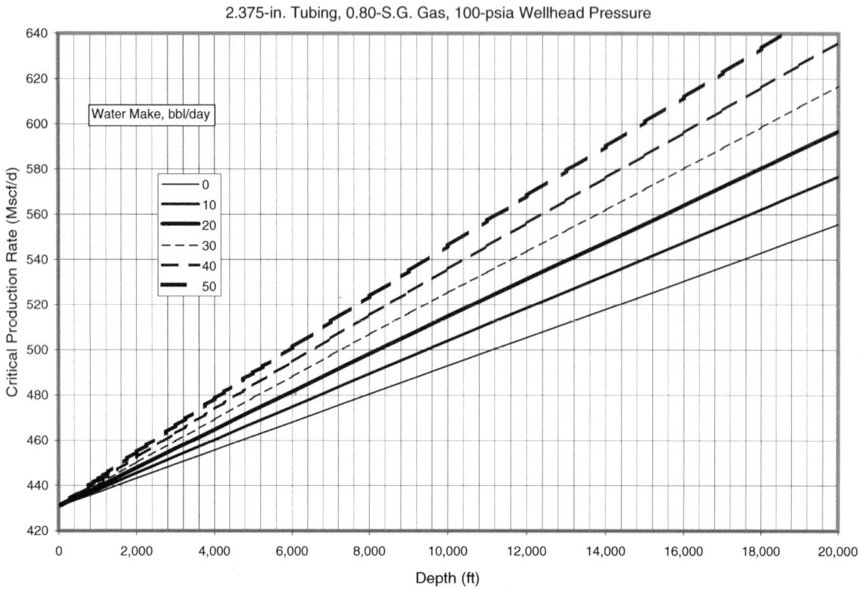

Figure D–61 **Critical gas production rate for water removal in 2.375-in tubing against 100 psia wellhead pressure, S.G. 0.80 gas.**

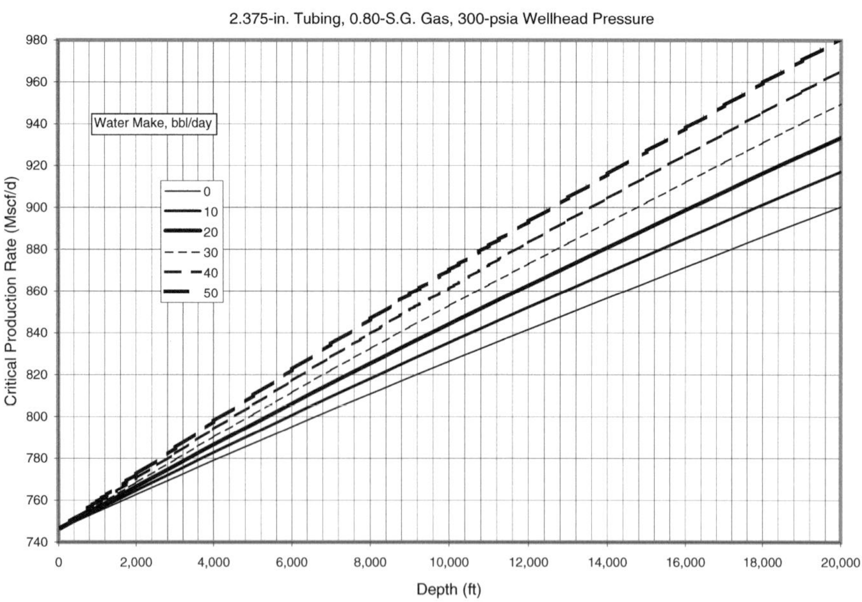

Figure D–62 **Critical gas production rate for water removal in 2.375-in tubing against 300 psia wellhead pressure, S.G. 0.80 gas.**

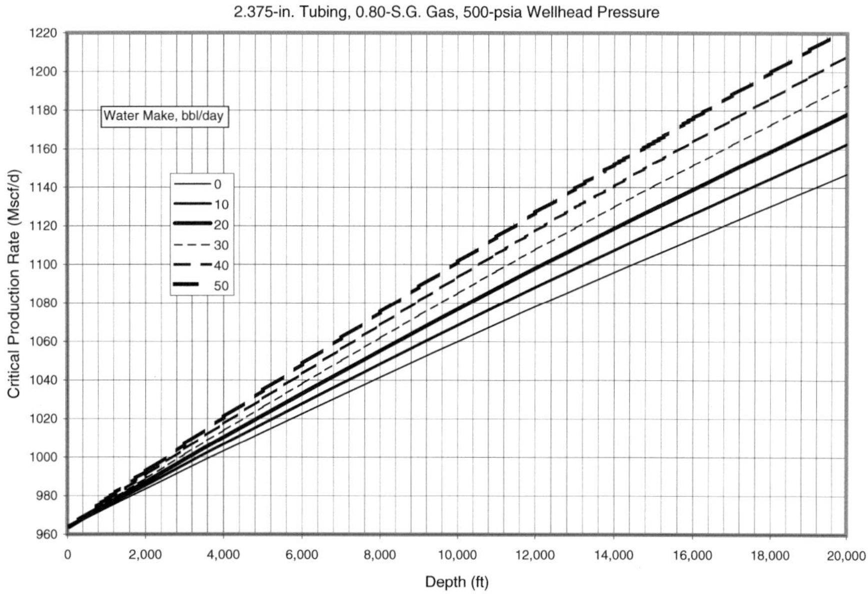

Figure D–63 **Critical gas production rate for water removal in 2.375-in tubing against 500 psia wellhead pressure, S.G. 0.80 gas.**

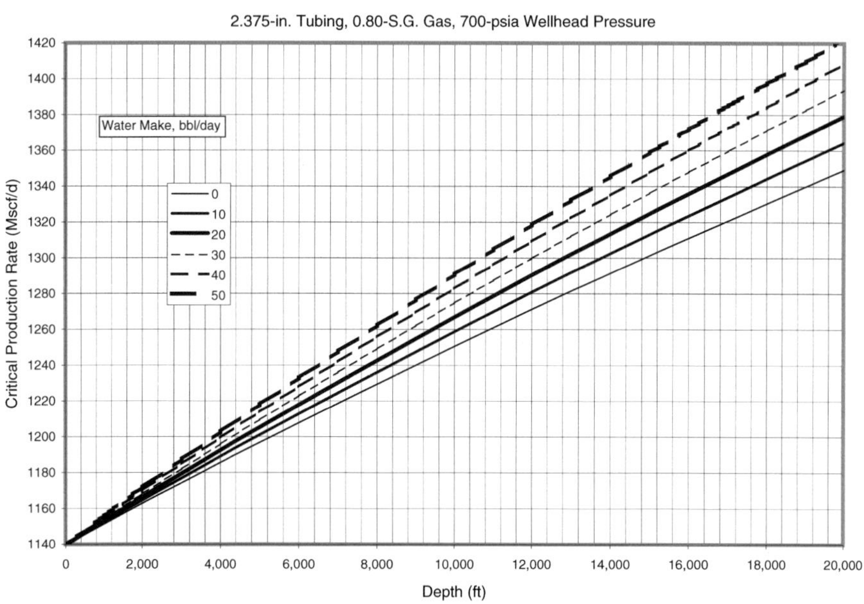

Figure D–64 **Critical gas production rate for water removal in 2.375-in tubing against 700 psia wellhead pressure, S.G. 0.80 gas.**

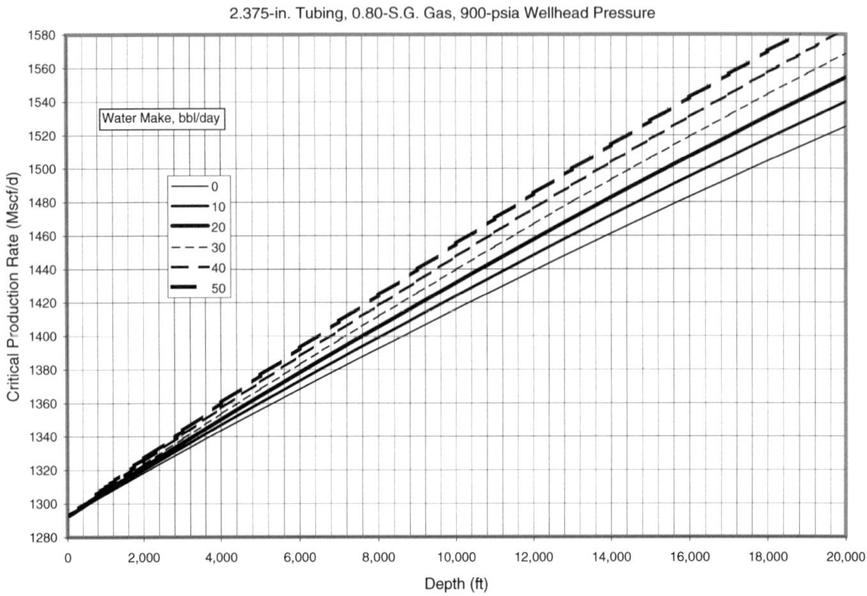

Figure D–65 **Critical gas production rate for water removal in 2.375-in tubing against 900 psia wellhead pressure, S.G. 0.80 gas.**

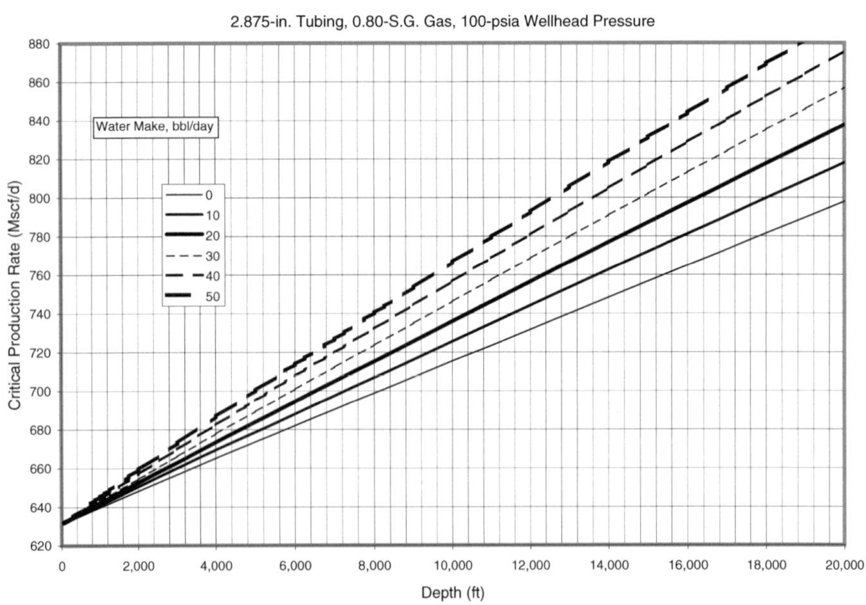

Figure D–66 **Critical gas production rate for water removal in 2.875-in tubing against 100 psia wellhead pressure, S.G. 0.80 gas.**

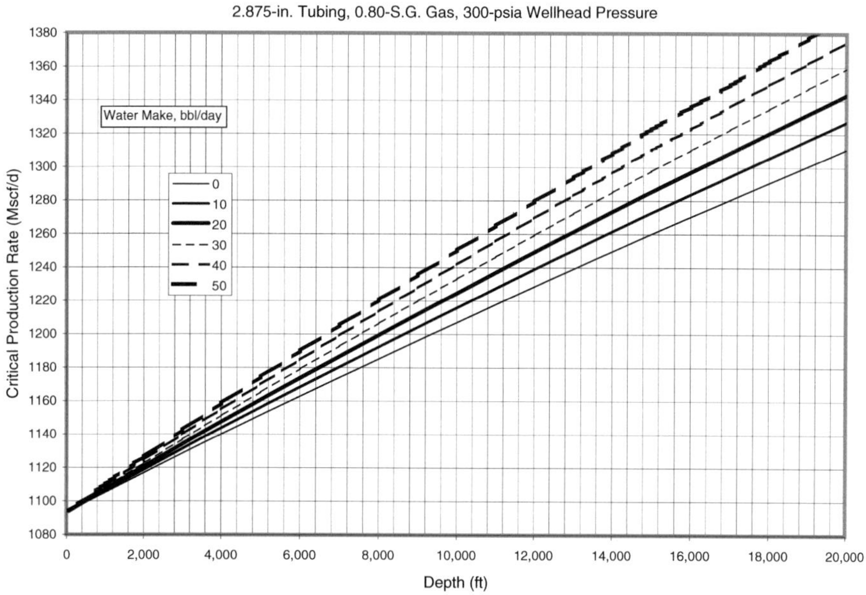

Figure D–67 Critical gas production rate for water removal in 2.875-in tubing against 300 psia wellhead pressure, S.G. 0.80 gas.

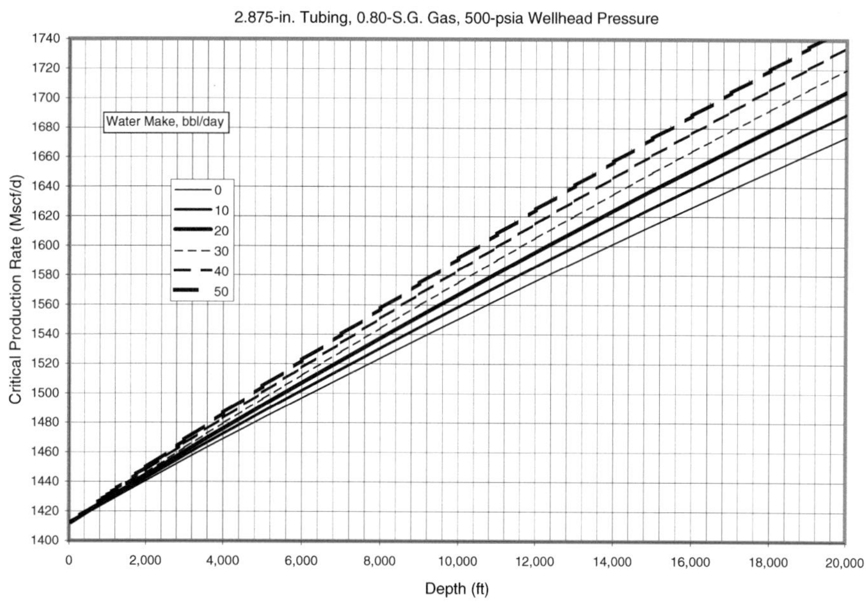

Figure D–68 Critical gas production rate for water removal in 2.875-in tubing against 500 psia wellhead pressure, S.G. 0.80 gas.

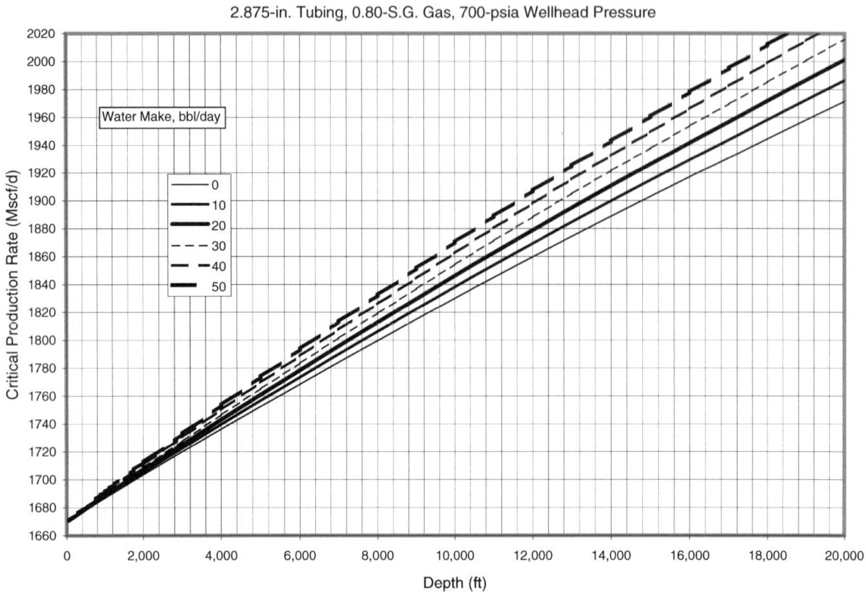

Figure D–69 **Critical gas production rate for water removal in 2.875-in tubing against 700 psia wellhead pressure, S.G. 0.80 gas.**

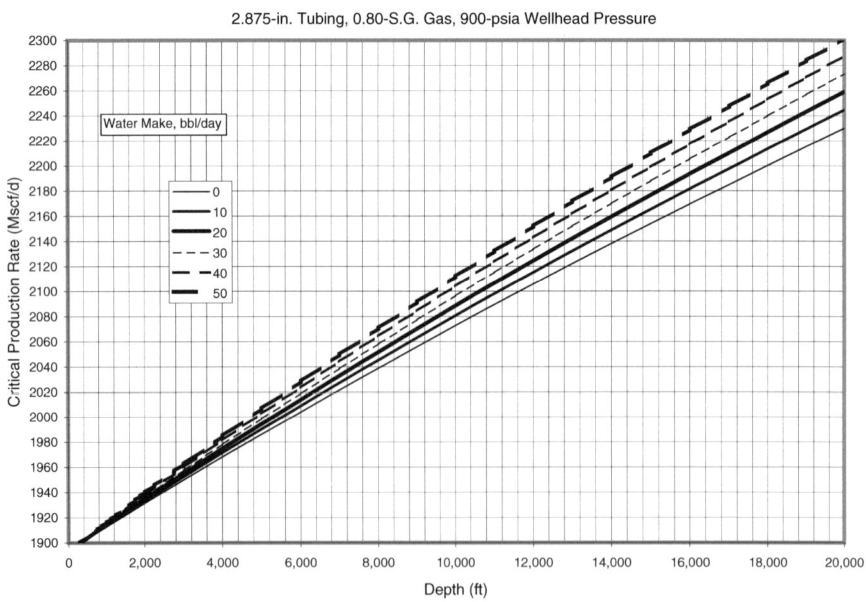

Figure D–70 **Critical gas production rate for water removal in 2.875-in tubing against 900 psia wellhead pressure, S.G. 0.80 gas.**

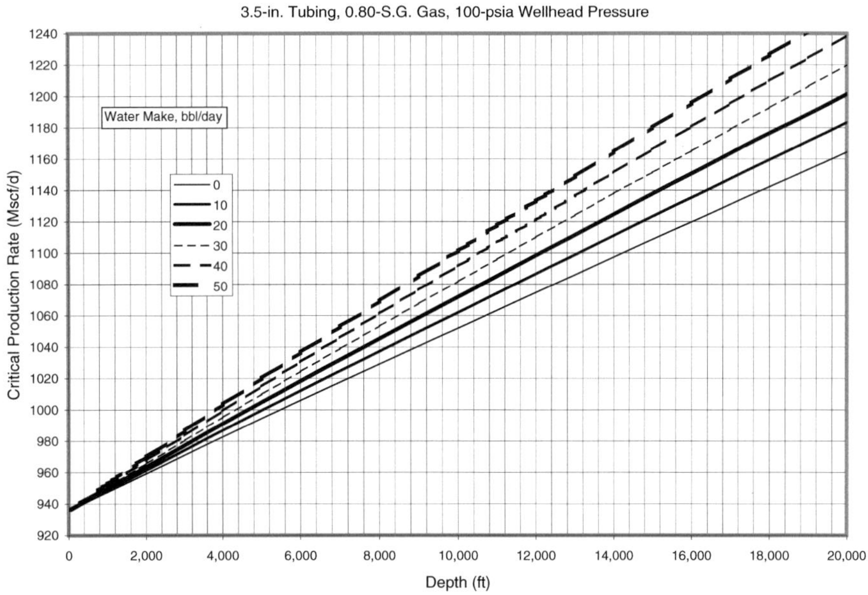

Figure D–71 **Critical gas production rate for water removal in 3.5-in tubing against 100 psia wellhead pressure, S.G. 0.80 gas.**

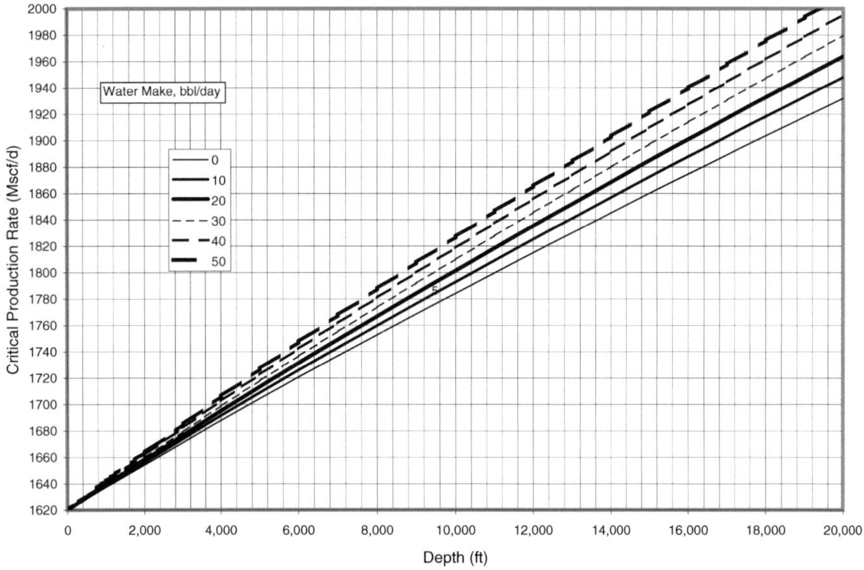

Figure D–72 **Critical gas production rate for water removal in 3.5-in tubing against 300 psia wellhead pressure, S.G. 0.80 gas.**

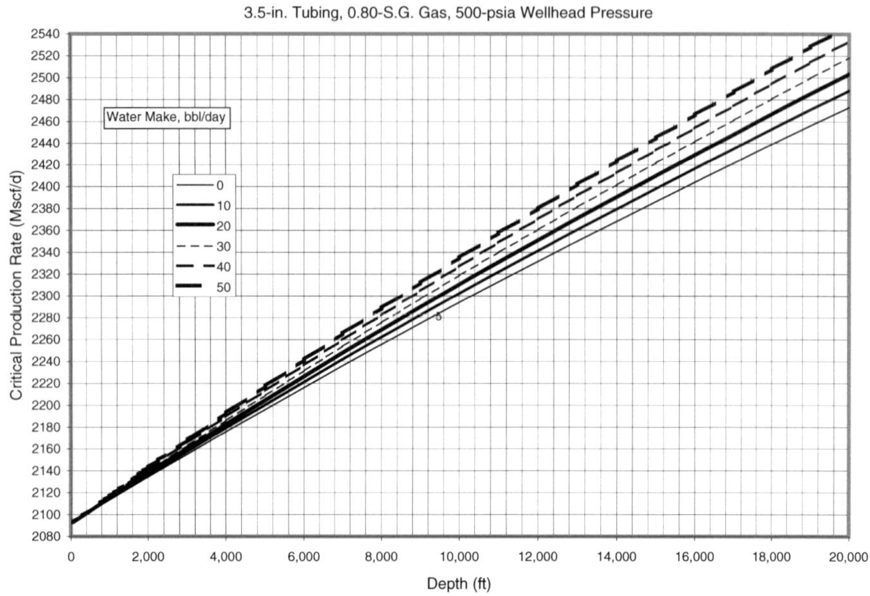

Figure D–73 **Critical gas production rate for water removal in 3.5-in tubing against 500 psia wellhead pressure, S.G. 0.80 gas.**

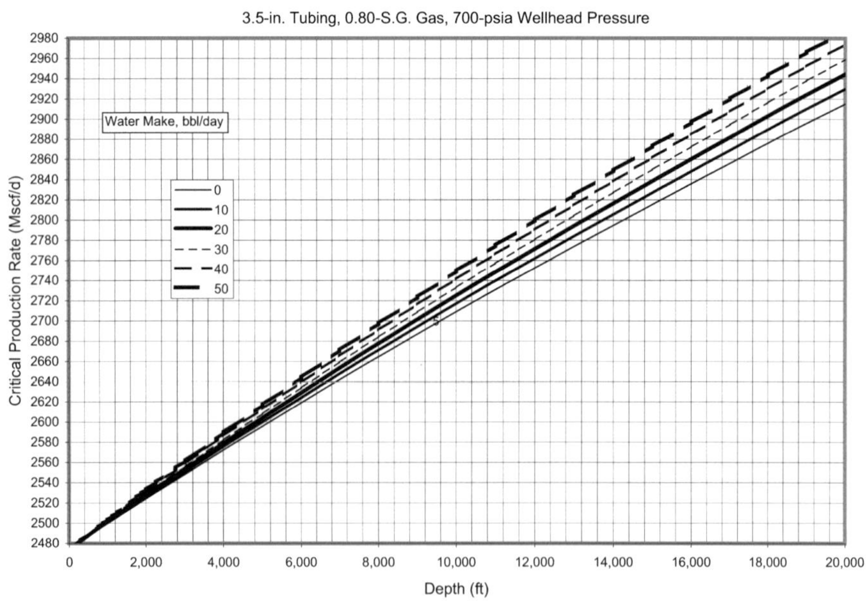

Figure D–74 **Critical gas production rate for water removal in 3.5-in tubing against 700 psia wellhead pressure, S.G. 0.80 gas.**

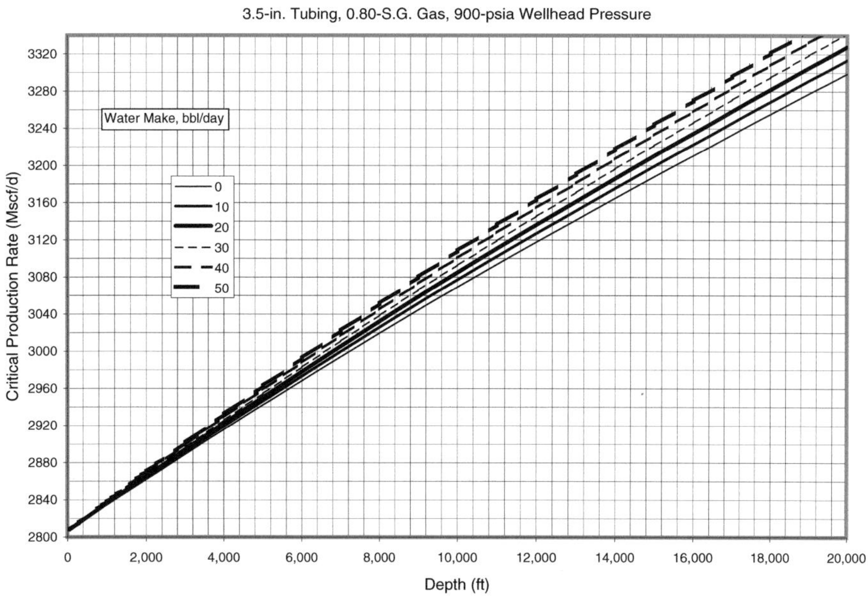

Figure D–75 **Critical gas production rate for water removal in 3.5-in tubing against 900 psia wellhead pressure, S.G. 0.80 gas.**

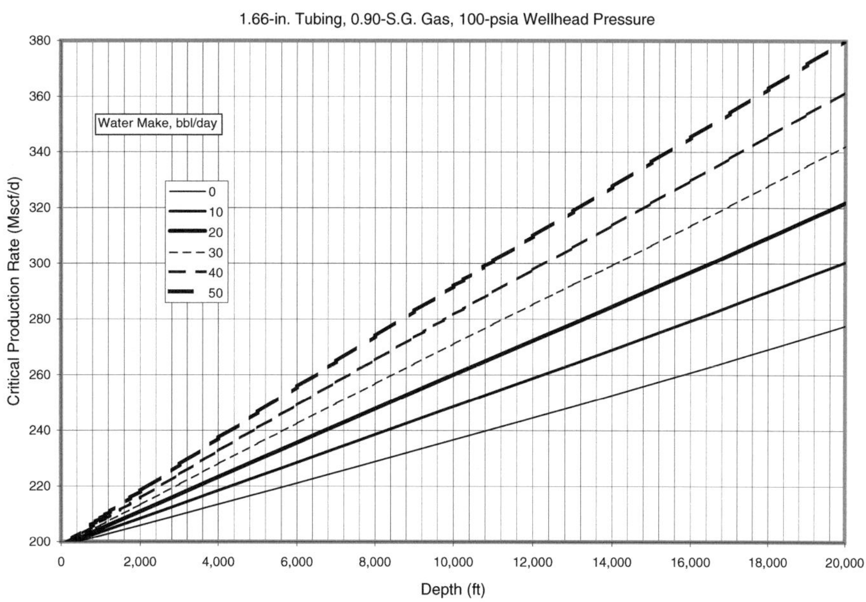

Figure D–76 **Critical gas production rate for water removal in 1.66-in tubing against 100 psia wellhead pressure, S.G. 0.90 gas.**

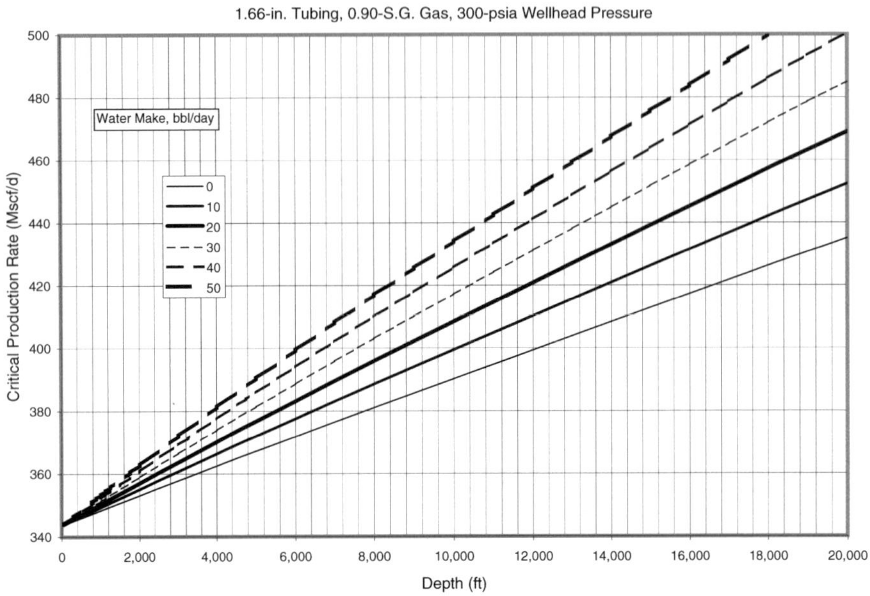

Figure D–77 Critical gas production rate for water removal in 1.66-in tubing against 300 psia wellhead pressure, S.G. 0.90 gas.

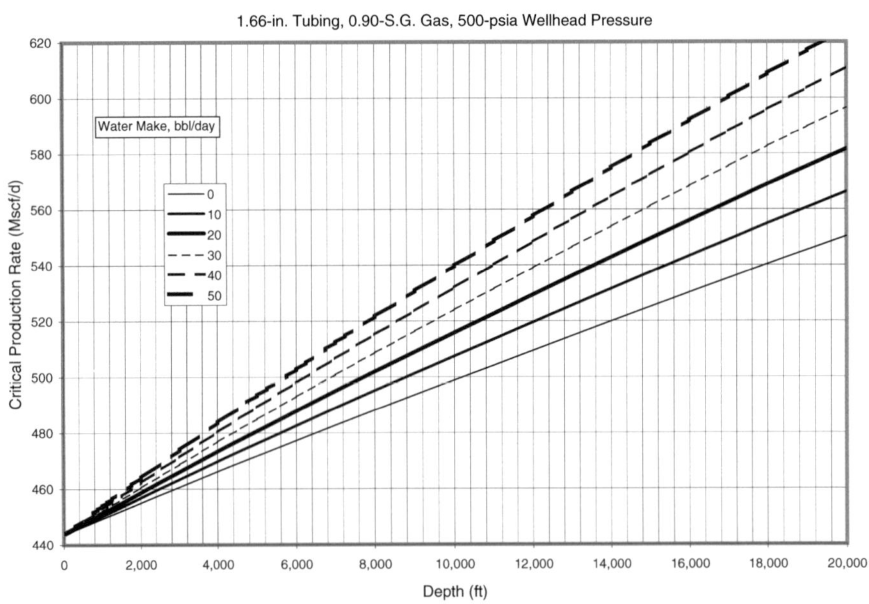

Figure D–78 Critical gas production rate for water removal in 1.66-in tubing against 500 psia wellhead pressure, S.G. 0.90 gas.

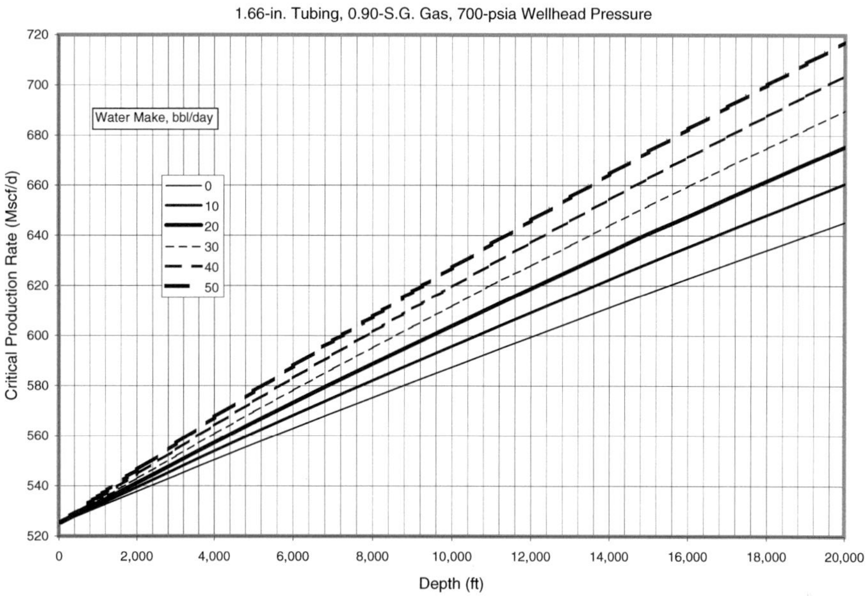

Figure D–79 **Critical gas production rate for water removal in 1.66-in tubing against 700 psia wellhead pressure, S.G. 0.90 gas.**

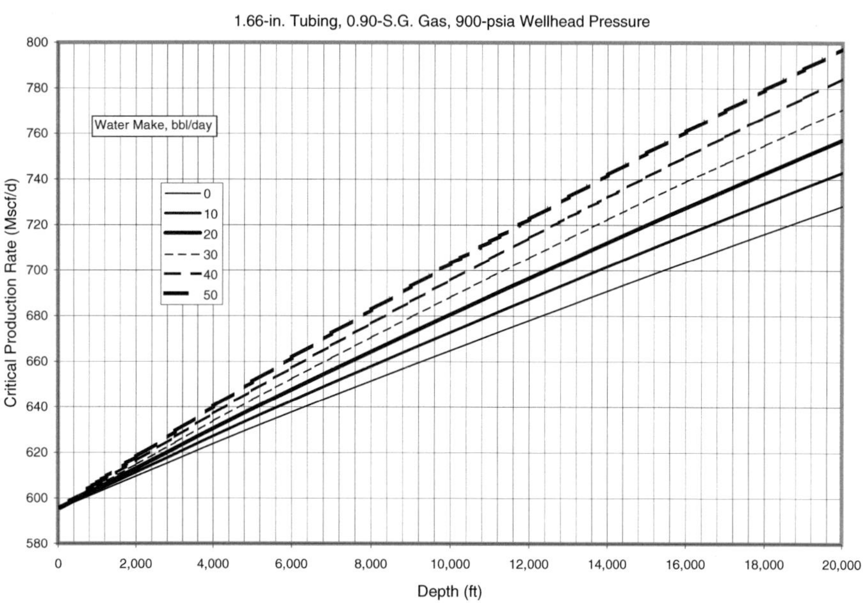

Figure D–80 **Critical gas production rate for water removal in 1.66-in tubing against 900 psia wellhead pressure, S.G. 0.90 gas.**

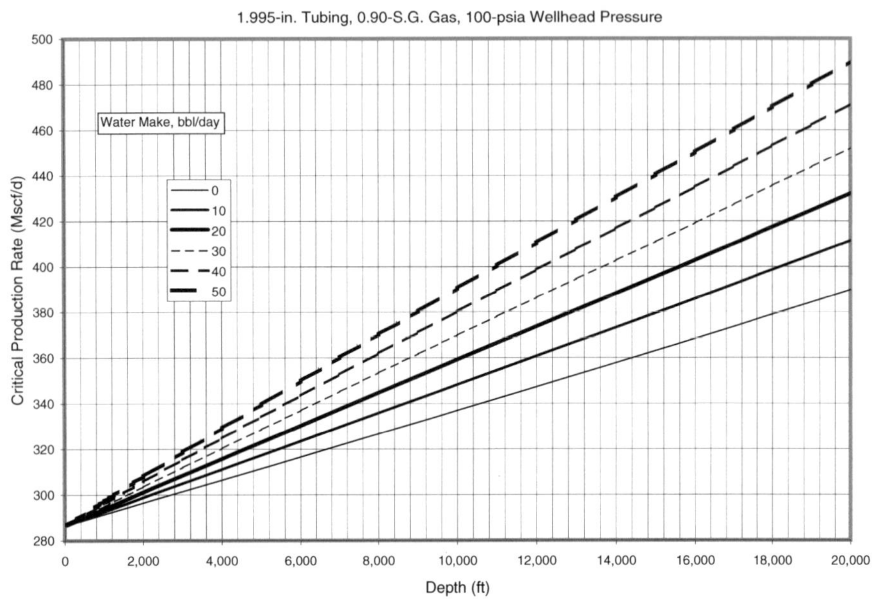

Figure D–81 **Critical gas production rate for water removal in 1.995-in tubing against 100 psia wellhead pressure, S.G. 0.90 gas.**

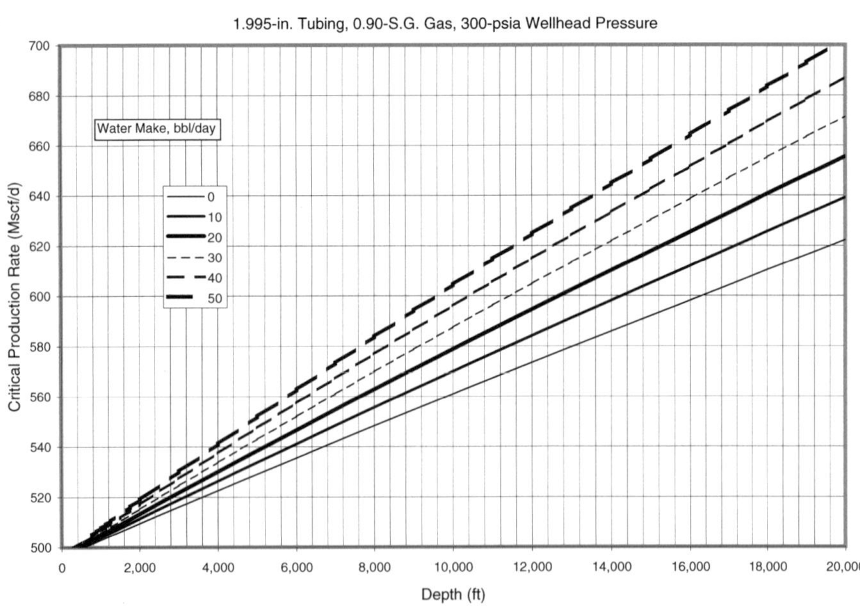

Figure D–82 **Critical gas production rate for water removal in 1.995-in tubing against 300 psia wellhead pressure, S.G. 0.90 gas.**

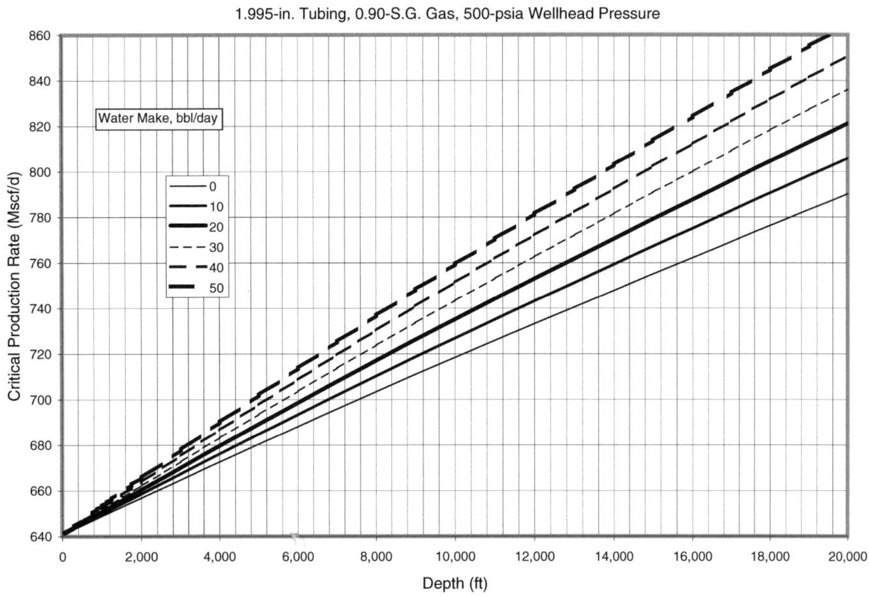

Figure D–83 Critical gas production rate for water removal in 1.995-in tubing against 500 psia wellhead pressure, S.G. 0.90 gas.

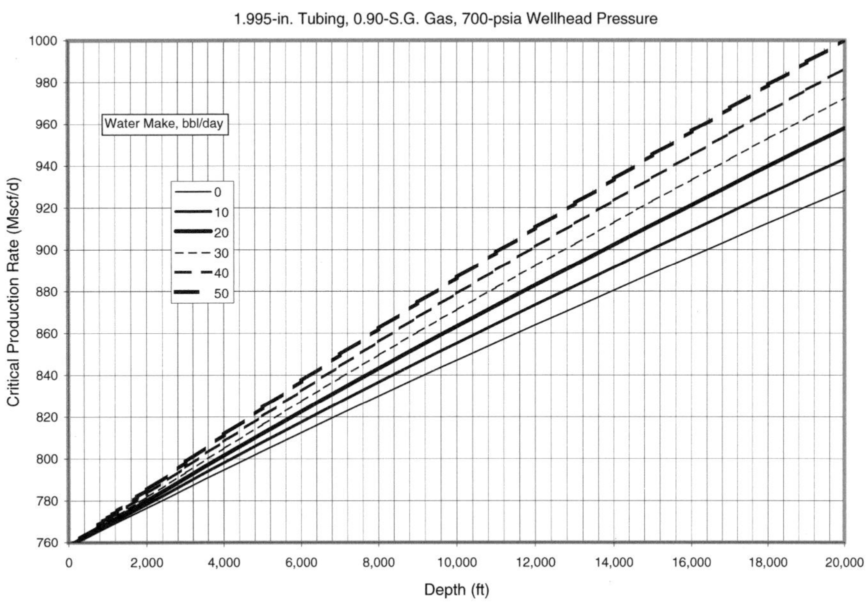

Figure D–84 Critical gas production rate for water removal in 1.995-in tubing against 700 psia wellhead pressure, S.G. 0.90 gas.

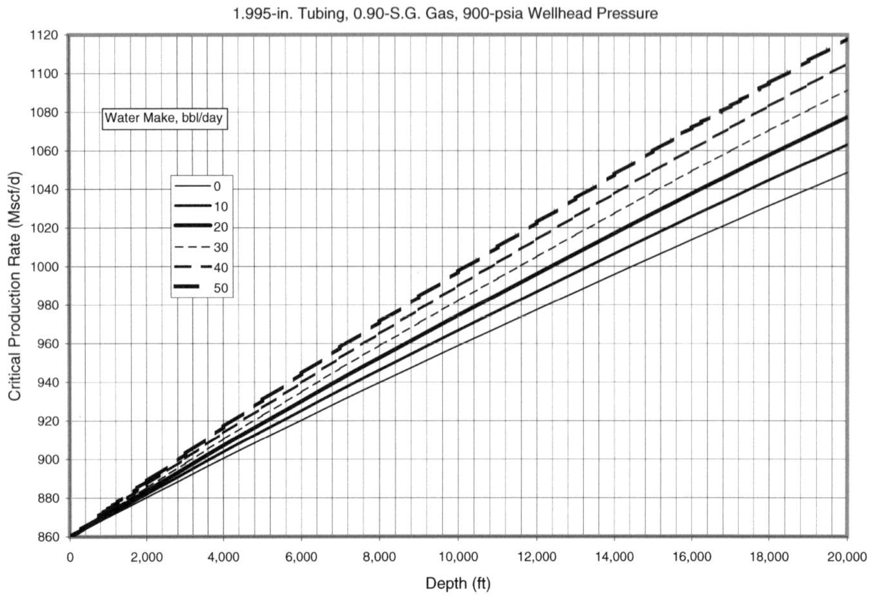

Figure D–85 **Critical gas production rate for water removal in 1.995-in tubing against 900 psia wellhead pressure, S.G. 0.90 gas.**

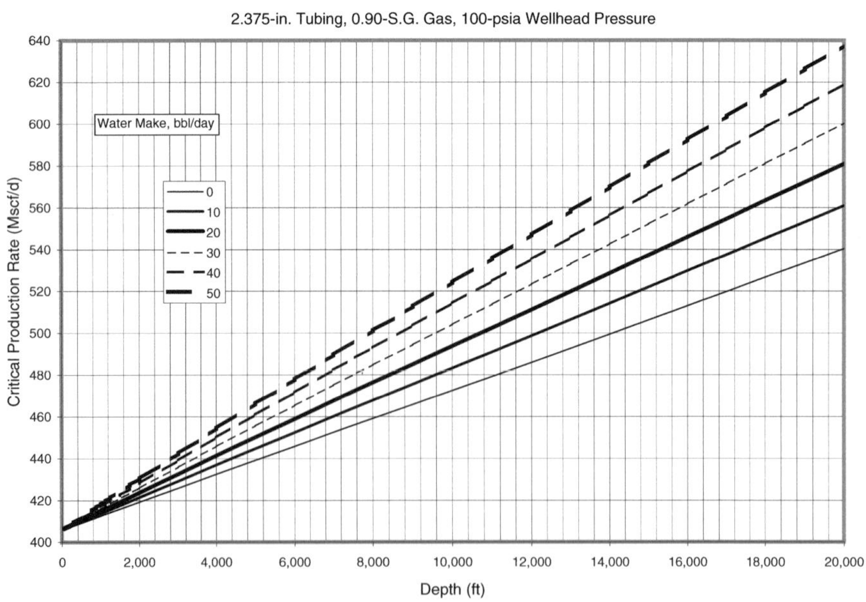

Figure D–86 **Critical gas production rate for water removal in 2.375-in tubing against 100 psia wellhead pressure, S.G. 0.90 gas.**

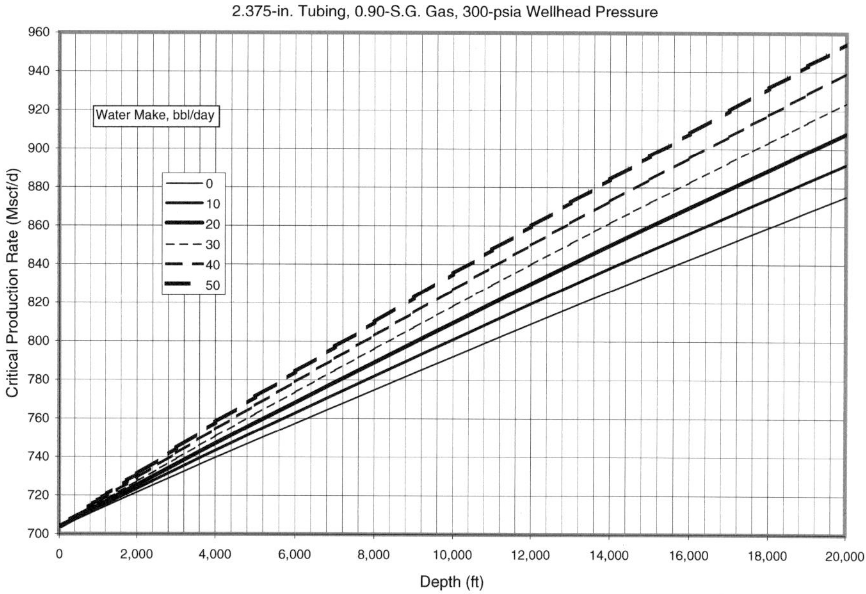

Figure D–87 **Critical gas production rate for water removal in 2.375-in tubing against 300 psia wellhead pressure, S.G. 0.90 gas.**

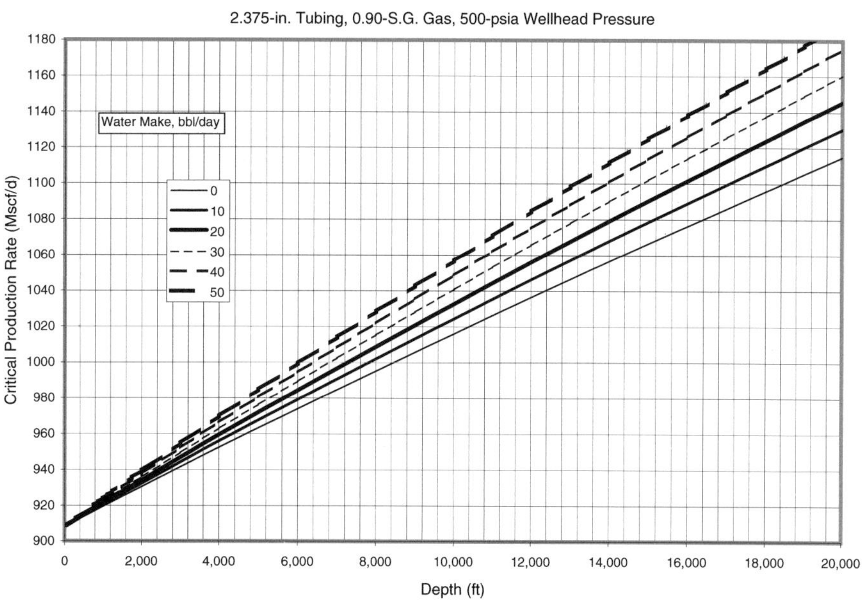

Figure D–88 **Critical gas production rate for water removal in 2.375-in tubing against 500 psia wellhead pressure, S.G. 0.90 gas.**

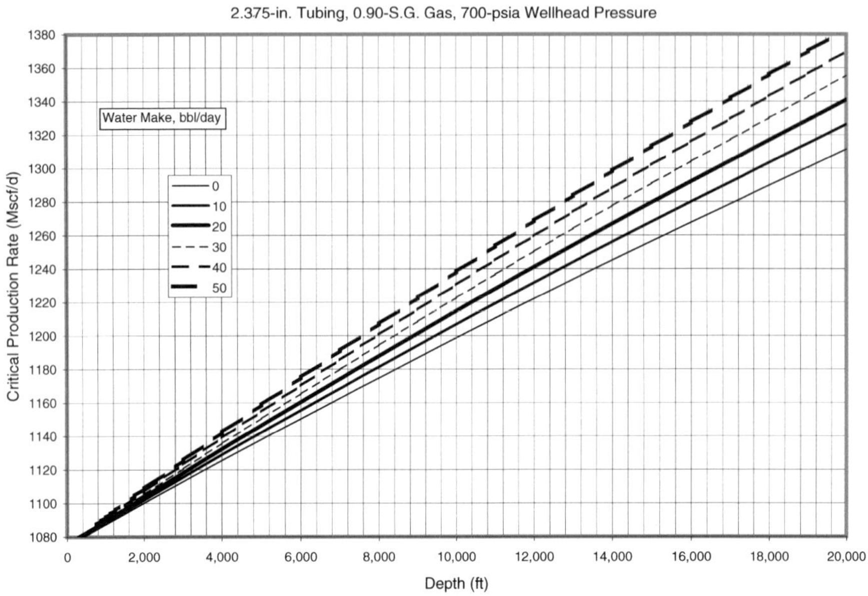

Figure D–89 **Critical gas production rate for water removal in 2.375-in tubing against 700 psia wellhead pressure, S.G. 0.90 gas.**

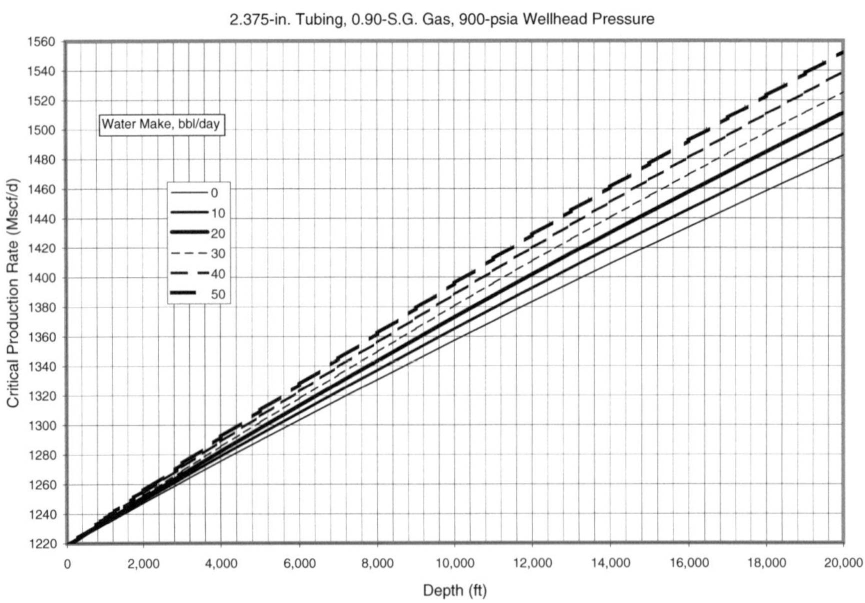

Figure D–90 **Critical gas production rate for water removal in 2.375-in tubing against 900 psia wellhead pressure, S.G. 0.90 gas.**

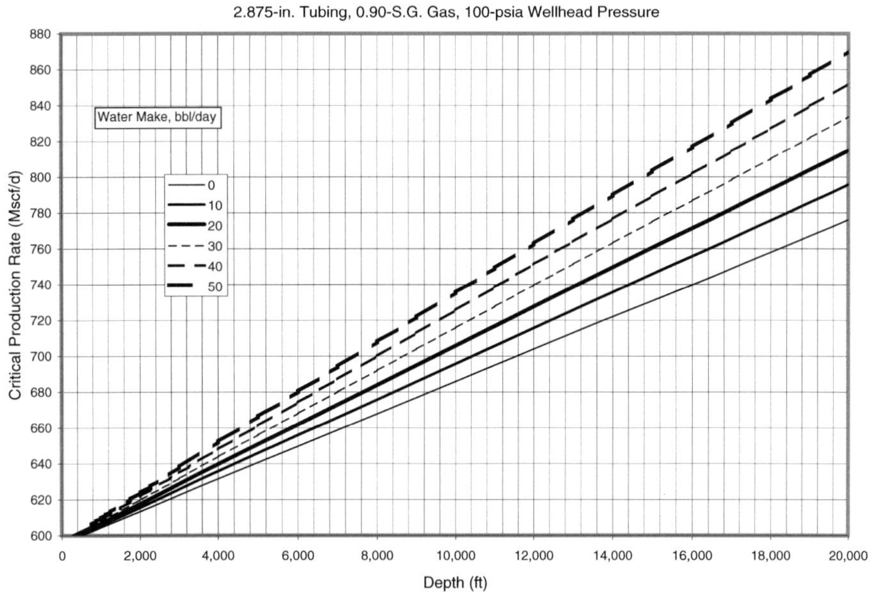

Figure D–91 **Critical gas production rate for water removal in 2.875-in tubing against 100 psia wellhead pressure, S.G. 0.90 gas.**

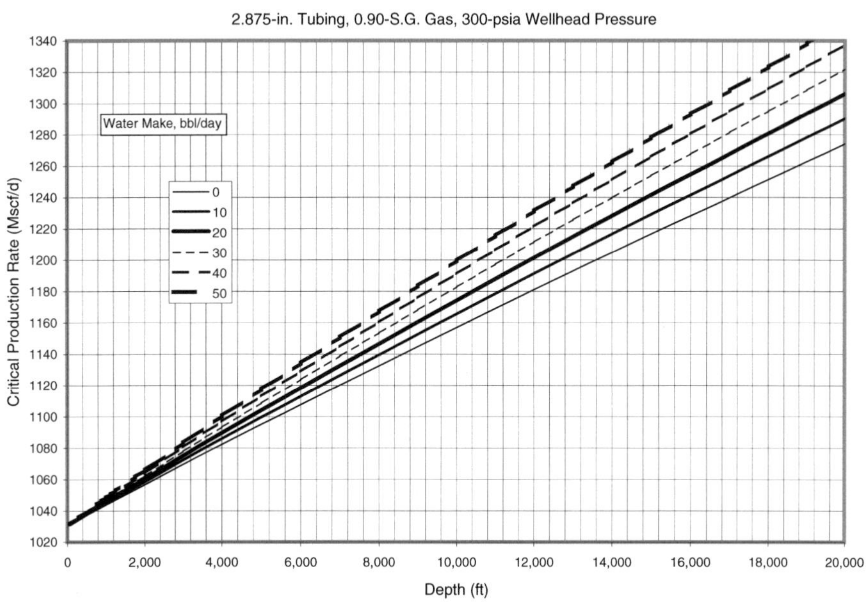

Figure D–92 **Critical gas production rate for water removal in 2.875-in tubing against 300 psia wellhead pressure, S.G. 0.90 gas.**

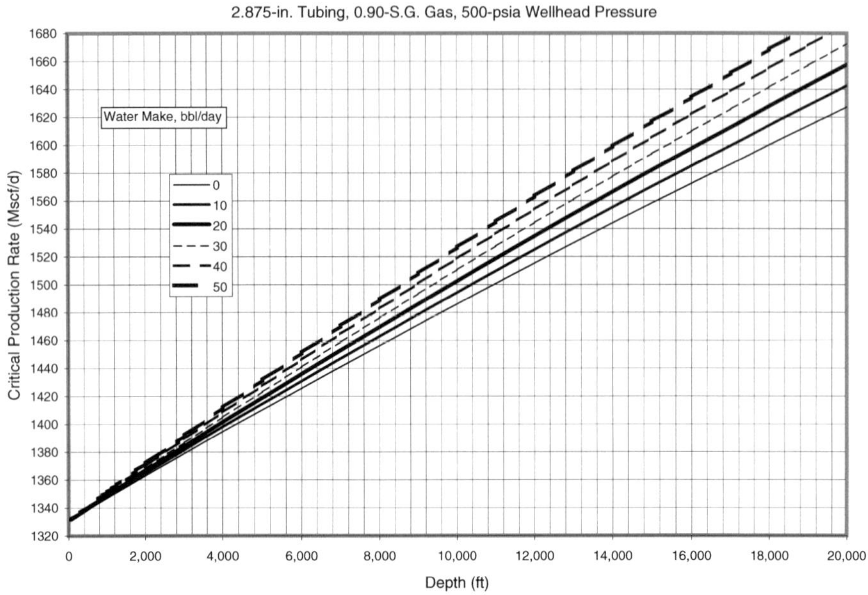

Figure D–93 **Critical gas production rate for water removal in 2.875-in tubing against 500 psia wellhead pressure, S.G. 0.90 gas.**

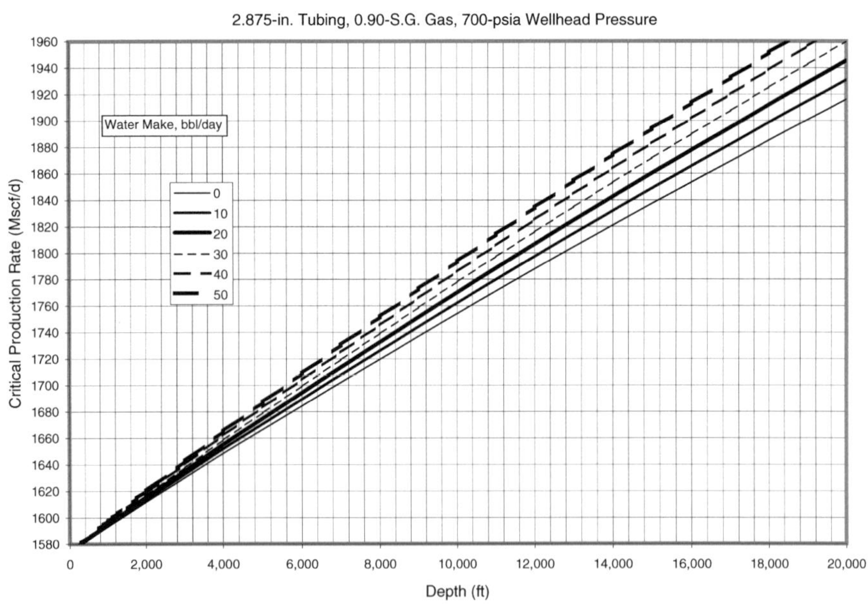

Figure D–94 **Critical gas production rate for water removal in 2.875-in tubing against 700 psia wellhead pressure, S.G. 0.90 gas.**

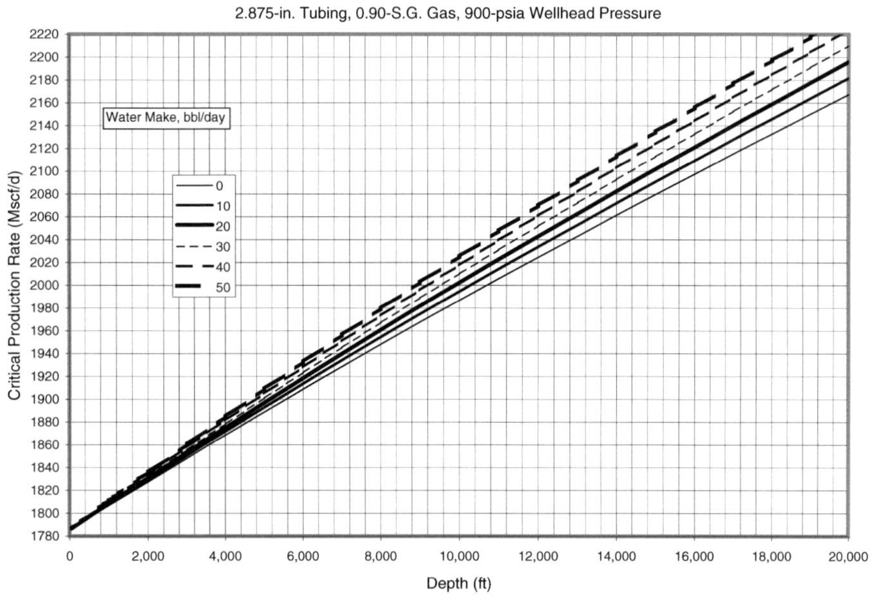

Figure D–95 **Critical gas production rate for water removal in 2.875-in tubing against 900 psia wellhead pressure, S.G. 0.90 gas.**

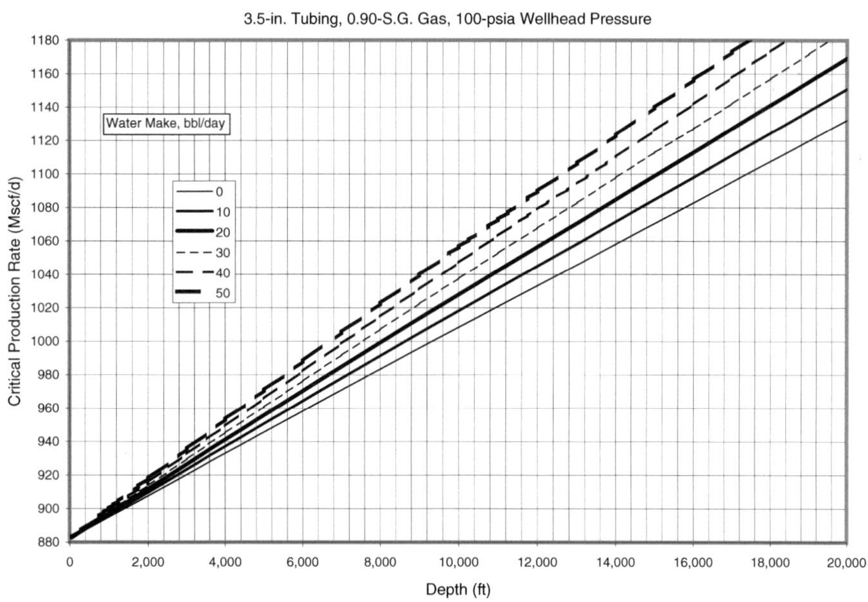

Figure D–96 **Critical gas production rate for water removal in 3.5-in tubing against 100 psia wellhead pressure, S.G. 0.90 gas.**

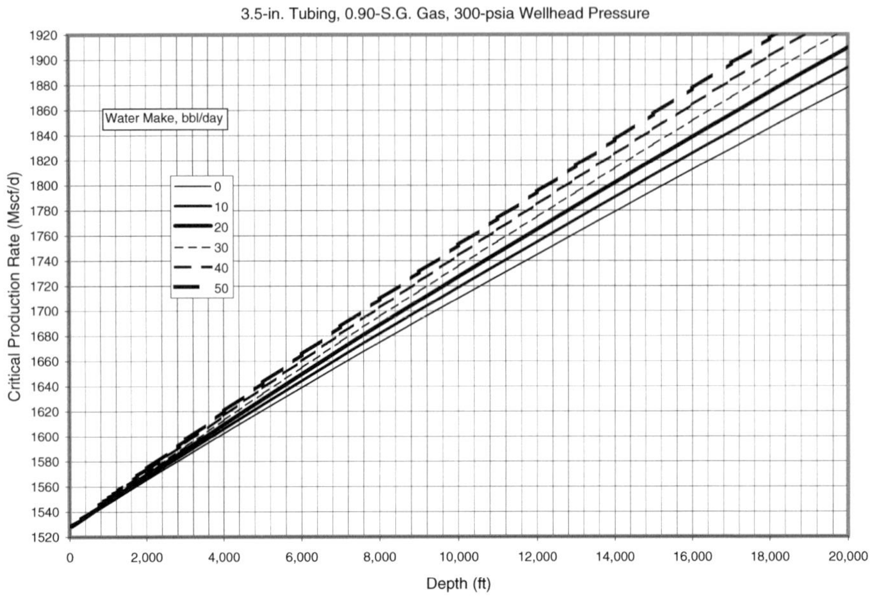

Figure D–97 **Critical gas production rate for water removal in 3.5-in tubing against 300 psia wellhead pressure, S.G. 0.90 gas.**

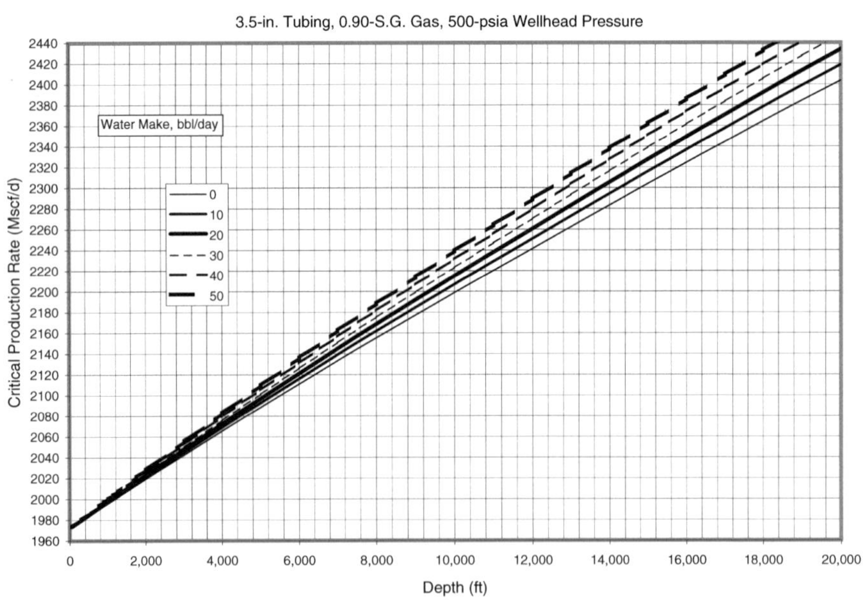

Figure D–98 **Critical gas production rate for water removal in 3.5-in tubing against 500 psia wellhead pressure, S.G. 0.90 gas.**

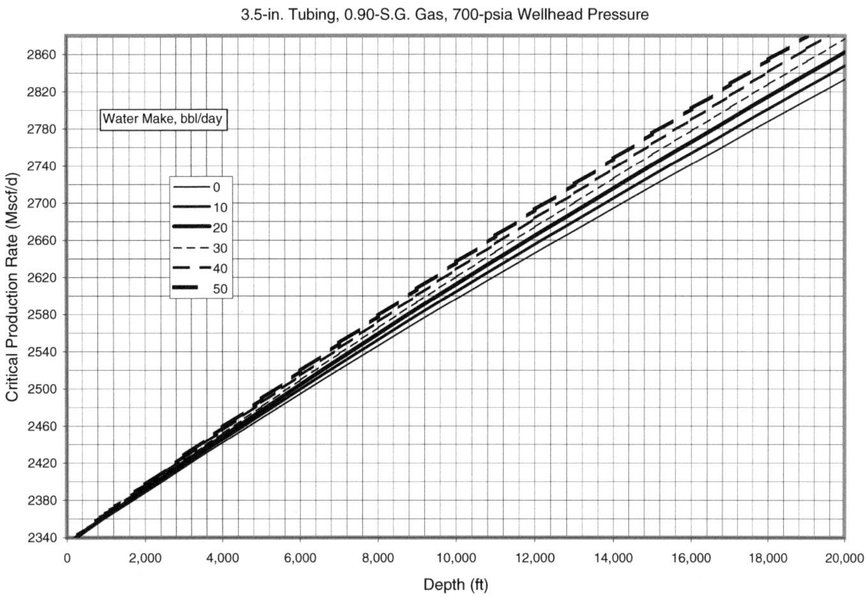

Figure D–99 **Critical gas production rate for water removal in 3.5-in tubing against 700 psia wellhead pressure, S.G. 0.90 gas.**

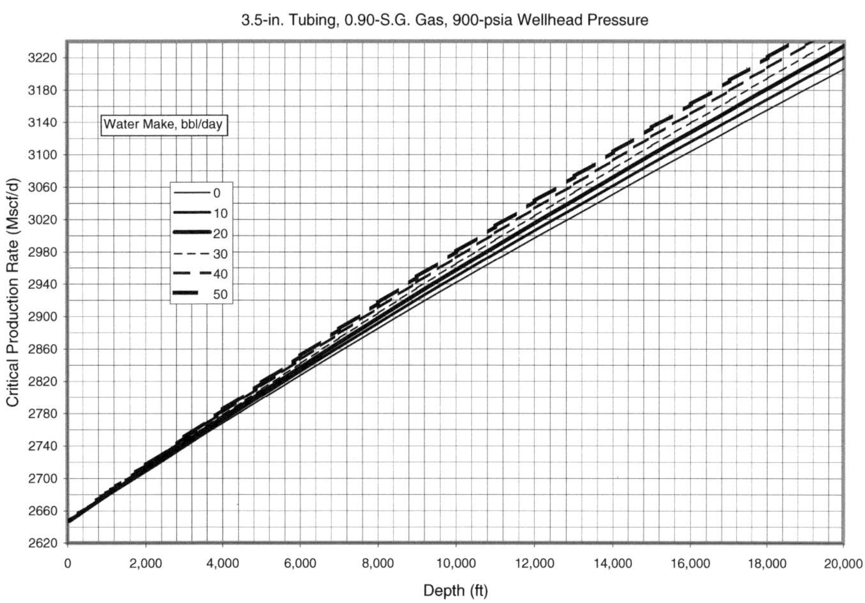

Figure D–100 **Critical gas production rate for water removal in 3.5-in tubing against 900 psia wellhead pressure, S.G. 0.90 gas.**

Appendix E

The Minimum Gas Production Rate for Condensate Removal in Gas Wells

Data presented in this section were generated with Guo's method on the basis of the following parameter values:

Gas specific gravity:	0.60 to 0.90 air =1
Hole inclination:	0°
Wellhead pressure:	100 to 900 psi
Wellhead temperature:	60 °F
Geothermal gradient:	0.01 °F/ft
Condensate gravity:	70 API
Condensate production rate:	0 to 50 bbl/day
Water specific gravity:	1.08 water = 1
Water production rate:	0
Solid make:	0
Tubing ID:	1.66 to 3.5 in
Conduit wall roughness:	0.000015 in
Maximum interfacial tension:	60 dyne/cm

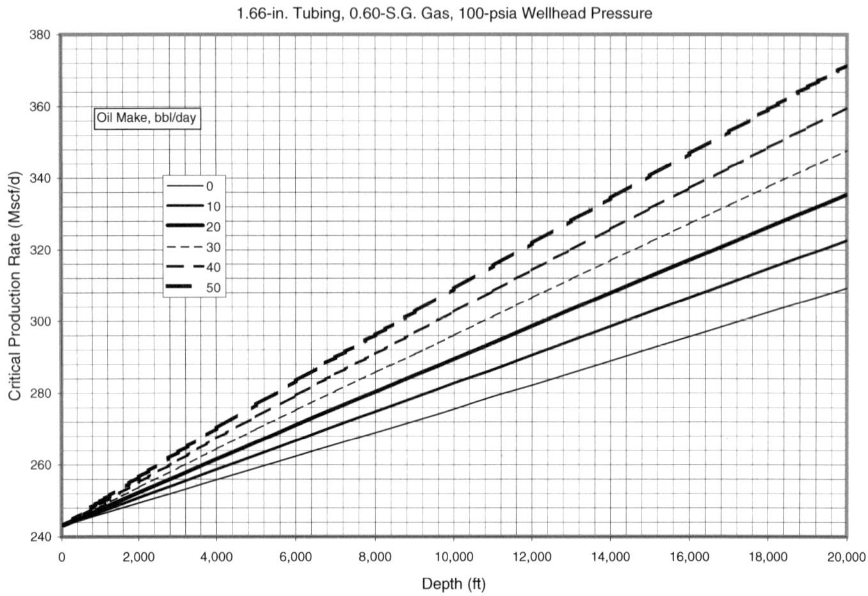

Figure E–1 **Critical gas production rate for condensate removal in 1.66-in tubing against 100 psia wellhead pressure, S.G. 0.60 gas.**

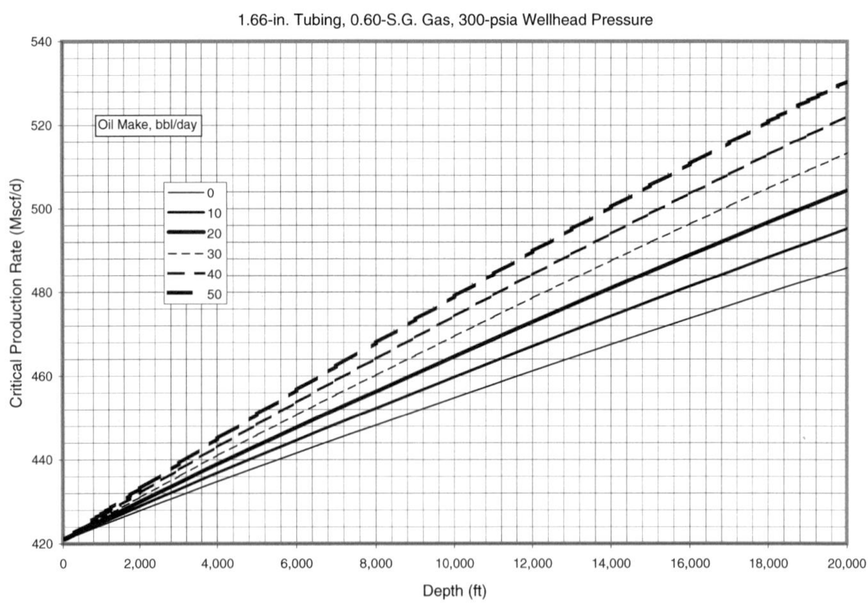

Figure E–2 **Critical gas production rate for condensate removal in 1.66-in tubing against 300 psia wellhead pressure, S.G. 0.60 gas.**

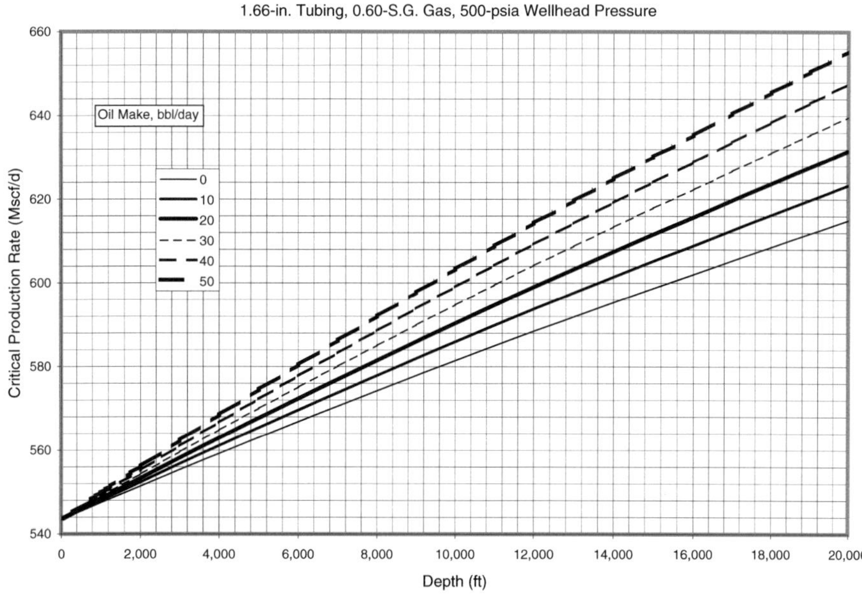

Figure E–3 **Critical gas production rate for condensate removal in 1.66-in tubing against 500 psia wellhead pressure, S.G. 0.60 gas.**

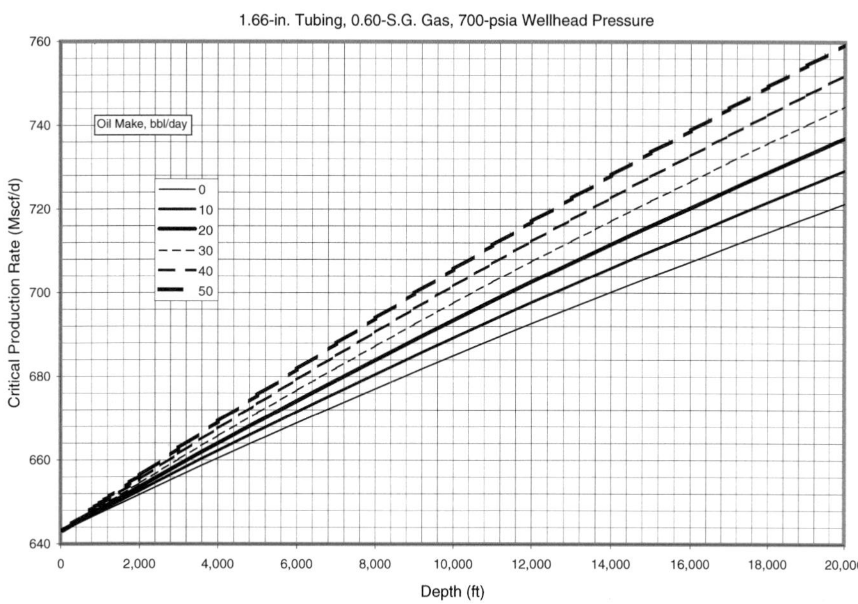

Figure E–4 **Critical gas production rate for condensate removal in 1.66-in tubing against 700 psia wellhead pressure, S.G. 0.60 gas.**

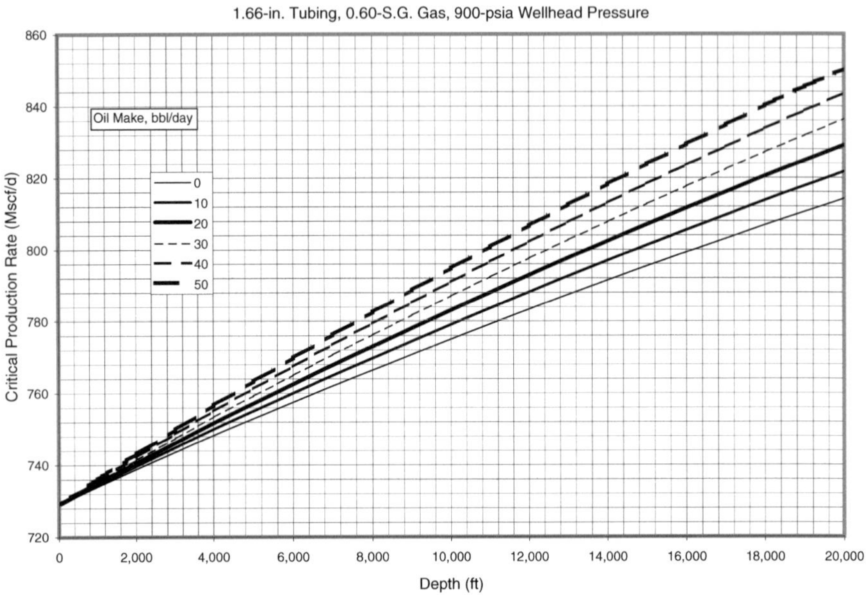

Figure E–5 **Critical gas production rate for condensate removal in 1.66-in tubing against 900 psia wellhead pressure, S.G. 0.60 gas.**

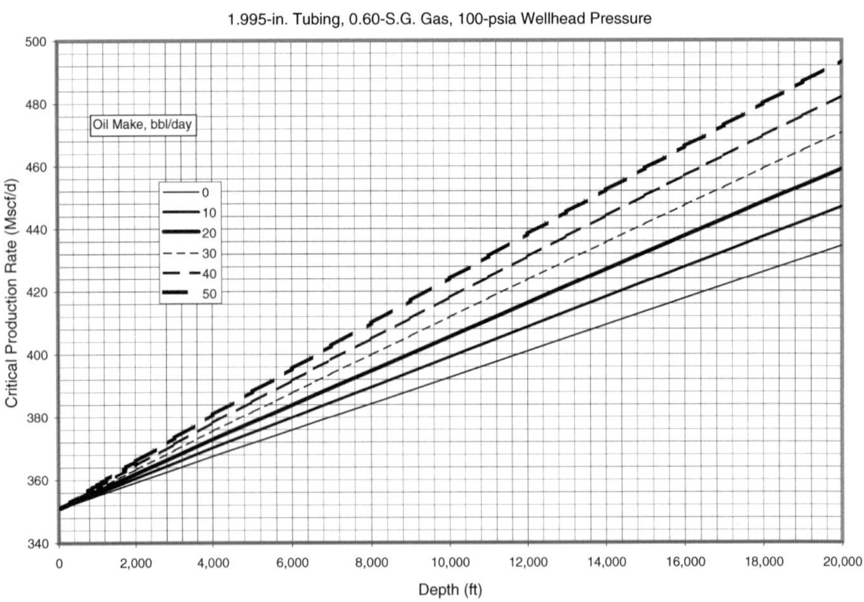

Figure E–6 **Critical gas production rate for condensate removal in 1.995-in tubing against 100 psia wellhead pressure, S.G. 0.60 gas.**

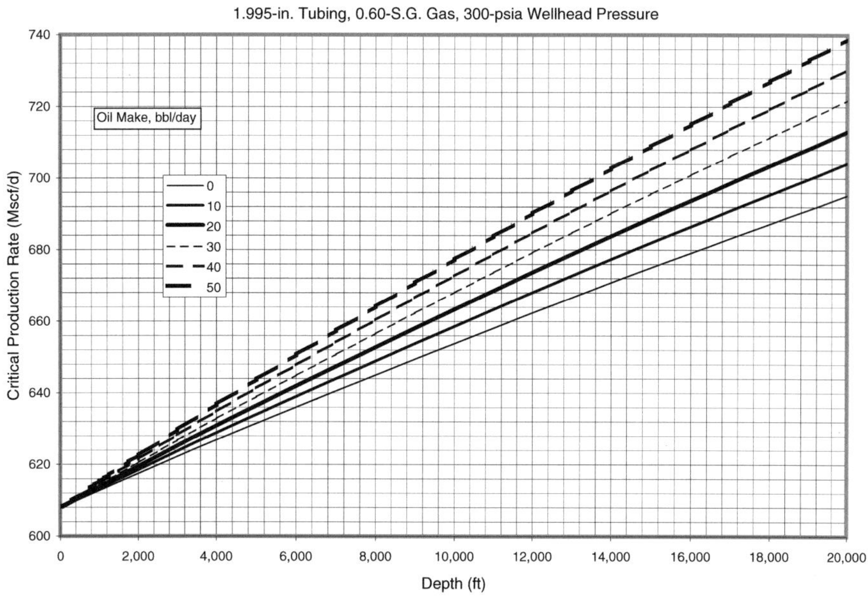

Figure E–7 Critical gas production rate for condensate removal in 1.995-in tubing against 300 psia wellhead pressure, S.G. 0.60 gas.

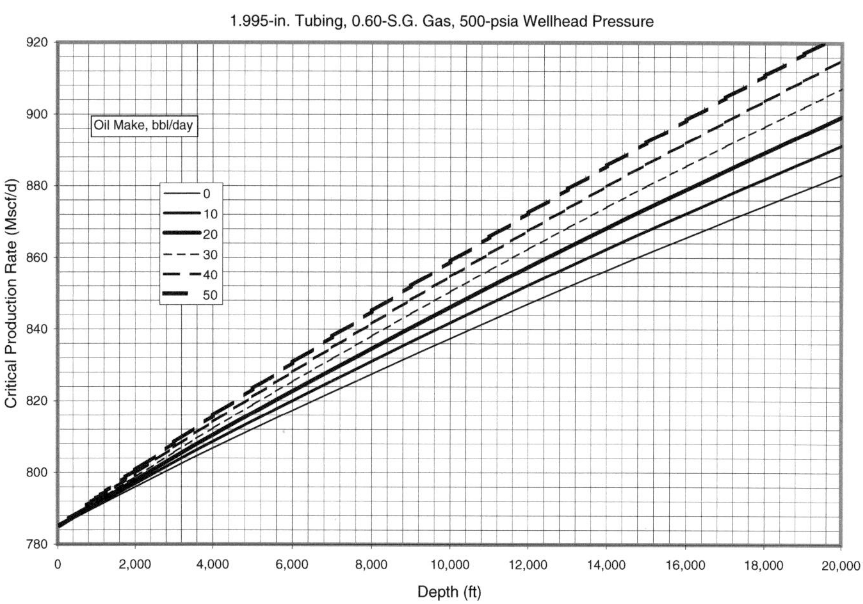

Figure E–8 Critical gas production rate for condensate removal in 1.995-in tubing against 500 psia wellhead pressure, S.G. 0.60 gas.

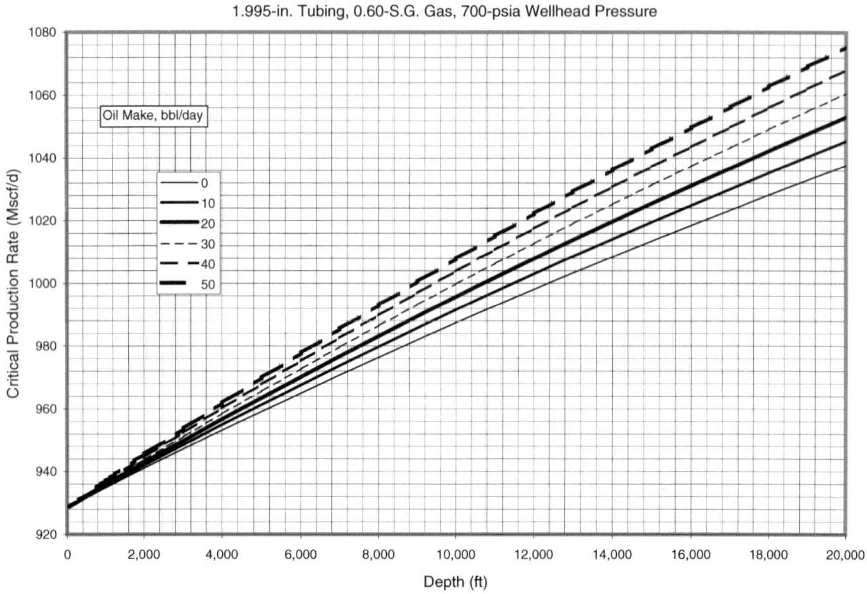

Figure E–9 Critical gas production rate for condensate removal in 1.995-in tubing against 700 psia wellhead pressure, S.G. 0.60 gas.

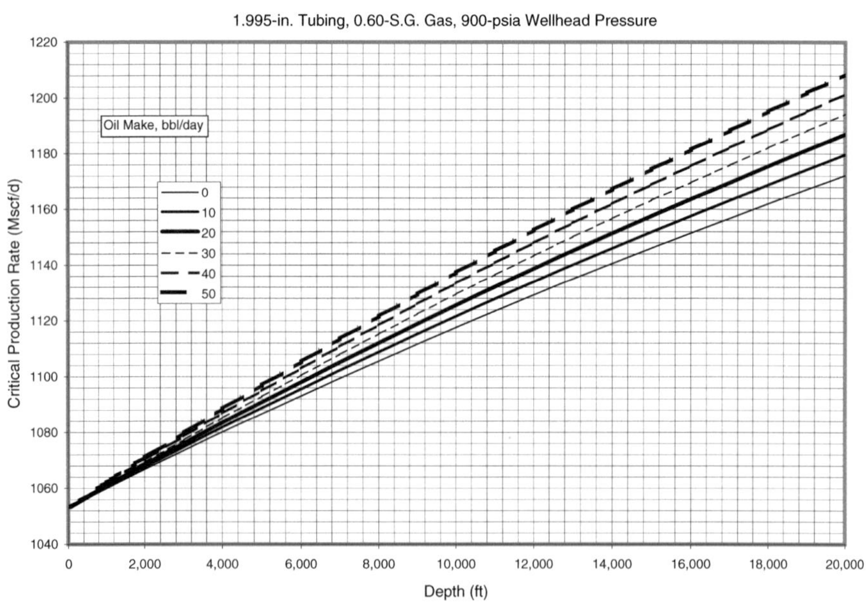

Figure E–10 Critical gas production rate for condensate removal in 1.995-in tubing against 900 psia wellhead pressure, S.G. 0.60 gas.

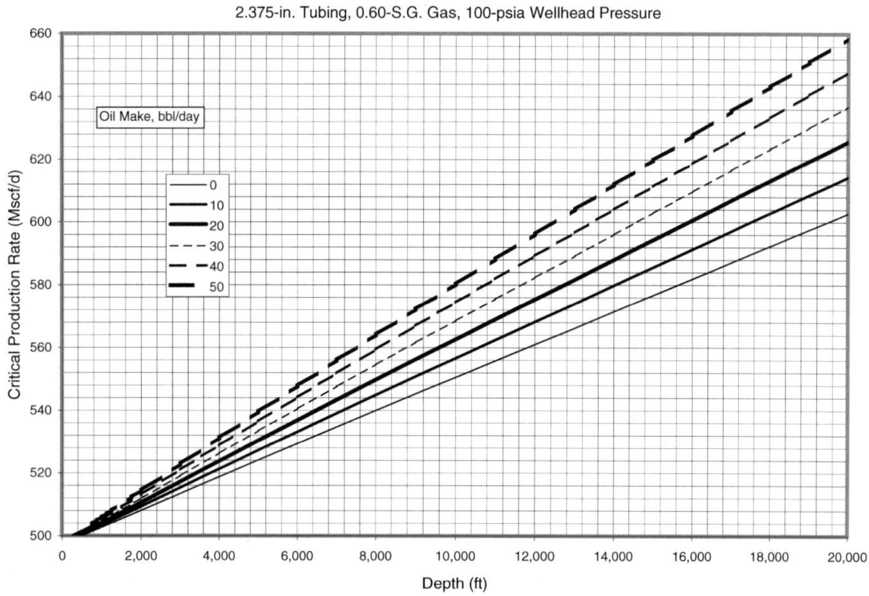

Figure E–11 Critical gas production rate for condensate removal in 2.375-in tubing against 100 psia wellhead pressure, S.G. 0.60 gas.

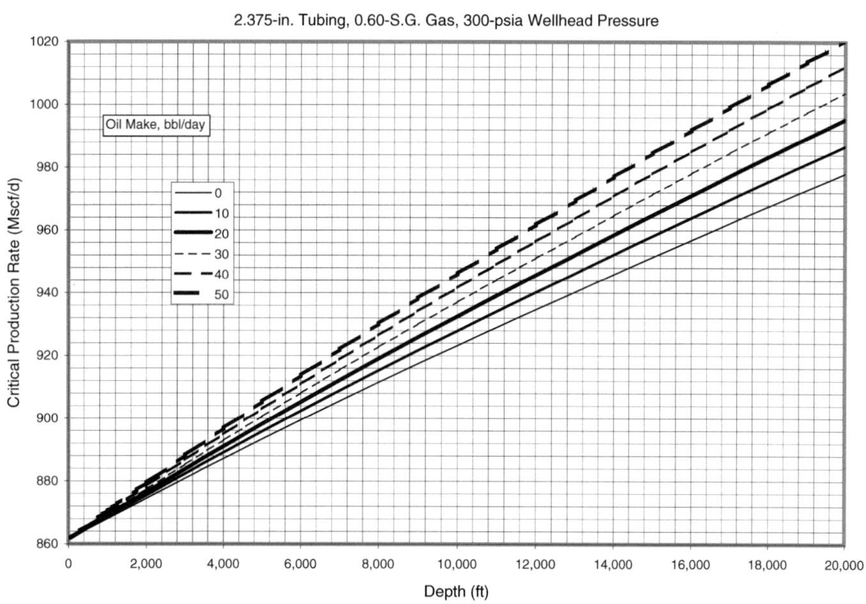

Figure E–12 Critical gas production rate for condensate removal in 2.375-in tubing against 300 psia wellhead pressure, S.G. 0.60 gas.

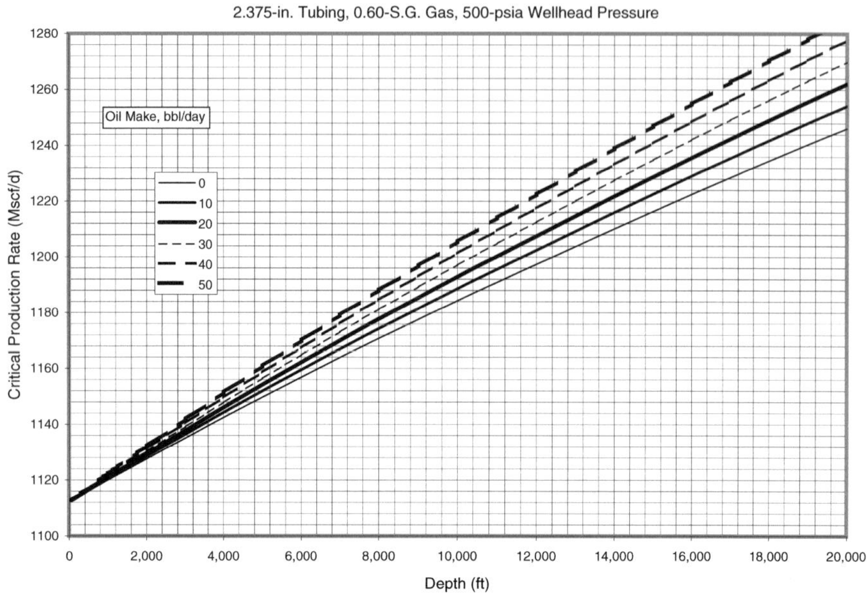

Figure E–13 Critical gas production rate for condensate removal in 2.375-in tubing against 500 psia wellhead pressure, S.G. 0.60 gas.

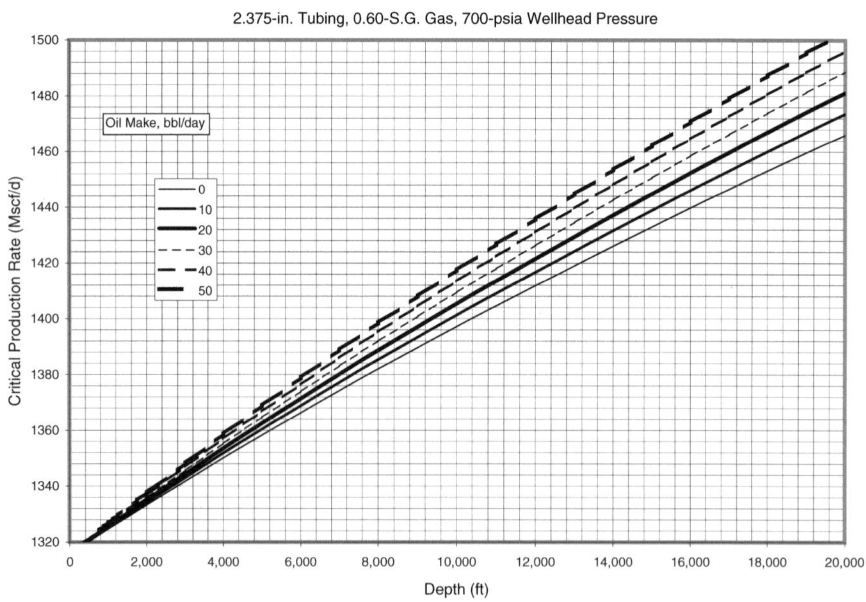

Figure E–14 Critical gas production rate for condensate removal in 2.375-in tubing against 700 psia wellhead pressure, S.G. 0.60 gas.

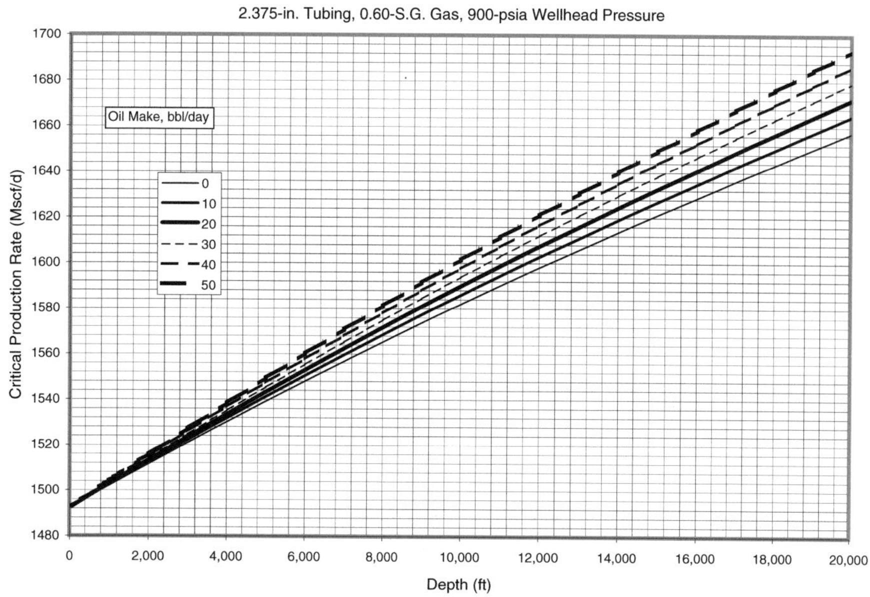

Figure E–15 Critical gas production rate for condensate removal in 2.375-in tubing against 900 psia wellhead pressure, S.G. 0.60 gas.

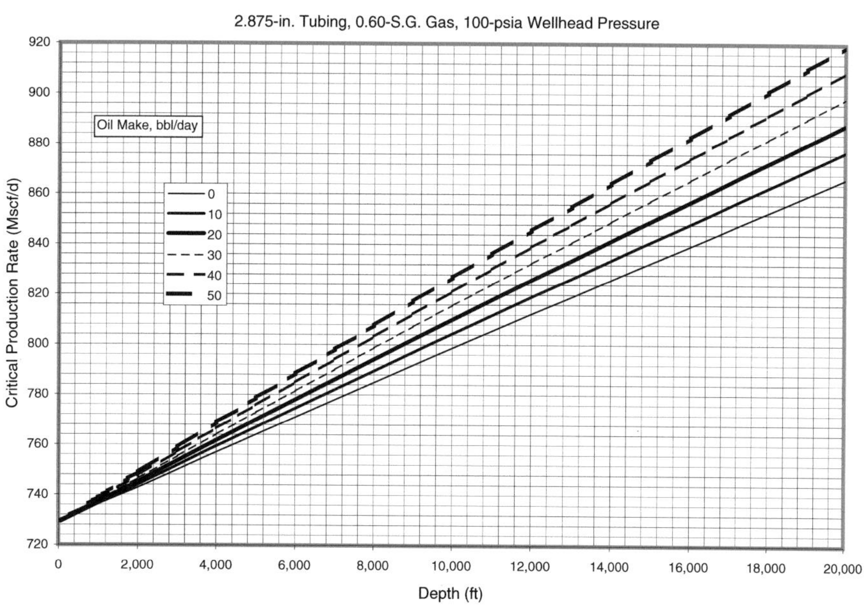

Figure E–16 Critical gas production rate for condensate removal in 2.875-in tubing against 100 psia wellhead pressure, S.G. 0.60 gas.

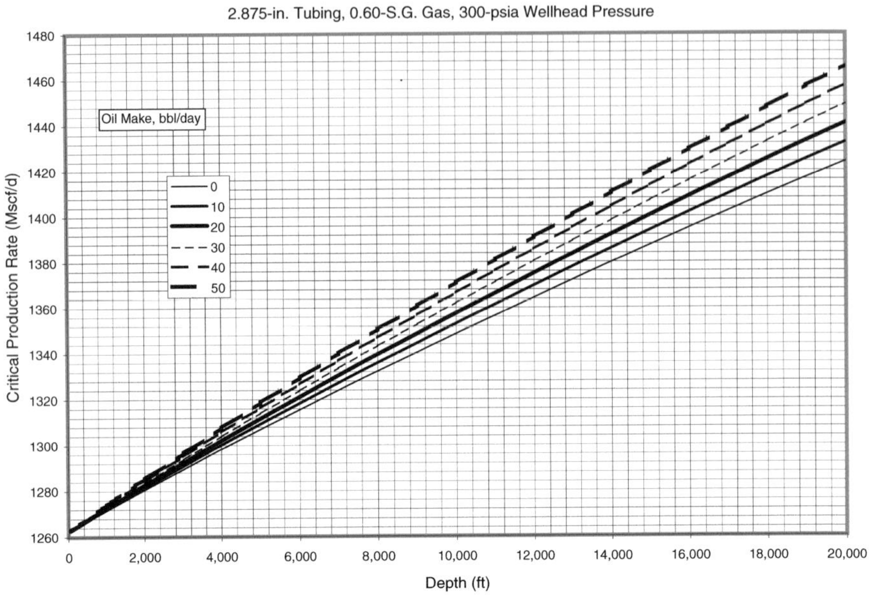

Figure E–17 Critical gas production rate for condensate removal in 2.875-in tubing against 300 psia wellhead pressure, S.G. 0.60 gas.

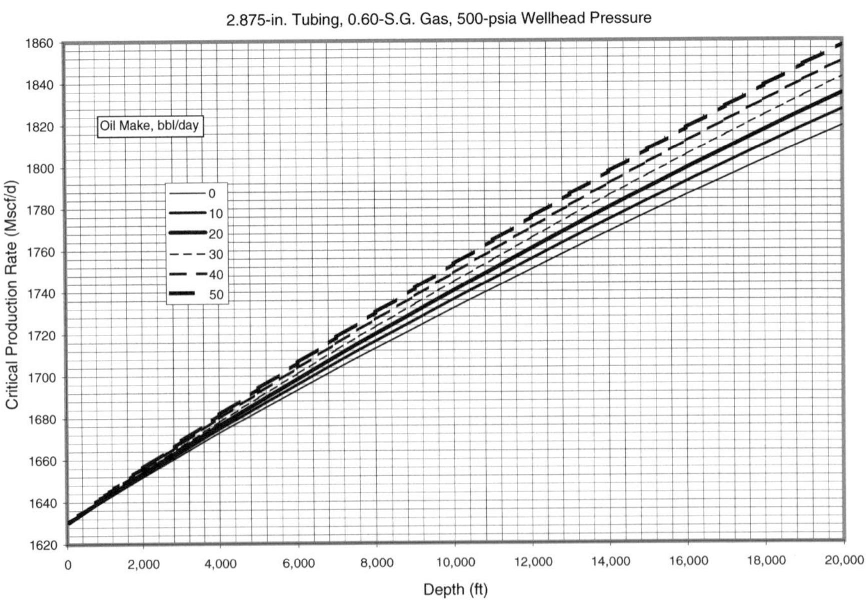

Figure E–18 Critical gas production rate for condensate removal in 2.875-in tubing against 500 psia wellhead pressure, S.G. 0.60 gas.

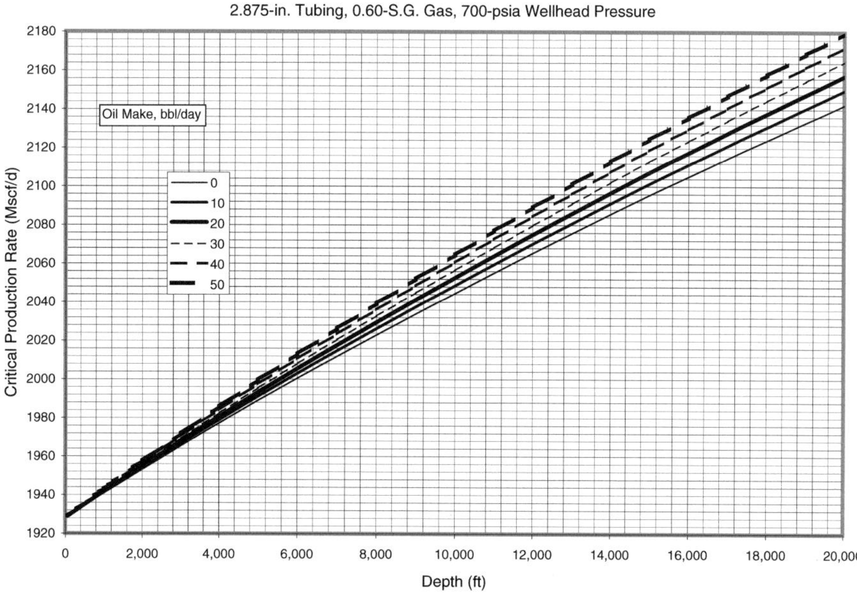

Figure E–19 **Critical gas production rate for condensate removal in 2.875-in tubing against 700 psia wellhead pressure, S.G. 0.60 gas.**

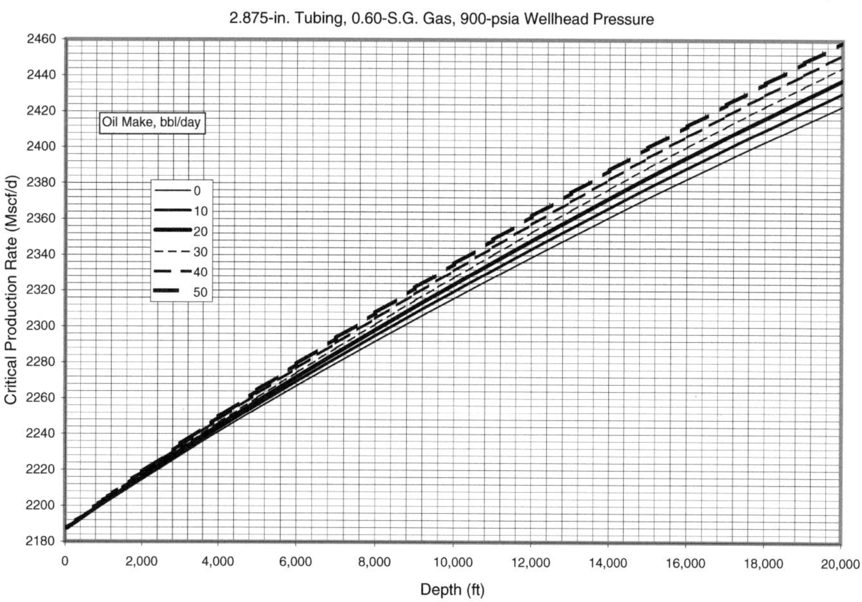

Figure E–20 **Critical gas production rate for condensate removal in 2.875-in tubing against 900 psia wellhead pressure, S.G. 0.60 gas.**

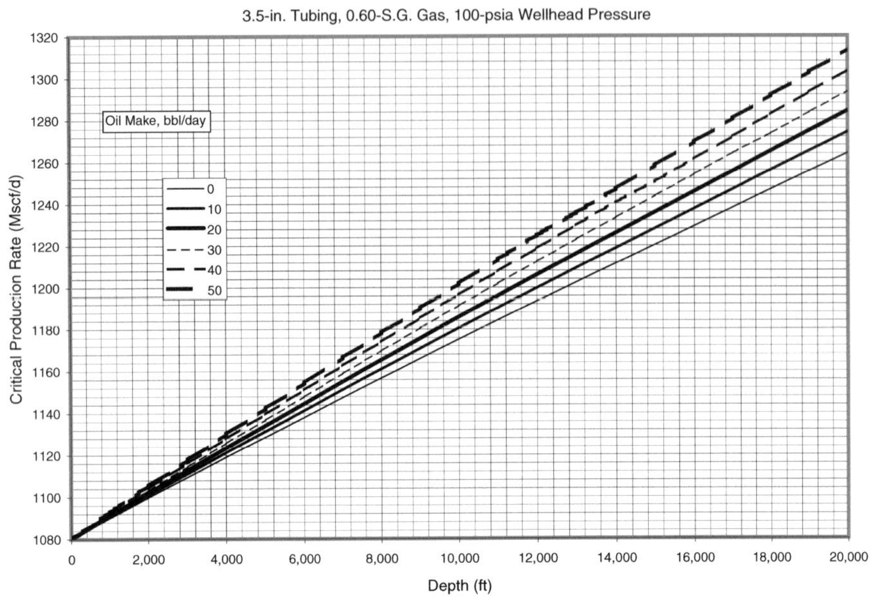

Figure E–21 **Critical gas production rate for condensate removal in 3.5-in tubing against 100 psia wellhead pressure, S.G. 0.60 gas.**

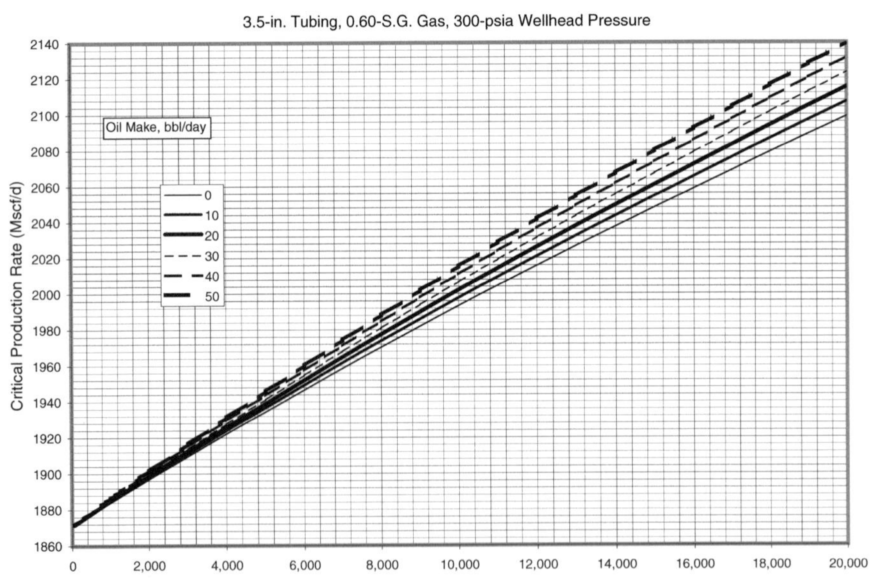

Figure E–22 **Critical gas production rate for condensate removal in 3.5-in tubing against 300 psia wellhead pressure, S.G. 0.60 gas.**

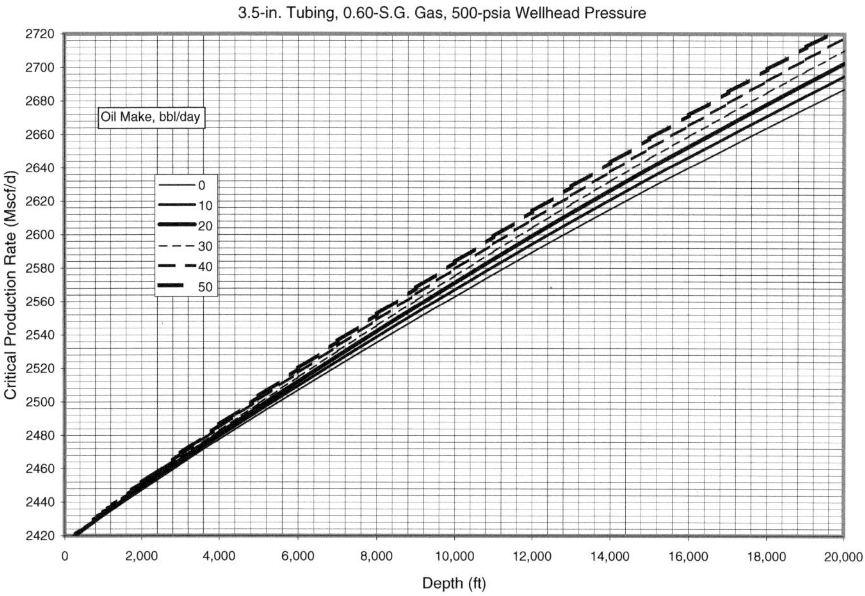

Figure E–23 **Critical gas production rate for condensate removal in 3.5-in tubing against 500 psia wellhead pressure, S.G. 0.60 gas.**

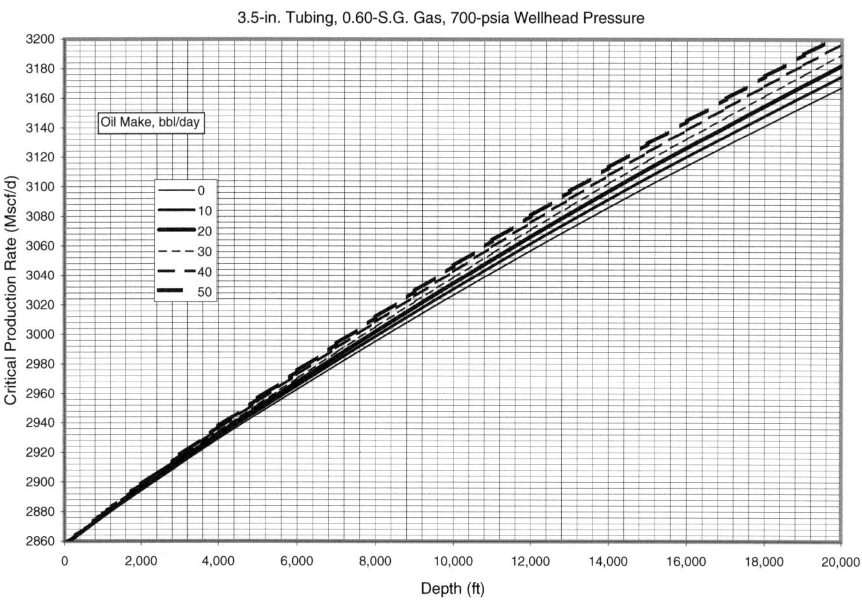

Figure E–24 **Critical gas production rate for condensate removal in 3.5-in tubing against 700 psia wellhead pressure, S.G. 0.60 gas.**

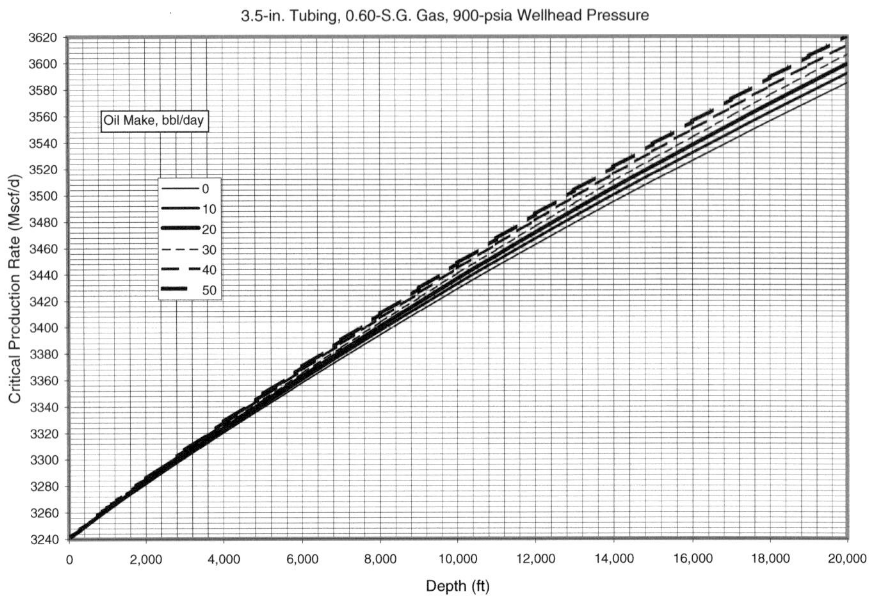

Figure E–25 **Critical gas production rate for condensate removal in 3.5-in tubing against 900 psia wellhead pressure, S.G. 0.60 gas.**

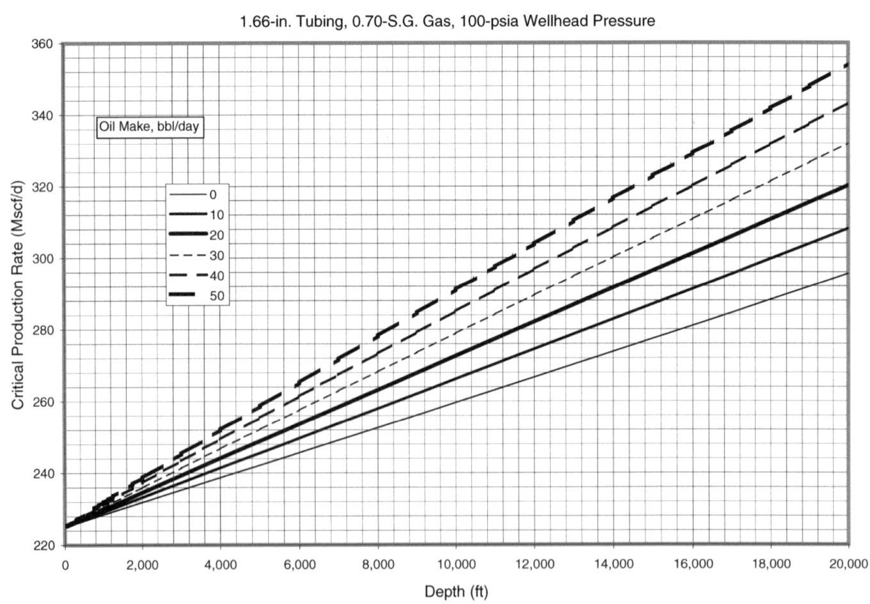

Figure E–26 **Critical gas production rate for condensate removal in 1.66-in tubing against 100 psia wellhead pressure, S.G. 0.70 gas.**

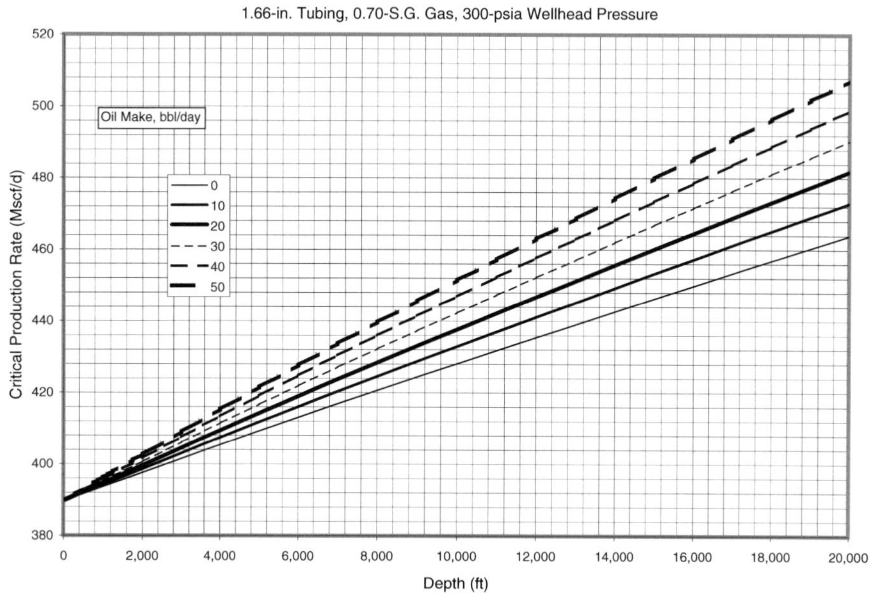

Figure E–27 **Critical gas production rate for condensate removal in 1.66-in tubing against 300 psia wellhead pressure, S.G. 0.70 gas.**

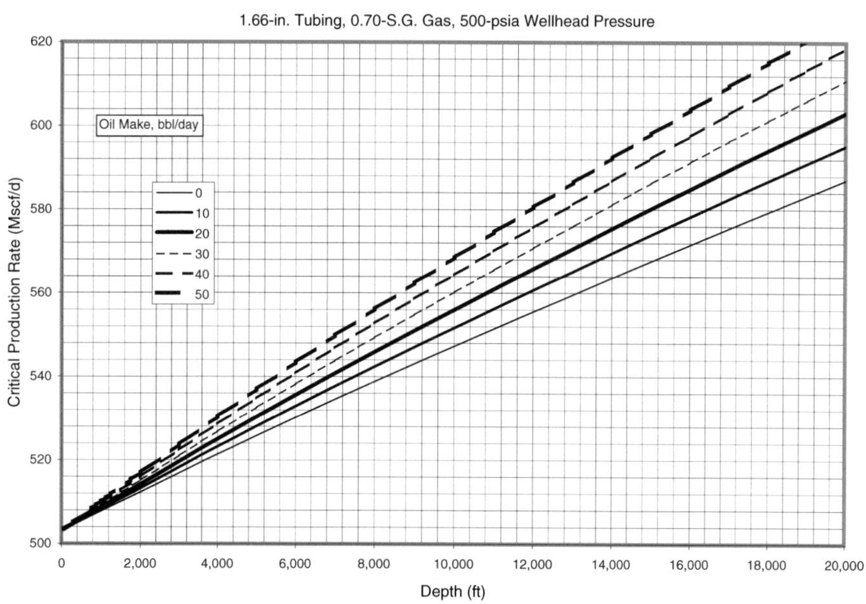

Figure E–28 **Critical gas production rate for condensate removal in 1.66-in tubing against 500 psia wellhead pressure, S.G. 0.70 gas.**

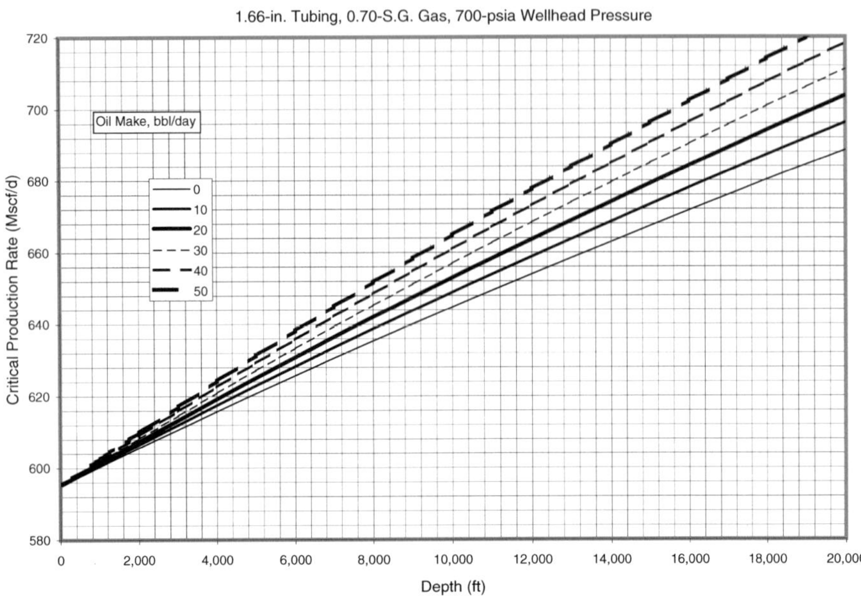

Figure E–29 **Critical gas production rate for condensate removal in 1.66-in tubing against 700 psia wellhead pressure, S.G. 0.70 gas.**

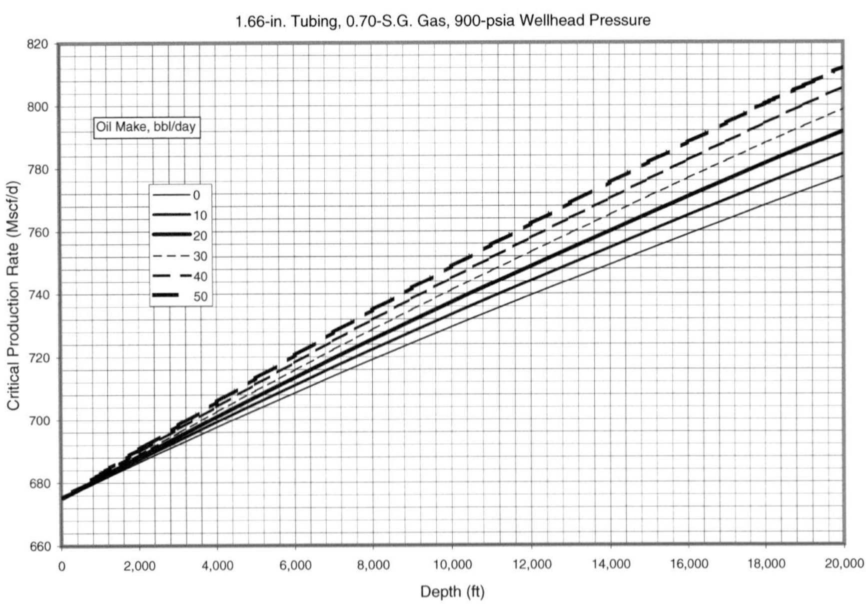

Figure E–30 **Critical gas production rate for condensate removal in 1.66-in tubing against 900 psia wellhead pressure, S.G. 0.70 gas.**

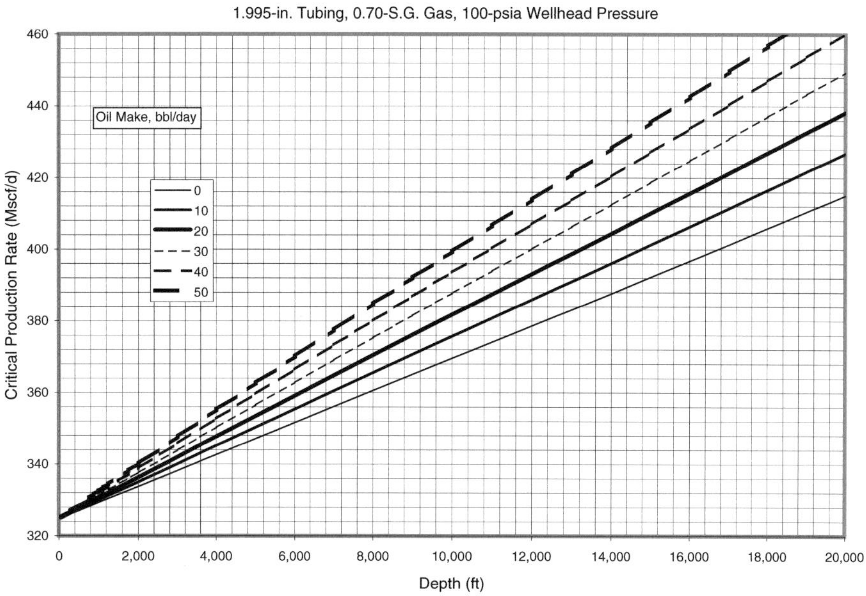

Figure E–31 **Critical gas production rate for condensate removal in 1.995-in tubing against 100 psia wellhead pressure, S.G. 0.70 gas.**

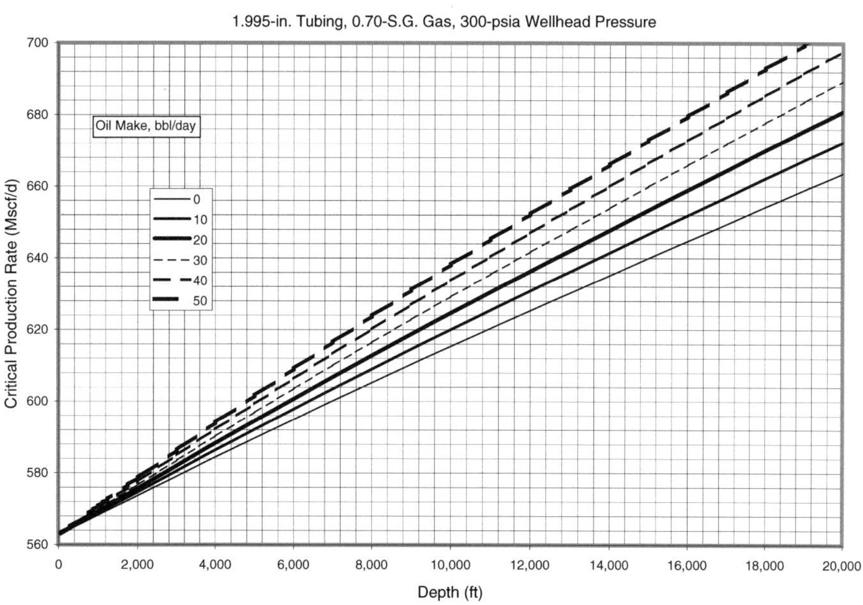

Figure E–32 **Critical gas production rate for condensate removal in 1.995-in tubing against 300 psia wellhead pressure, S.G. 0.70 gas.**

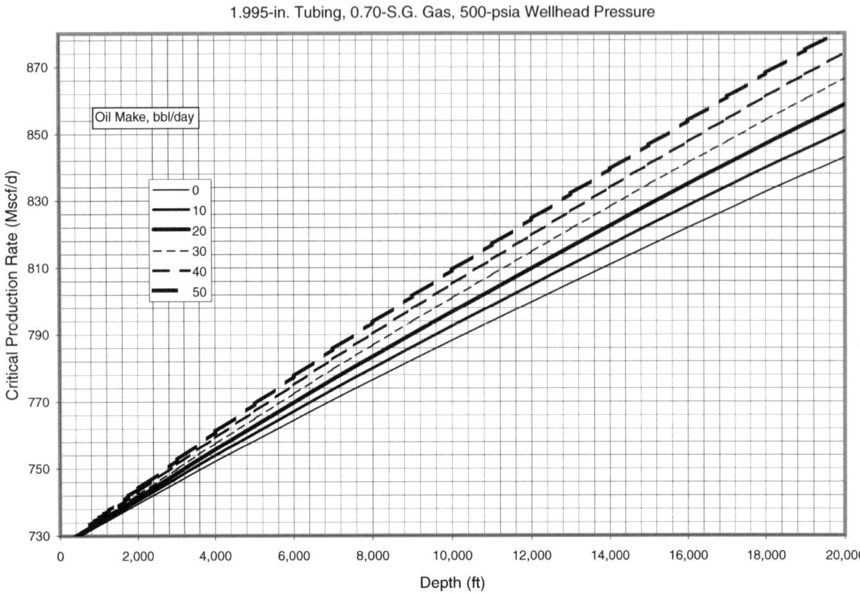

Figure E–33 **Critical gas production rate for condensate removal in 1.995-in tubing against 500 psia wellhead pressure, S.G. 0.70 gas.**

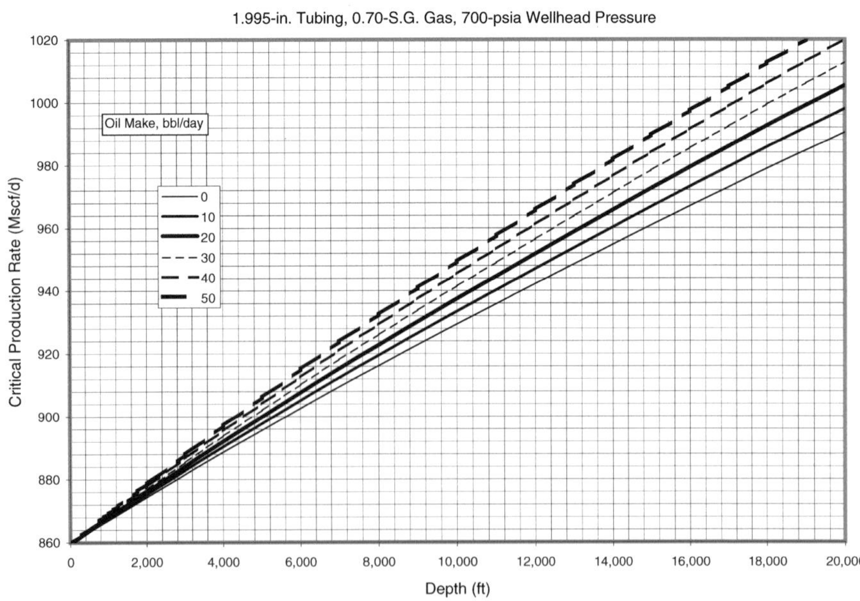

Figure E–34 **Critical gas production rate for condensate removal in 1.995-in tubing against 700 psia wellhead pressure, S.G. 0.70 gas.**

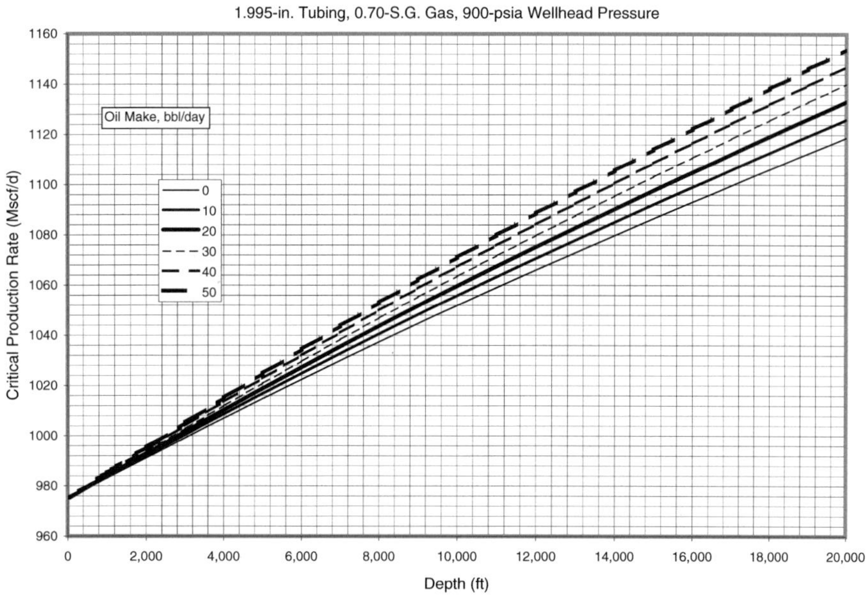

Figure E–35 Critical gas production rate for condensate removal in 1.995-in tubing against 900 psia wellhead pressure, S.G. 0.70 gas.

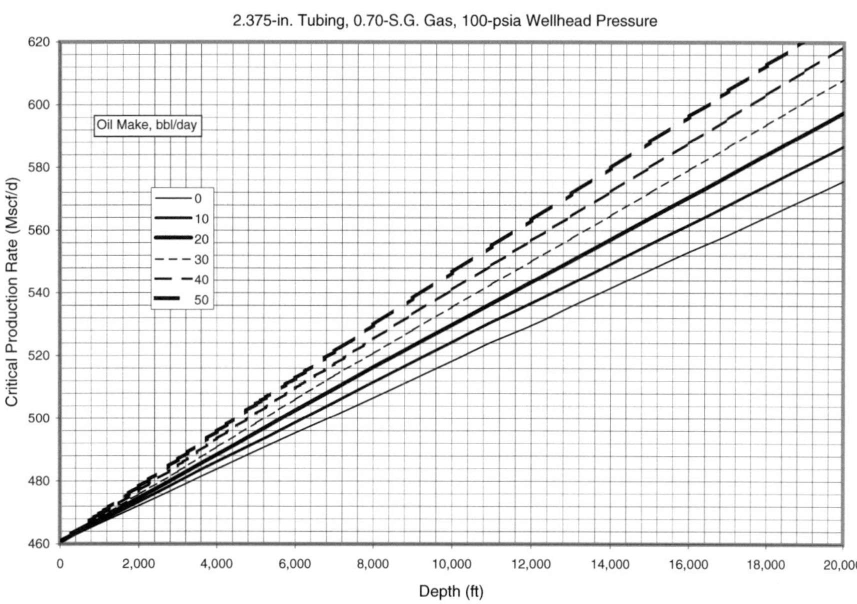

Figure E–36 Critical gas production rate for condensate removal in 2.375-in tubing against 100 psia wellhead pressure, S.G. 0.70 gas.

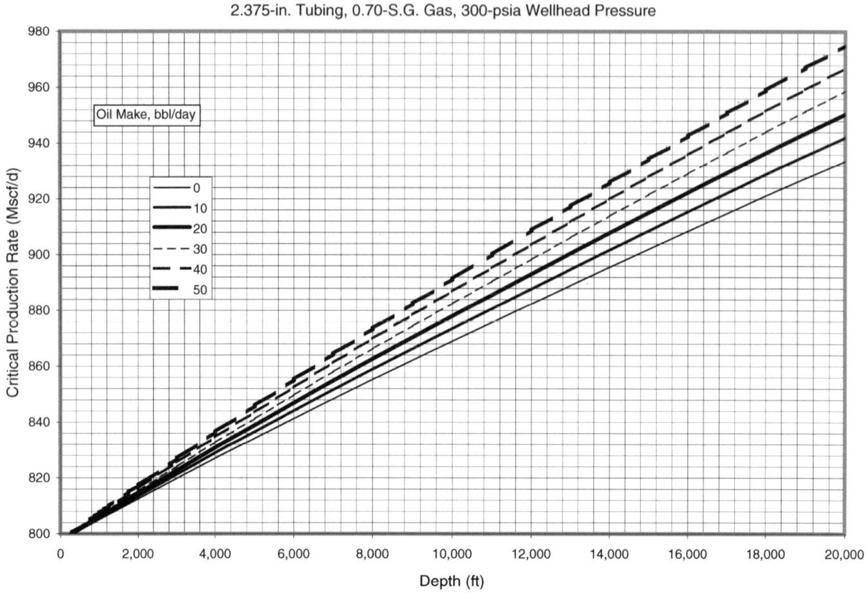

Figure E–37 **Critical gas production rate for condensate removal in 2.375-in tubing against 300 psia wellhead pressure, S.G. 0.70 gas.**

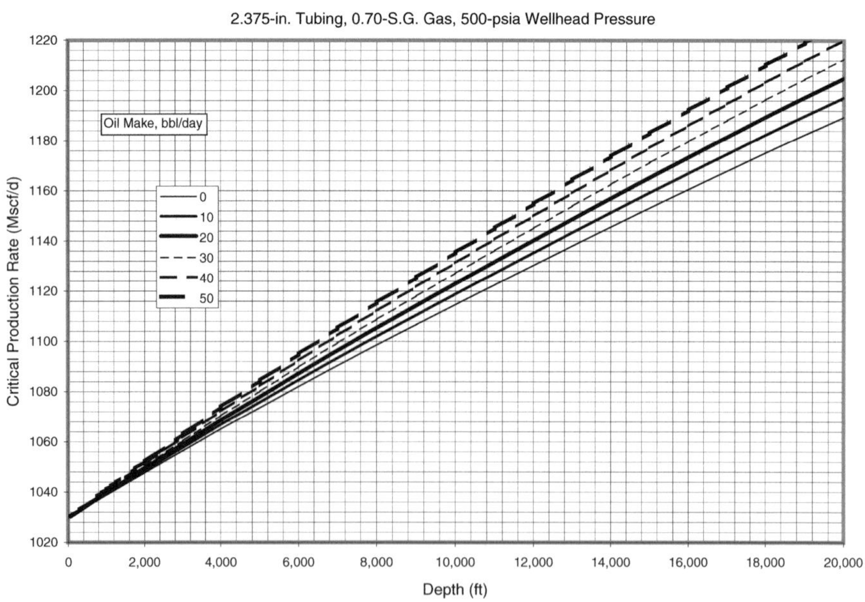

Figure E–38 **Critical gas production rate for condensate removal in 2.375-in tubing against 500 psia wellhead pressure, S.G. 0.70 gas.**

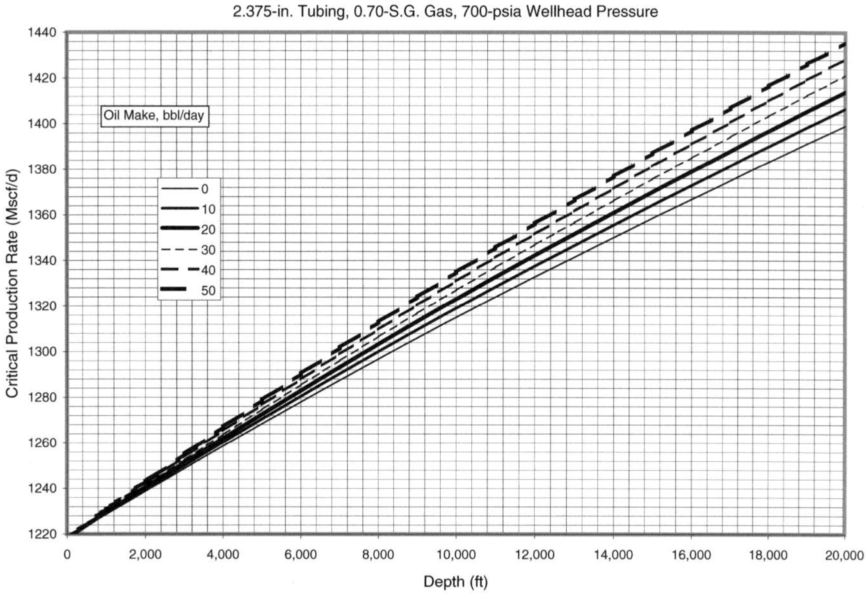

Figure E–39 **Critical gas production rate for condensate removal in 2.375-in tubing against 700 psia wellhead pressure, S.G. 0.70 gas.**

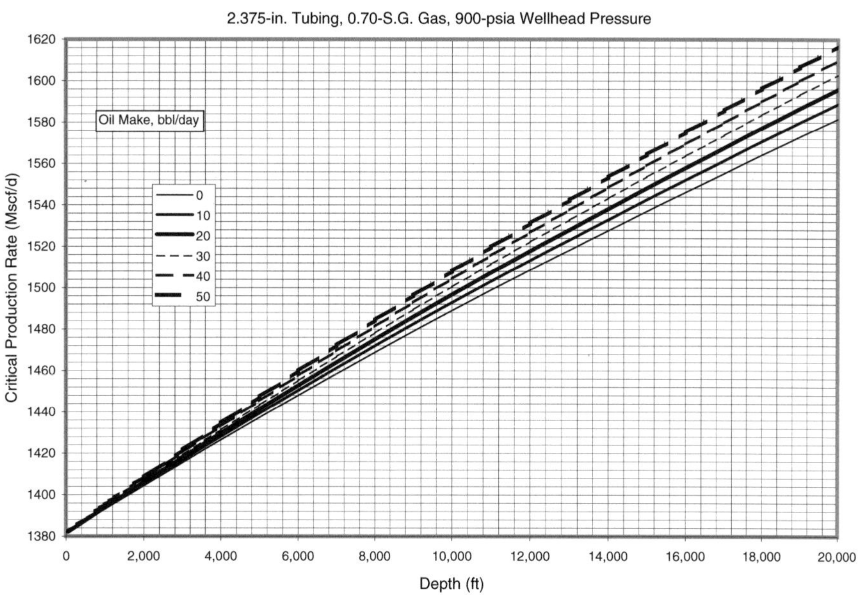

Figure E–40 **Critical gas production rate for condensate removal in 2.375-in tubing against 900 psia wellhead pressure, S.G. 0.70 gas.**

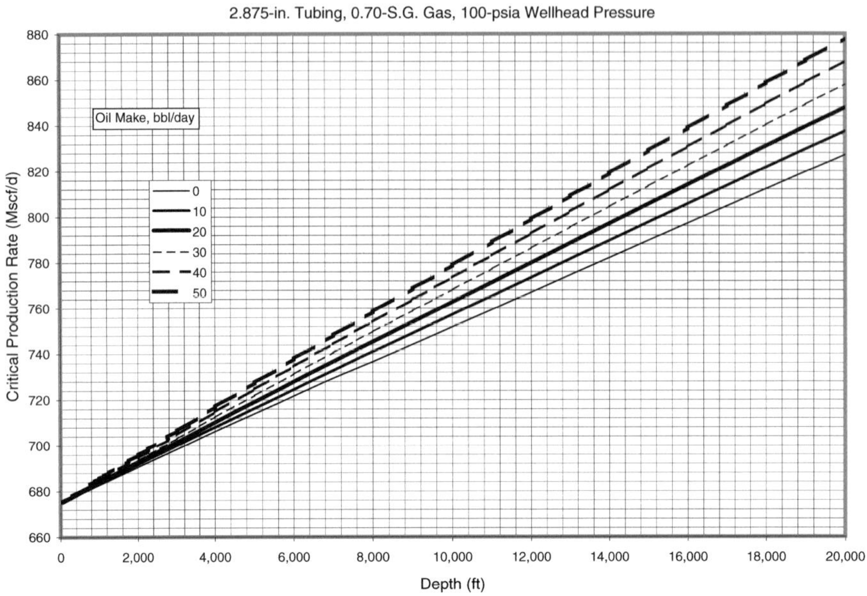

Figure E–41 Critical gas production rate for condensate removal in 2.875-in tubing against 100 psia wellhead pressure, S.G. 0.70 gas.

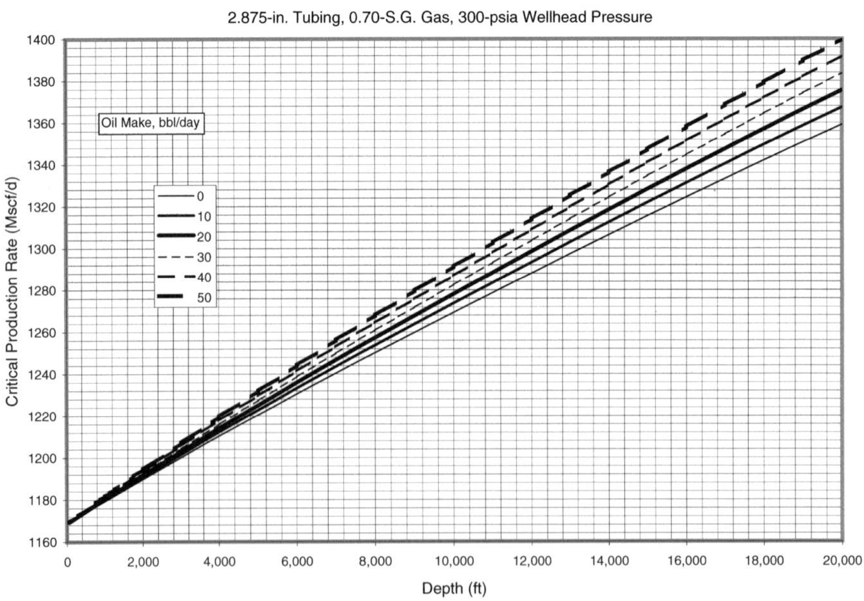

Figure E–42 Critical gas production rate for condensate removal in 2.875-in tubing against 300 psia wellhead pressure, S.G. 0.70 gas.

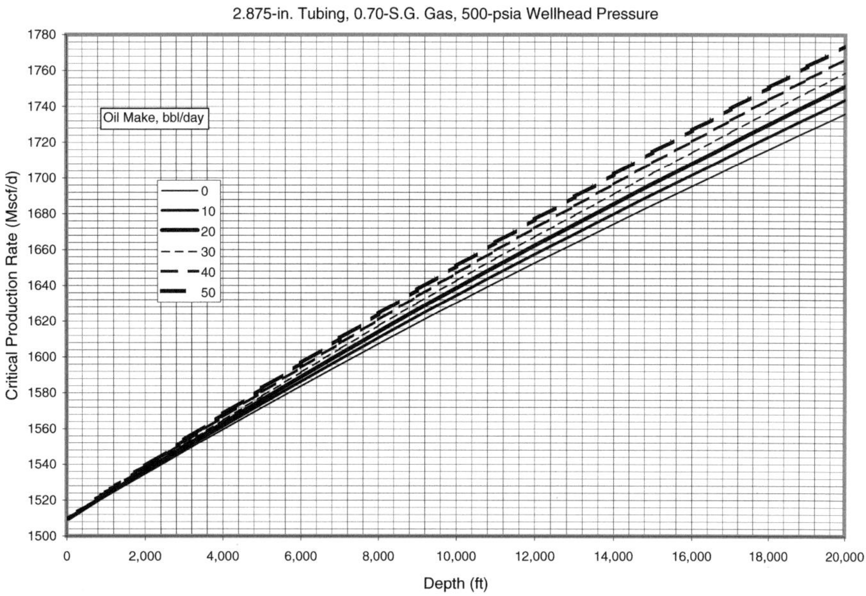

Figure E–43 **Critical gas production rate for condensate removal in 2.875-in tubing against 500 psia wellhead pressure, S.G. 0.70 gas.**

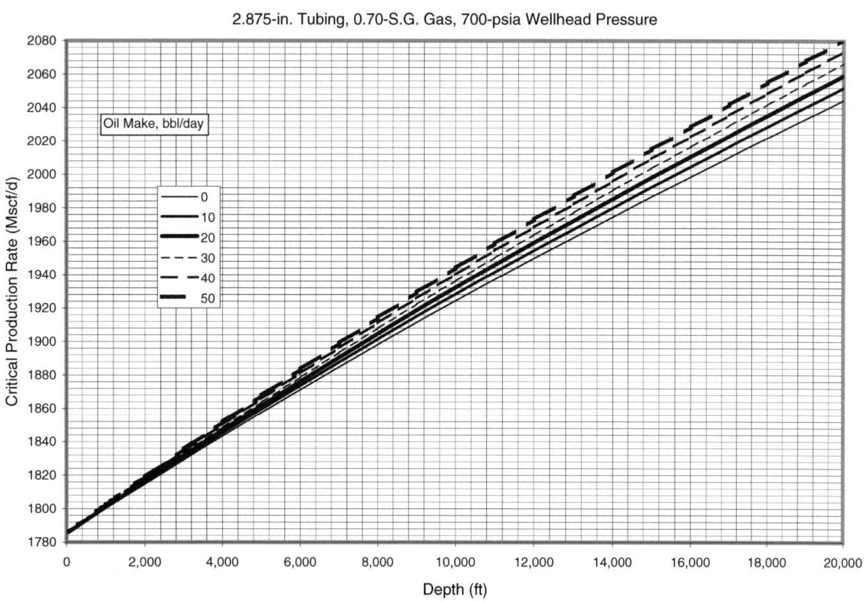

Figure E–44 **Critical gas production rate for condensate removal in 2.875-in tubing against 700 psia wellhead pressure, S.G. 0.70 gas.**

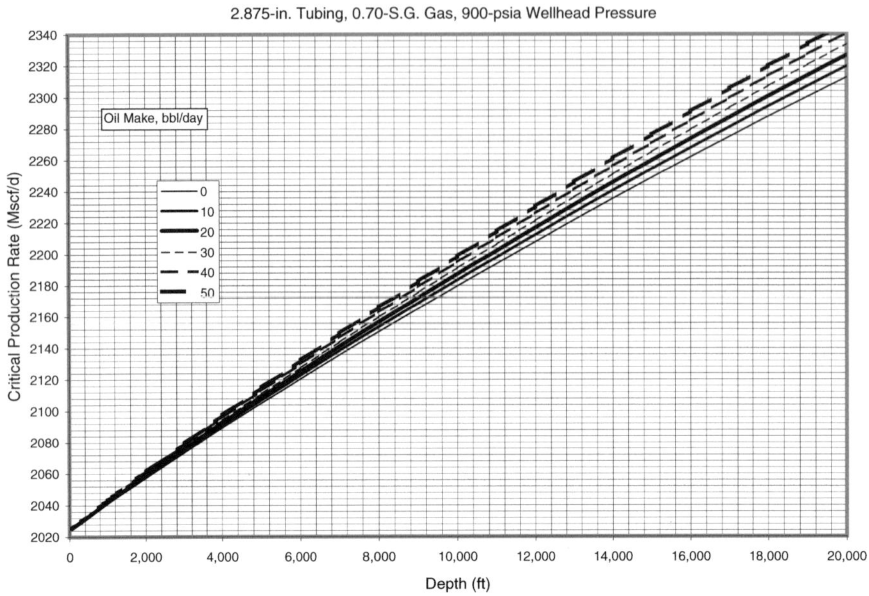

Figure E–45 Critical gas production rate for condensate removal in 2.875-in tubing against 900 psia wellhead pressure, S.G. 0.70 gas.

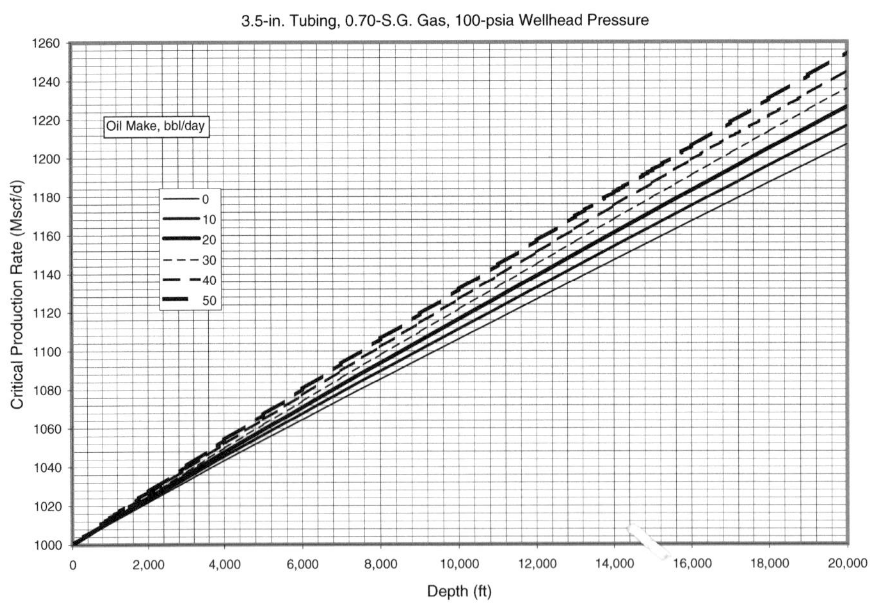

Figure E–46 Critical gas production rate for condensate removal in 3.5-in tubing against 100 psia wellhead pressure, S.G. 0.70 gas.

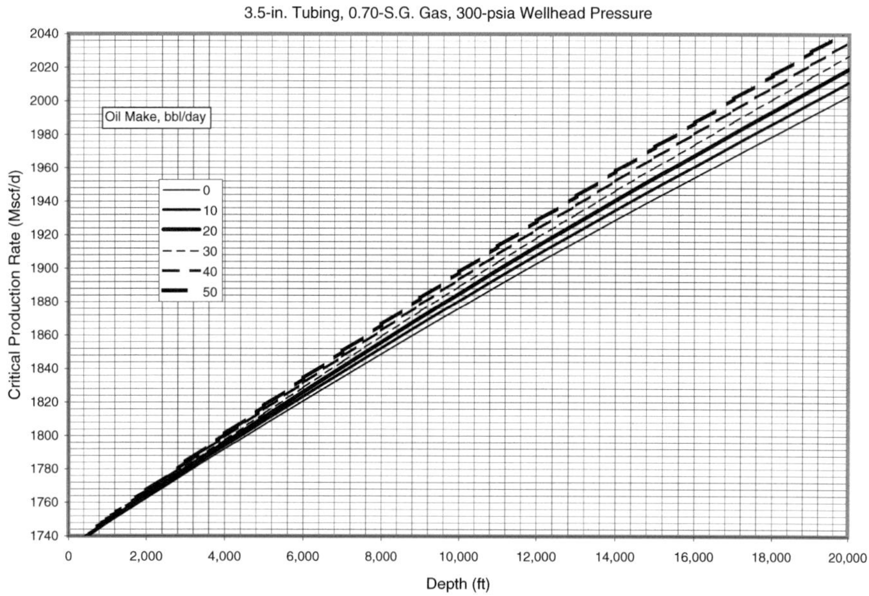

Figure E–47 Critical gas production rate for condensate removal in 3.5-in tubing against 300 psia wellhead pressure, S.G. 0.70 gas.

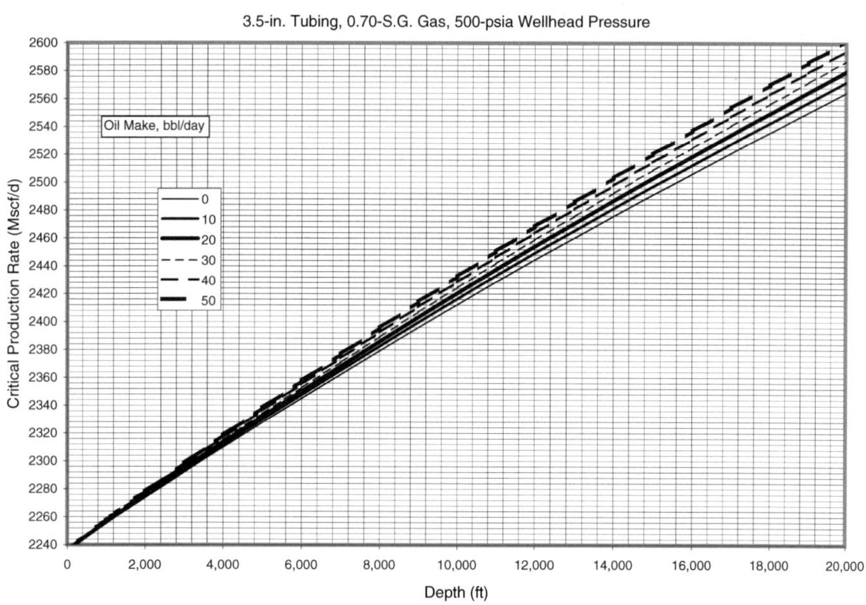

Figure E–48 Critical gas production rate for condensate removal in 3.5-in tubing against 500 psia wellhead pressure, S.G. 0.70 gas.

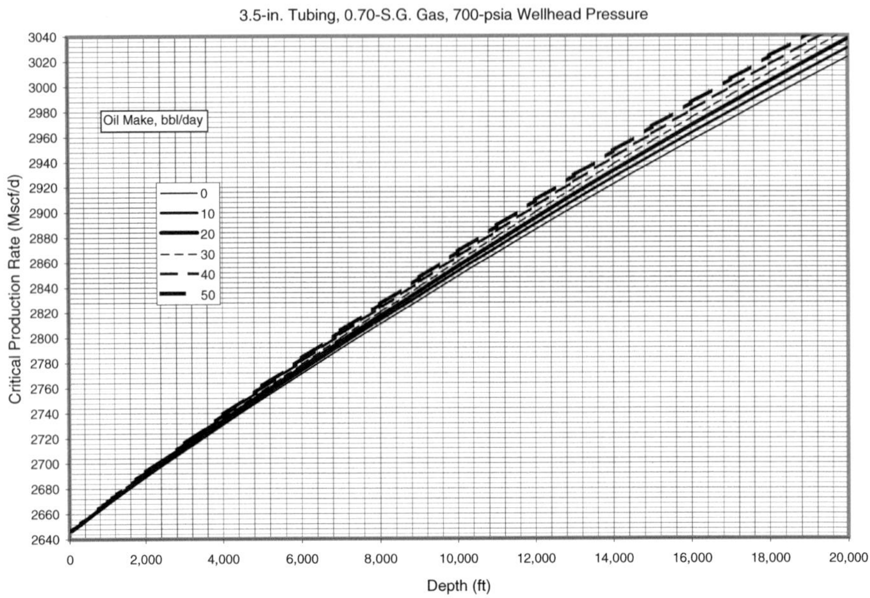

Figure E–49 Critical gas production rate for condensate removal in 3.5-in tubing against 700 psia wellhead pressure, S.G. 0.70 gas.

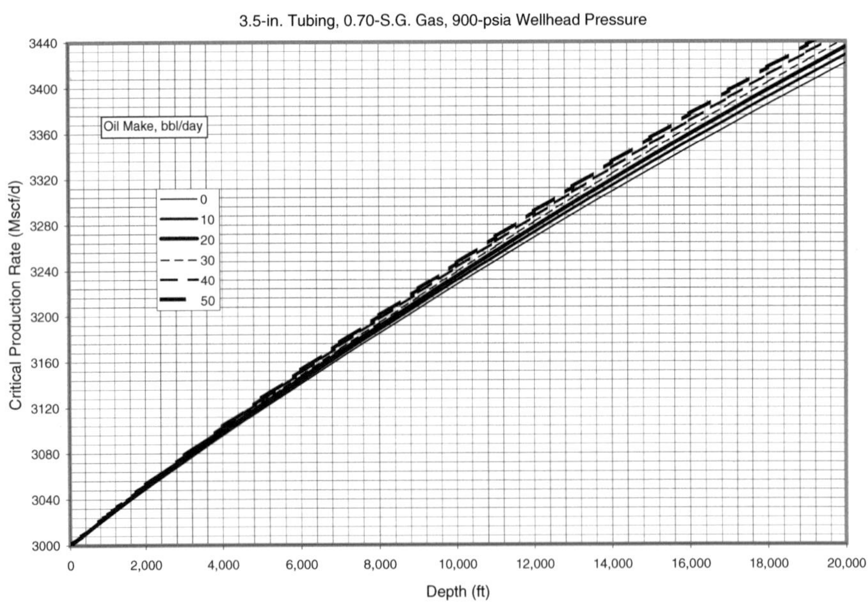

Figure E–50 Critical gas production rate for condensate removal in 3.5-in tubing against 900 psia wellhead pressure, S.G. 0.70 gas.

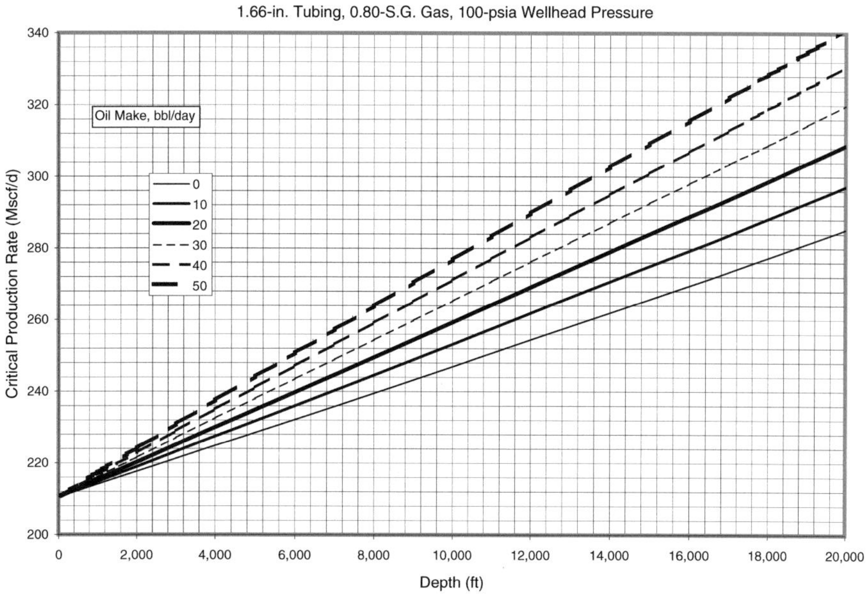

Figure E–51 **Critical gas production rate for condensate removal in 1.66-in tubing against 100 psia wellhead pressure, S.G. 0.80 gas.**

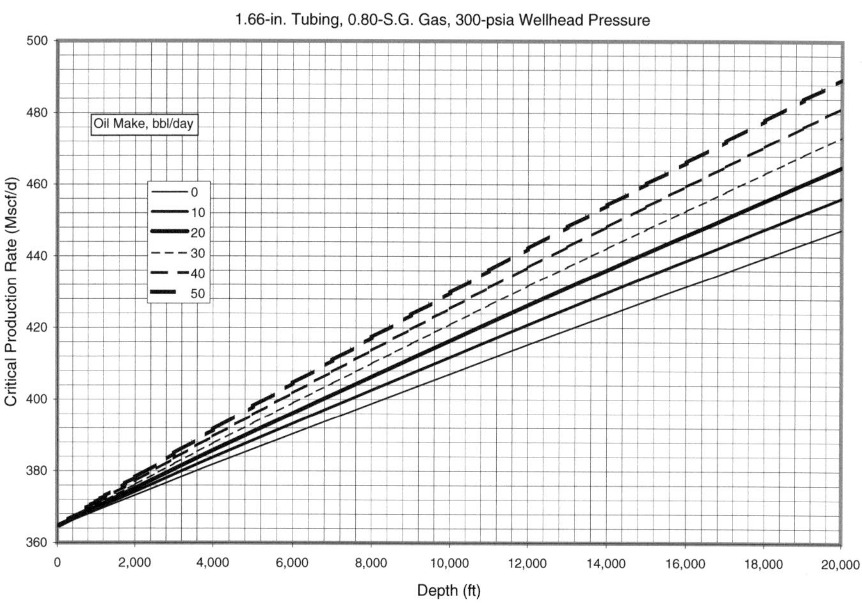

Figure E–52 **Critical gas production rate for condensate removal in 1.66-in tubing against 300 psia wellhead pressure, S.G. 0.80 gas.**

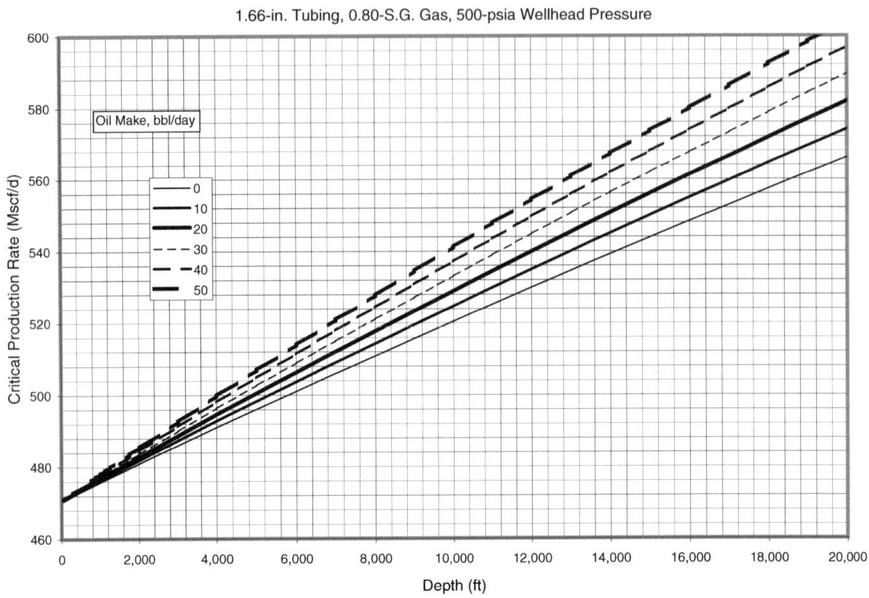

Figure E–53 Critical gas production rate for condensate removal in 1.66-in tubing against 500 psia wellhead pressure, S.G. 0.80 gas.

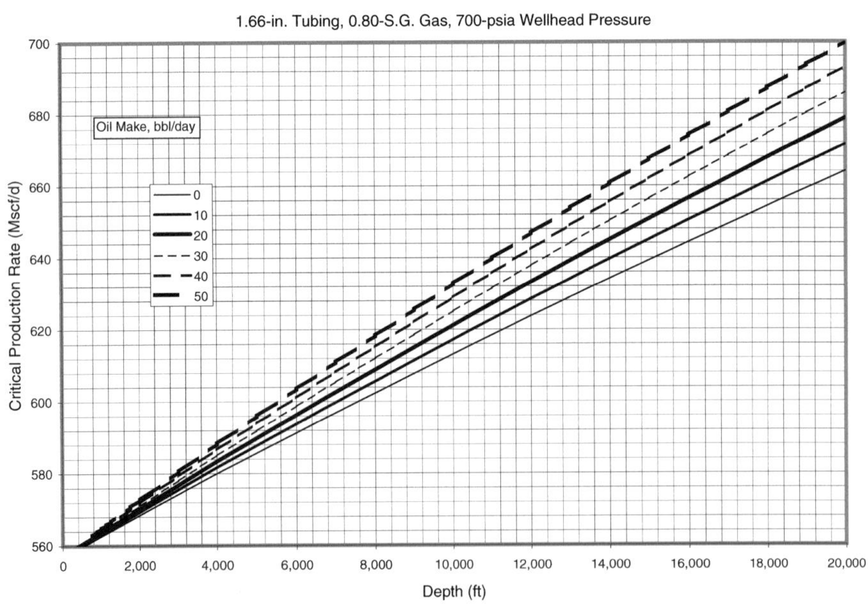

Figure E–54 Critical gas production rate for condensate removal in 1.66-in tubing against 700 psia wellhead pressure, S.G. 0.80 gas.

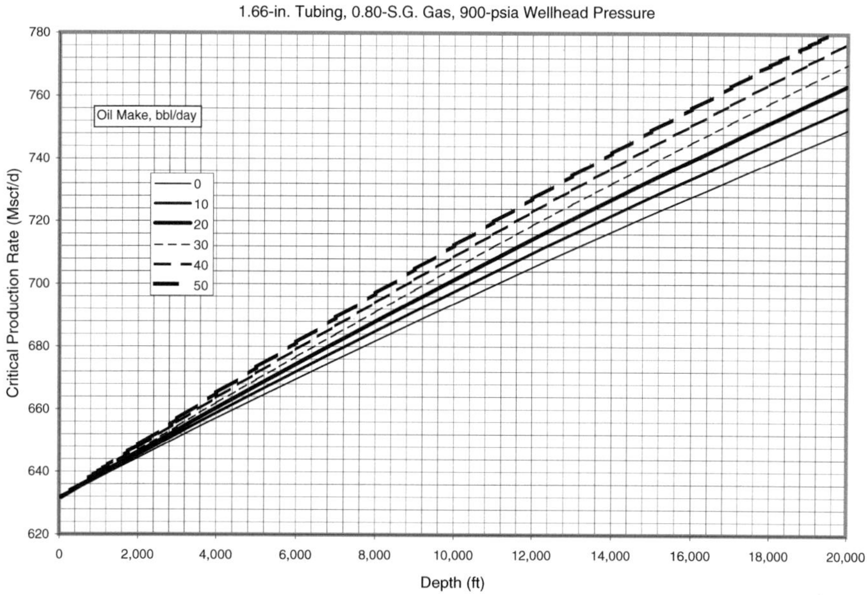

Figure E–55 Critical gas production rate for condensate removal in 1.66-in tubing against 900 psia wellhead pressure, S.G. 0.80 gas.

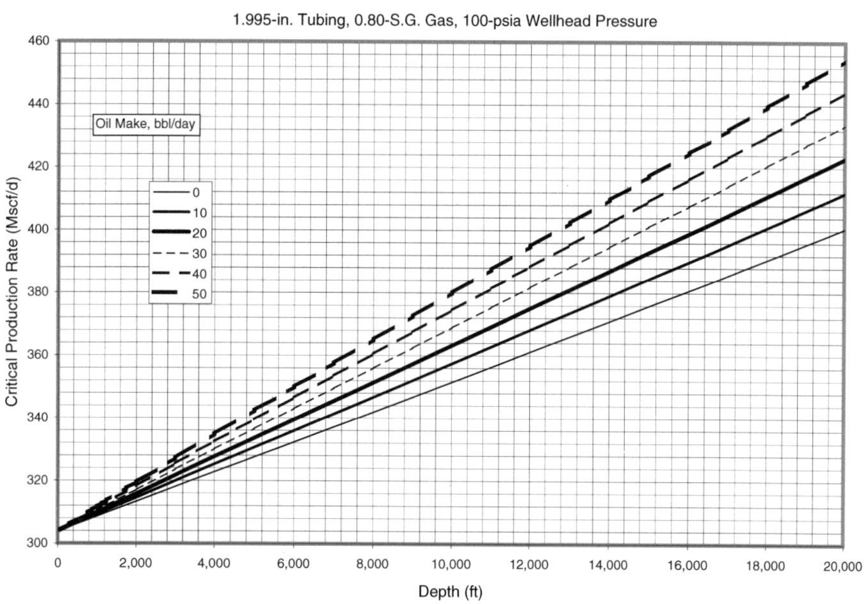

Figure E–56 Critical gas production rate for condensate removal in 1.995-in tubing against 100 psia wellhead pressure, S.G. 0.80 gas.

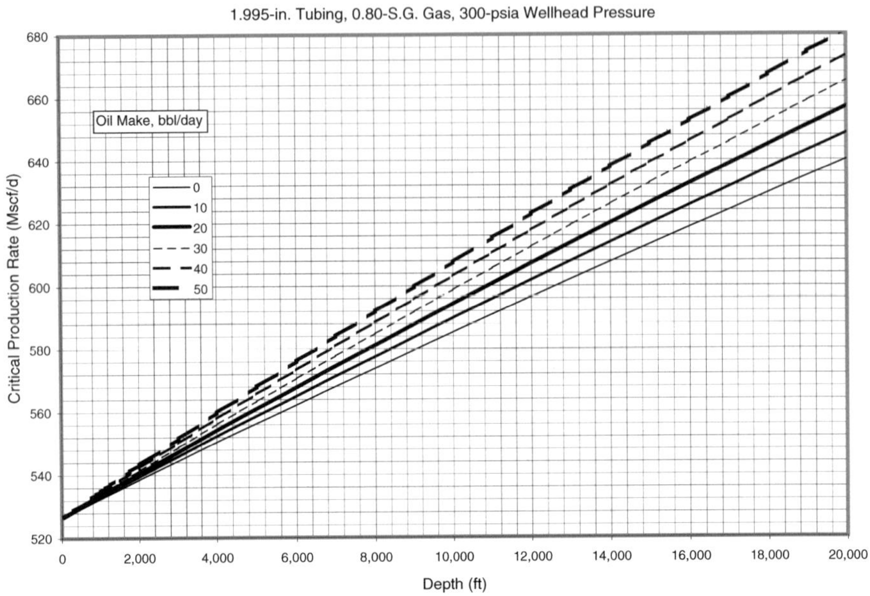

Figure E–57 Critical gas production rate for condensate removal in 1.995-in tubing against 300 psia wellhead pressure, S.G. 0.80 gas.

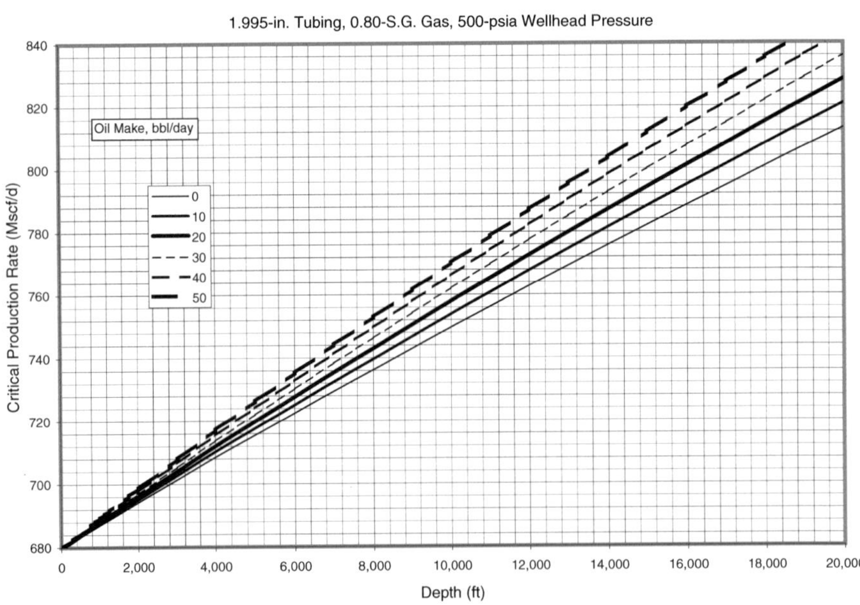

Figure E–58 Critical gas production rate for condensate removal in 1.995-in tubing against 500 psia wellhead pressure, S.G. 0.80 gas.

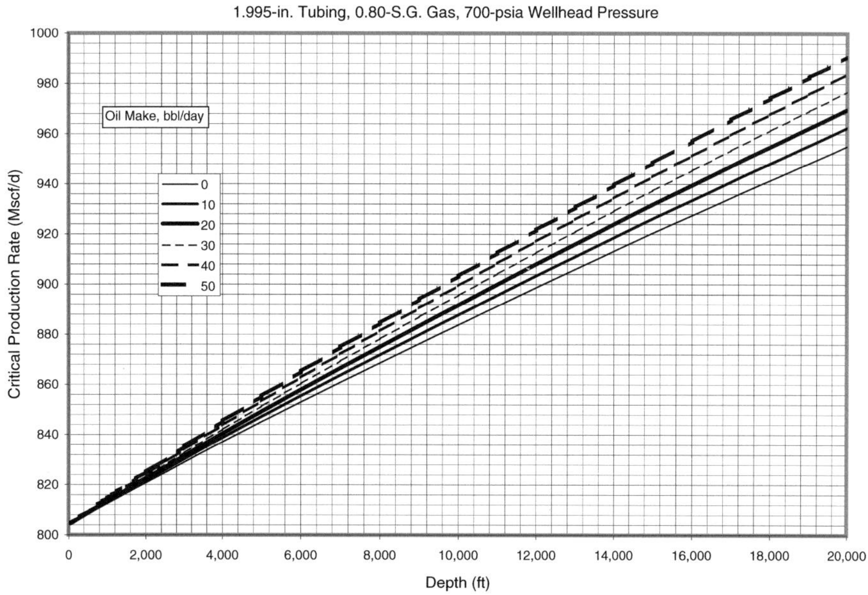

Figure E–59 **Critical gas production rate for condensate removal in 1.995-in tubing against 700 psia wellhead pressure, S.G. 0.80 gas.**

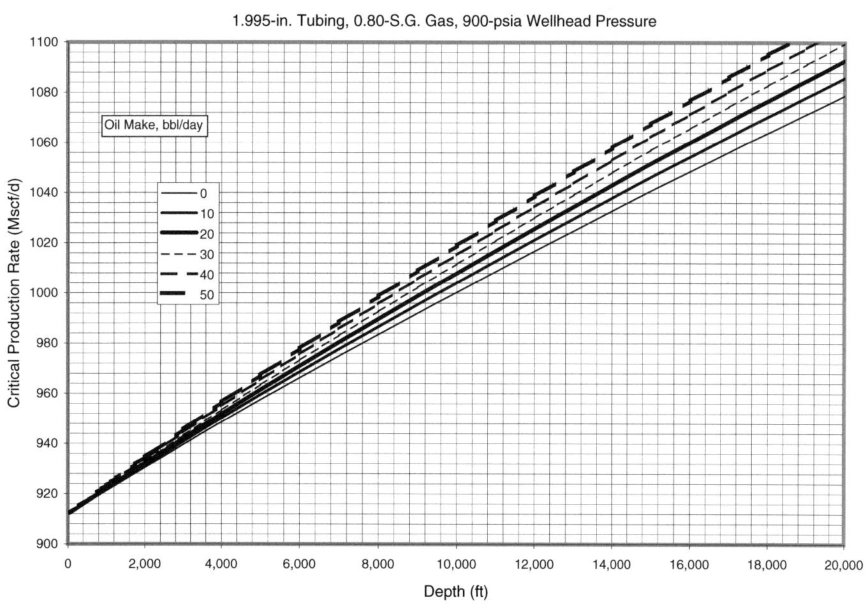

Figure E–60 **Critical gas production rate for condensate removal in 1.995-in tubing against 900 psia wellhead pressure, S.G. 0.80 gas.**

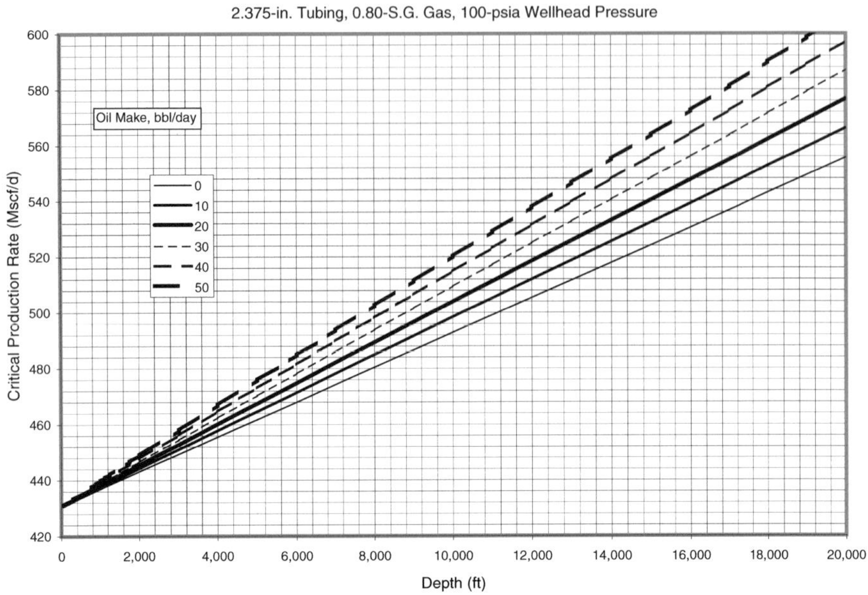

Figure E–61 Critical gas production rate for condensate removal in 2.375-in tubing against 100 psia wellhead pressure, S.G. 0.80 gas.

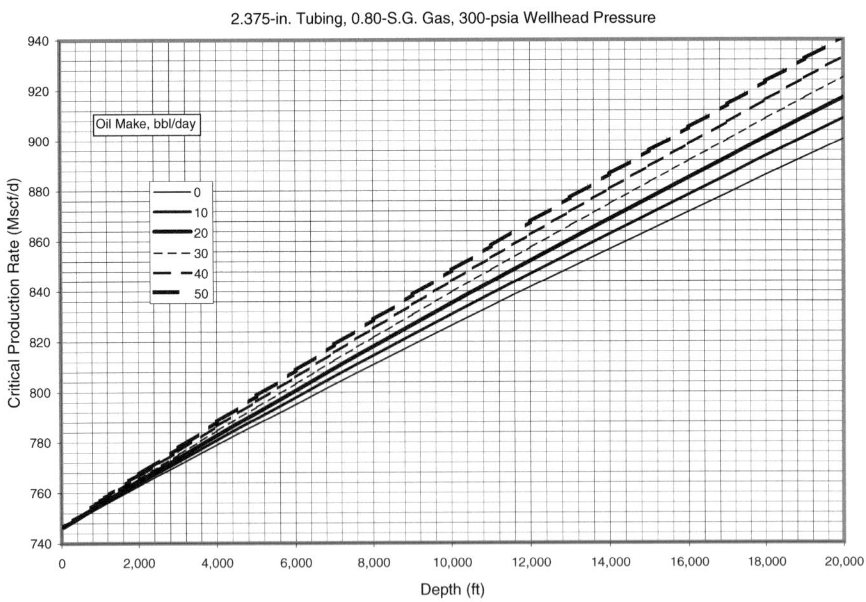

Figure E–62 Critical gas production rate for condensate removal in 2.375-in tubing against 300 psia wellhead pressure, S.G. 0.80 gas.

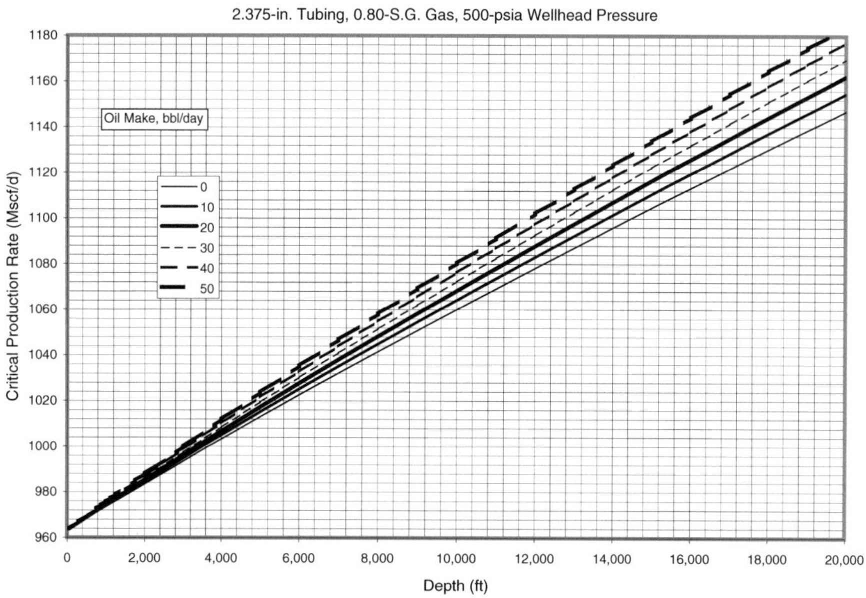

Figure E–63 **Critical gas production rate for condensate removal in 2.375-in tubing against 500 psia wellhead pressure, S.G. 0.80 gas.**

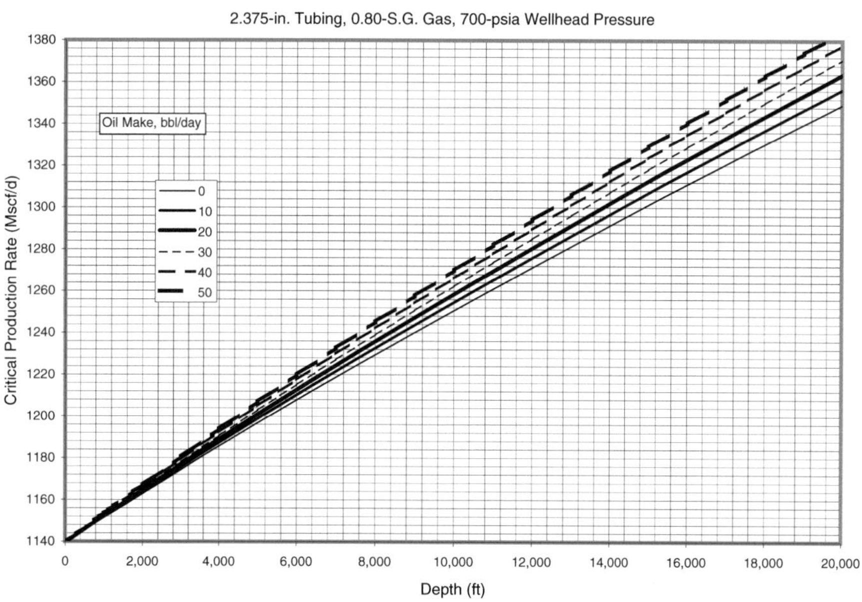

Figure E–64 **Critical gas production rate for condensate removal in 2.375-in tubing against 700 psia wellhead pressure, S.G. 0.80 gas.**

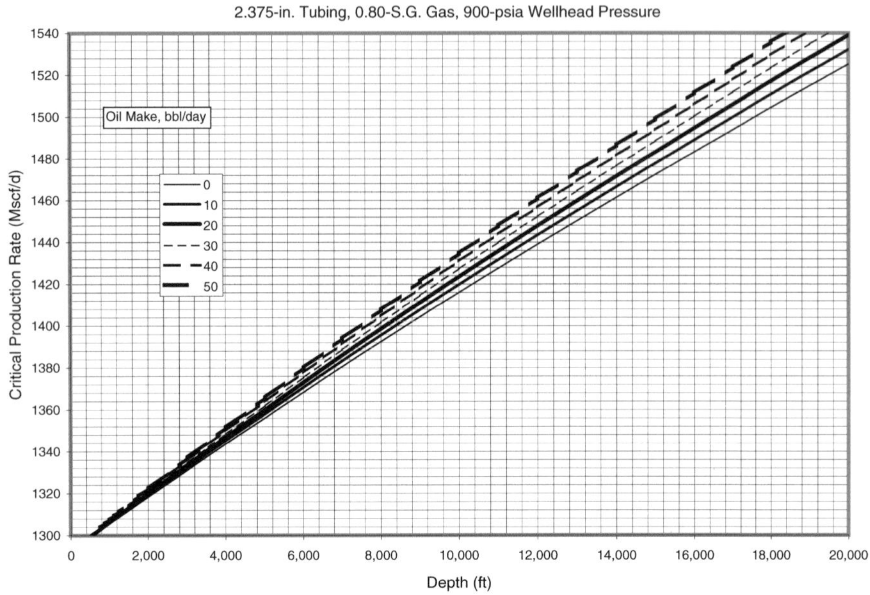

Figure E–65 **Critical gas production rate for condensate removal in 2.375-in tubing against 900 psia wellhead pressure, S.G. 0.80 gas.**

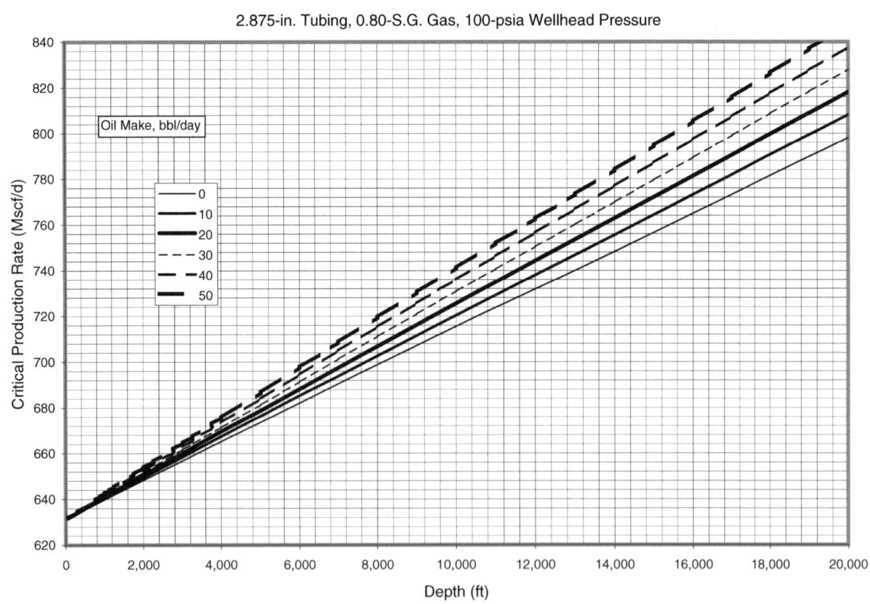

Figure E–66 **Critical gas production rate for condensate removal in 2.875-in tubing against 100 psia wellhead pressure, S.G. 0.80 gas.**

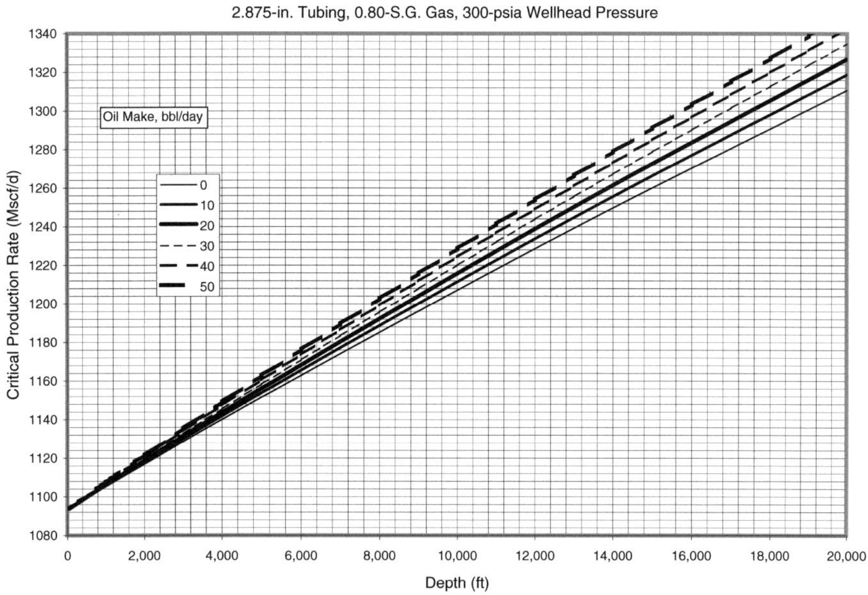

Figure E–67 Critical gas production rate for condensate removal in 2.875-in tubing against 300 psia wellhead pressure, S.G. 0.80 gas.

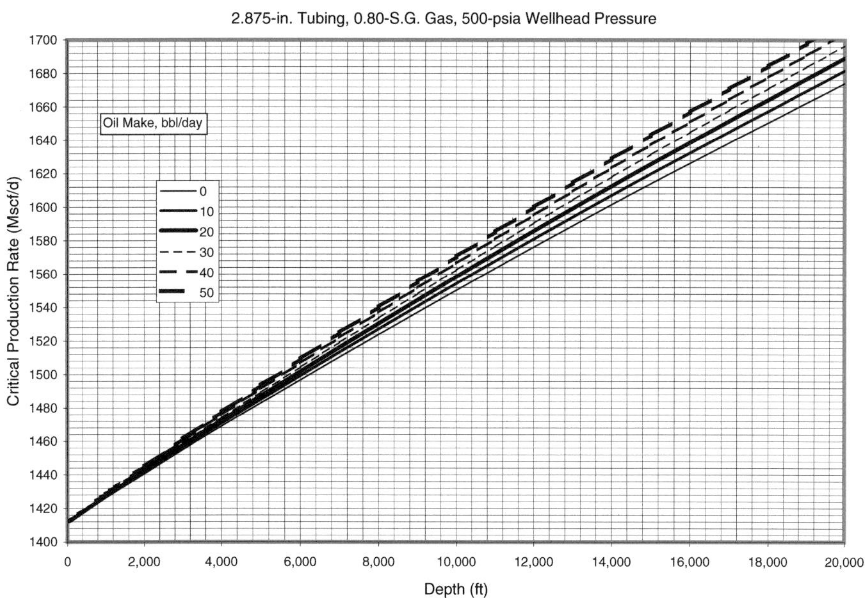

Figure E–68 Critical gas production rate for condensate removal in 2.875-in tubing against 500 psia wellhead pressure, S.G. 0.80 gas.

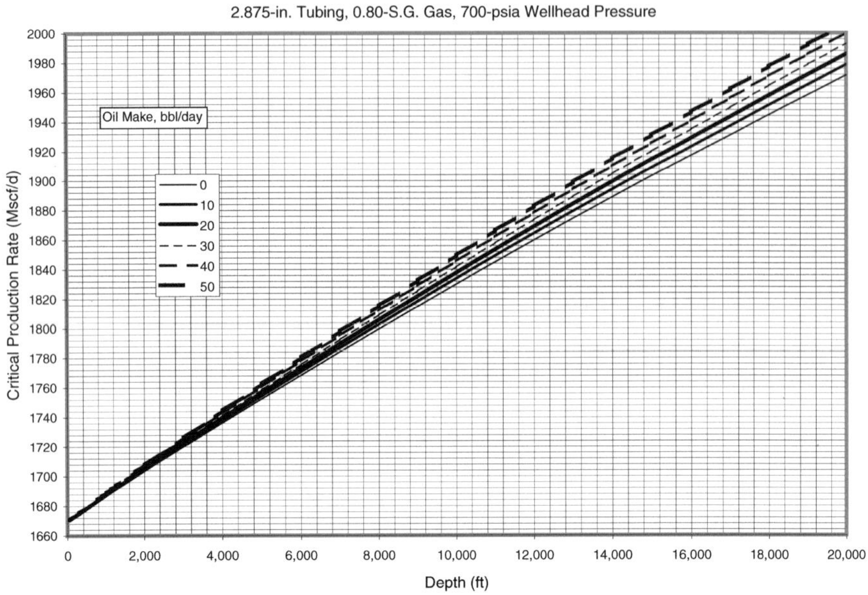

Figure E–69 Critical gas production rate for condensate removal in 2.875-in tubing against 700 psia wellhead pressure, S.G. 0.80 gas.

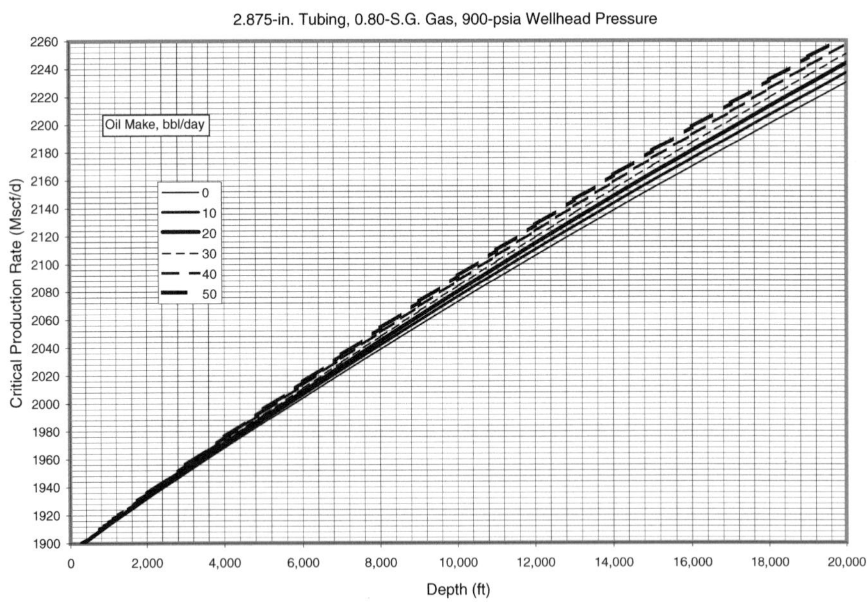

Figure E–70 Critical gas production rate for condensate removal in 2.875-in tubing against 900 psia wellhead pressure, S.G. 0.80 gas.

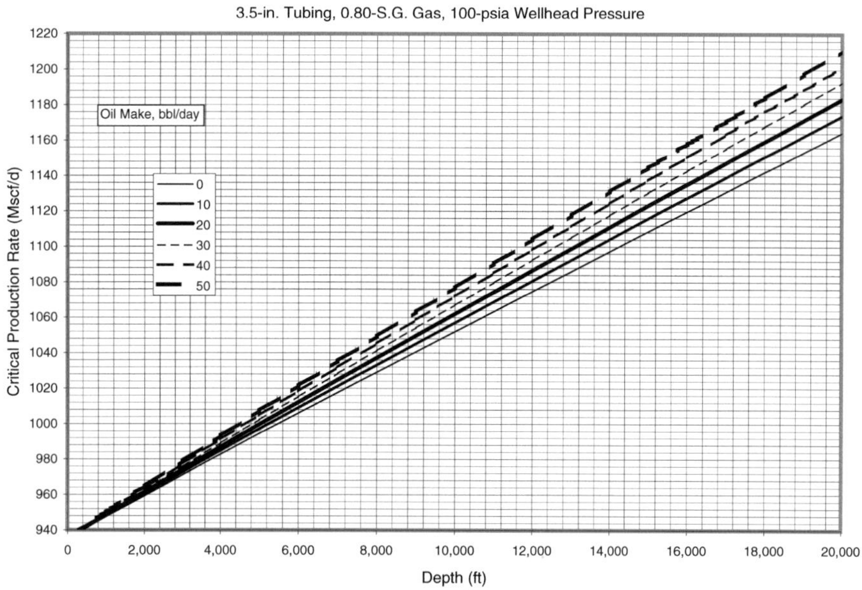

Figure E–71 **Critical gas production rate for condensate removal in 3.5-in tubing against 100 psia wellhead pressure, S.G. 0.80 gas.**

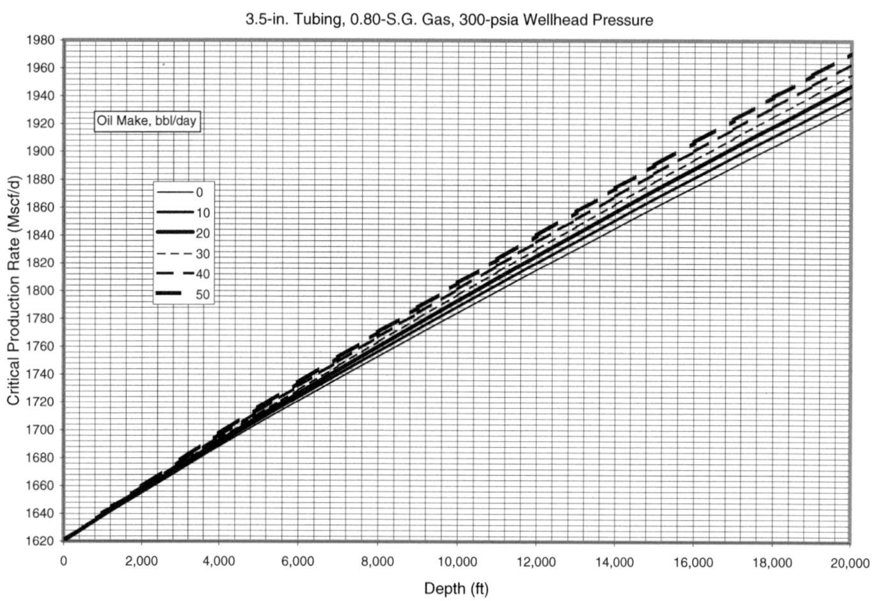

Figure E–72 **Critical gas production rate for condensate removal in 3.5-in tubing against 300 psia wellhead pressure, S.G. 0.80 gas.**

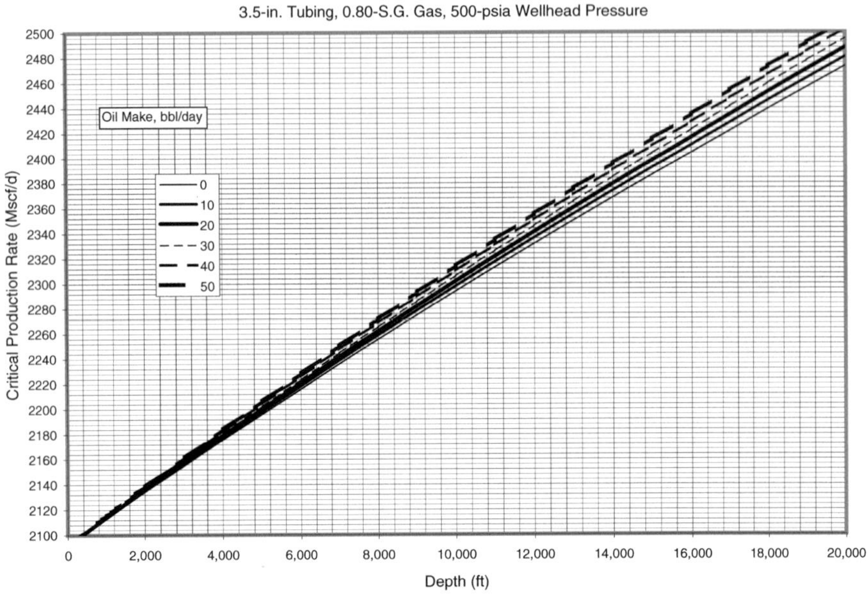

Figure E–73 **Critical gas production rate for condensate removal in 3.5-in tubing against 500 psia wellhead pressure, S.G. 0.80 gas.**

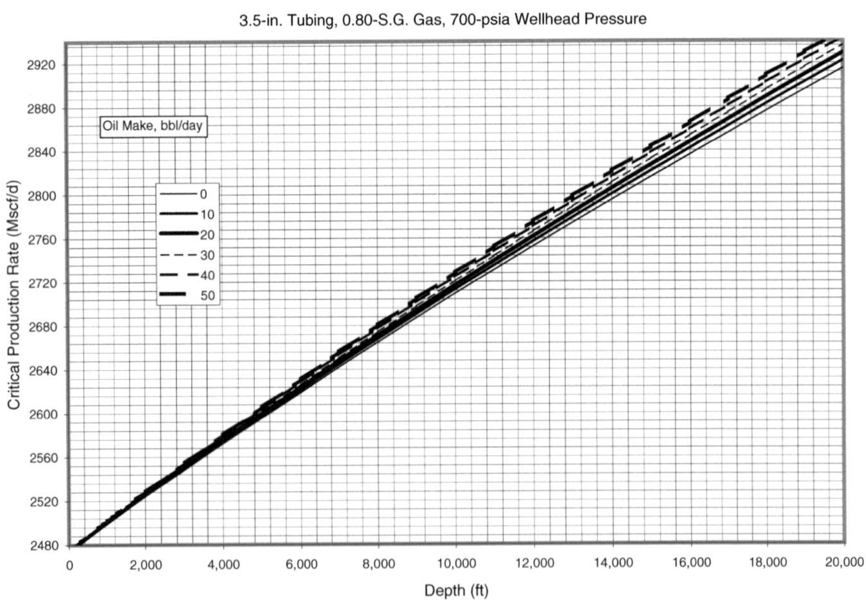

Figure E–74 **Critical gas production rate for condensate removal in 3.5-in tubing against 700 psia wellhead pressure, S.G. 0.80 gas.**

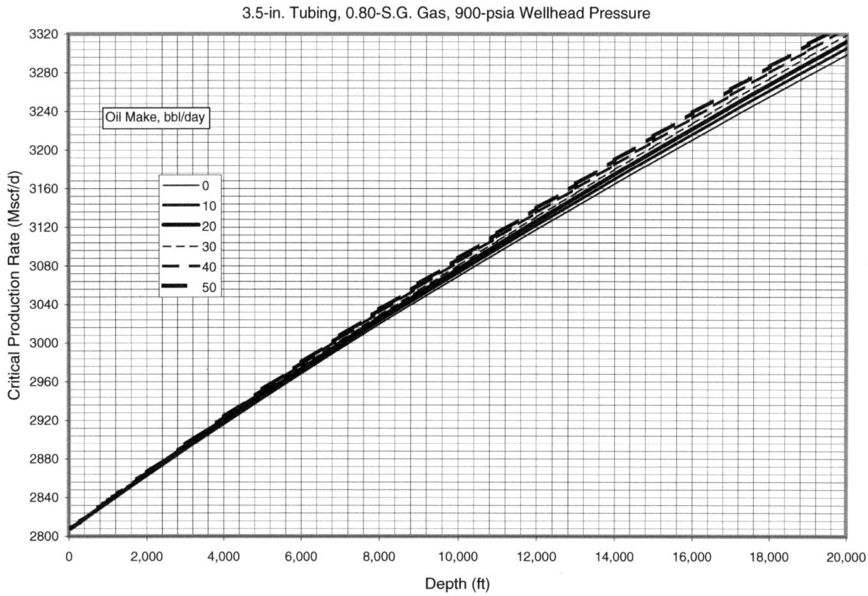

Figure E–75 **Critical gas production rate for condensate removal in 3.5-in tubing against 900 psia wellhead pressure, S.G. 0.80 gas.**

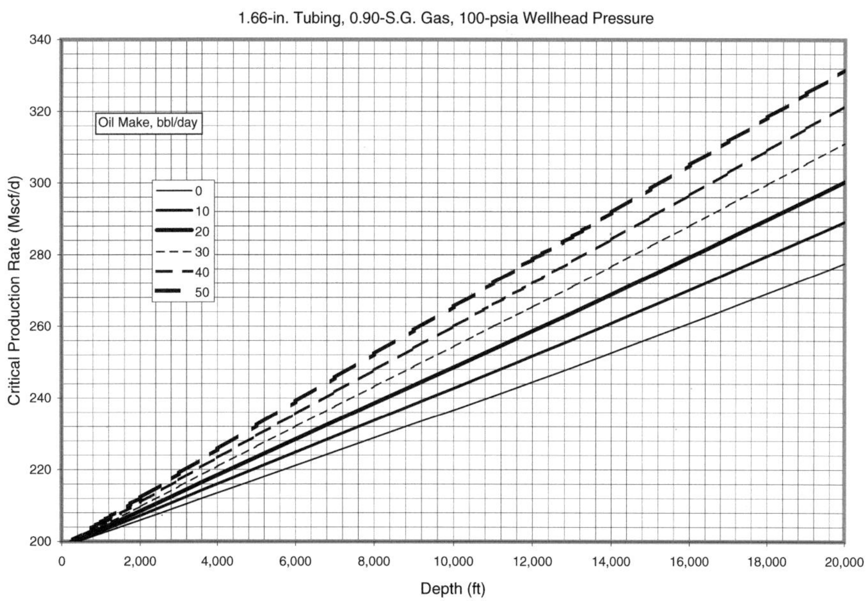

Figure E–76 **Critical gas production rate for condensate removal in 1.66-in tubing against 100 psia wellhead pressure, S.G. 0.90 gas.**

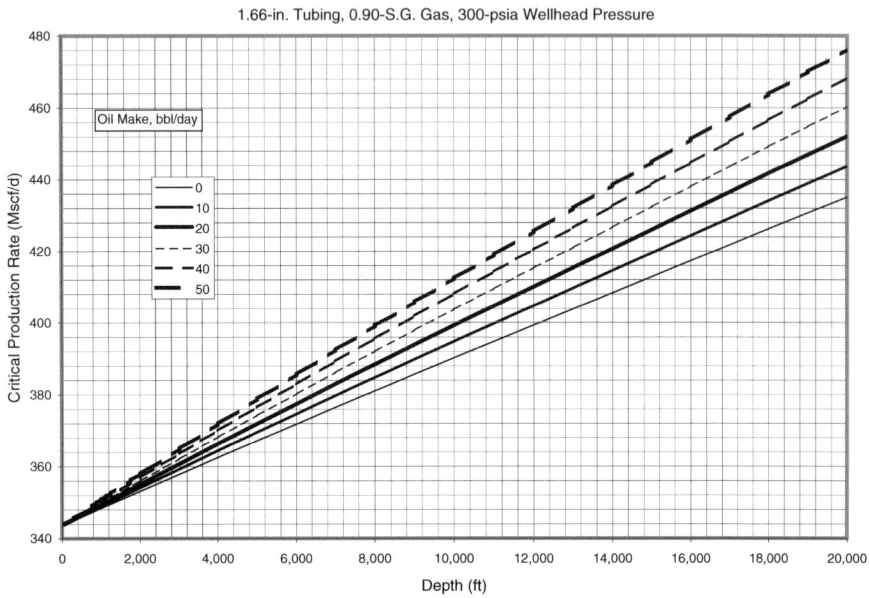

Figure E–77 **Critical gas production rate for condensate removal in 1.66-in tubing against 300 psia wellhead pressure, S.G. 0.90 gas.**

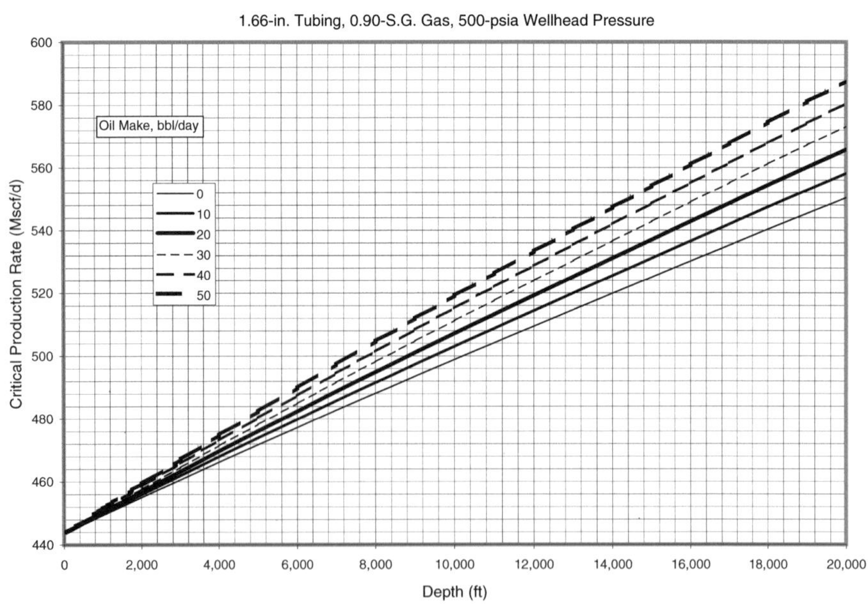

Figure E–78 **Critical gas production rate for condensate removal in 1.66-in tubing against 500 psia wellhead pressure, S.G. 0.90 gas.**

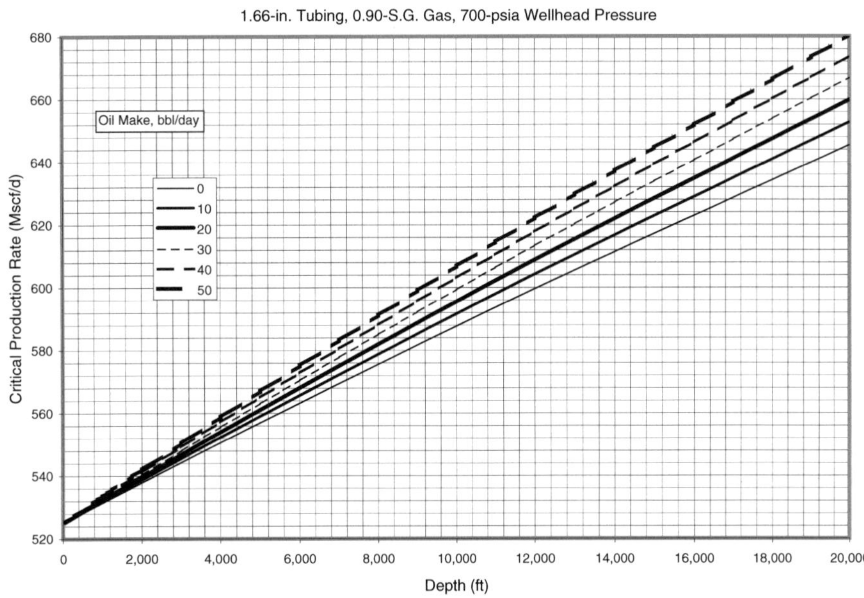

Figure E–79 **Critical gas production rate for condensate removal in 1.66-in tubing against 700 psia wellhead pressure, S.G. 0.90 gas.**

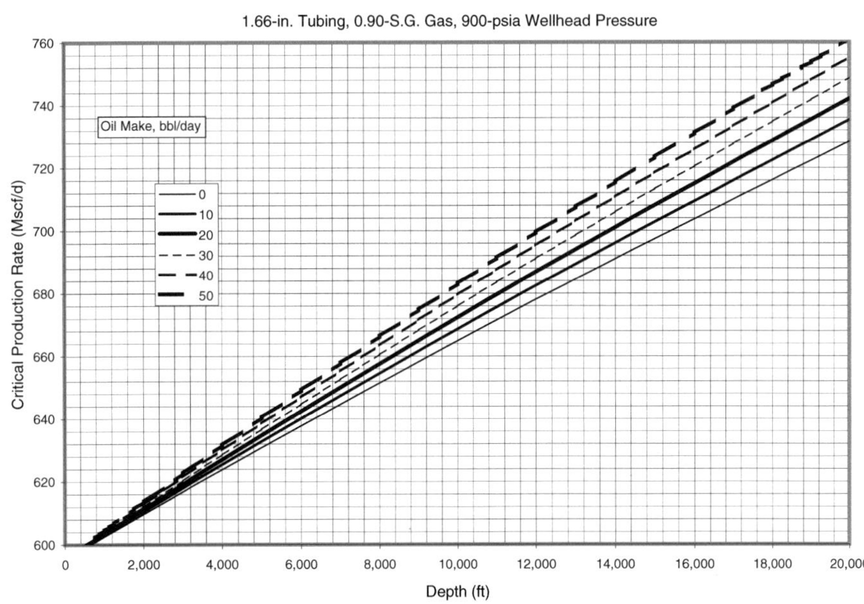

Figure E–80 **Critical gas production rate for condensate removal in 1.66-in tubing against 900 psia wellhead pressure, S.G. 0.90 gas.**

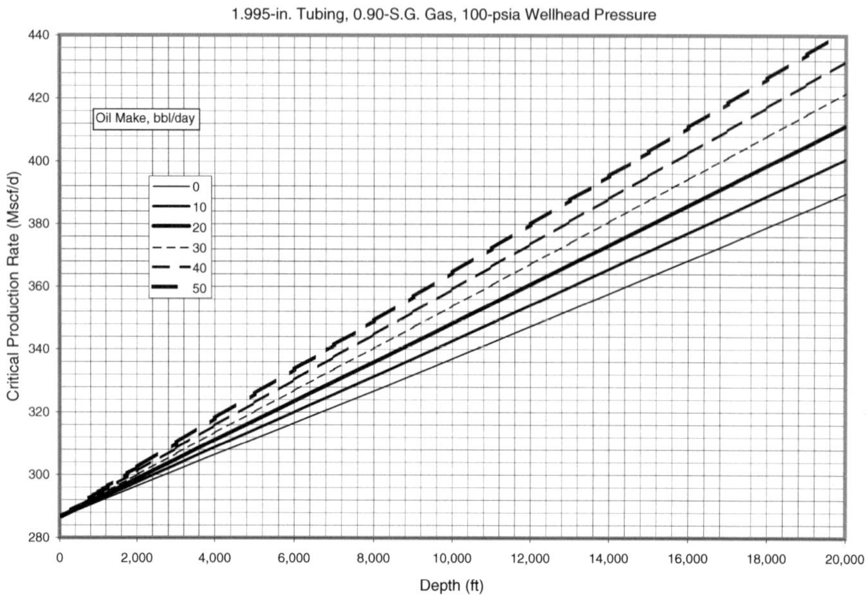

Figure E–81 **Critical gas production rate for condensate removal in 1.995-in tubing against 100 psia wellhead pressure, S.G. 0.90 gas.**

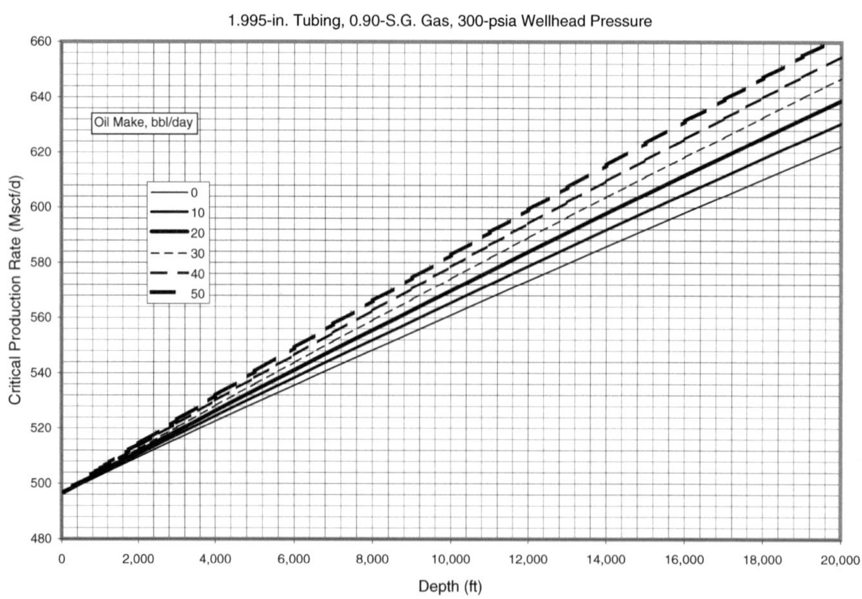

Figure E–82 **Critical gas production rate for condensate removal in 1.995-in tubing against 300 psia wellhead pressure, S.G. 0.90 gas.**

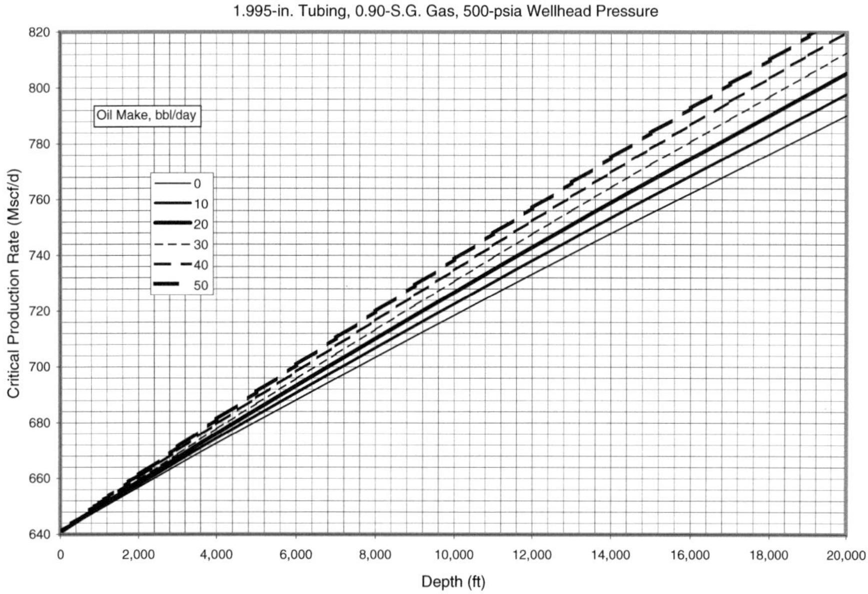

Figure E–83 **Critical gas production rate for condensate removal in 1.995-in tubing against 500 psia wellhead pressure, S.G. 0.90 gas.**

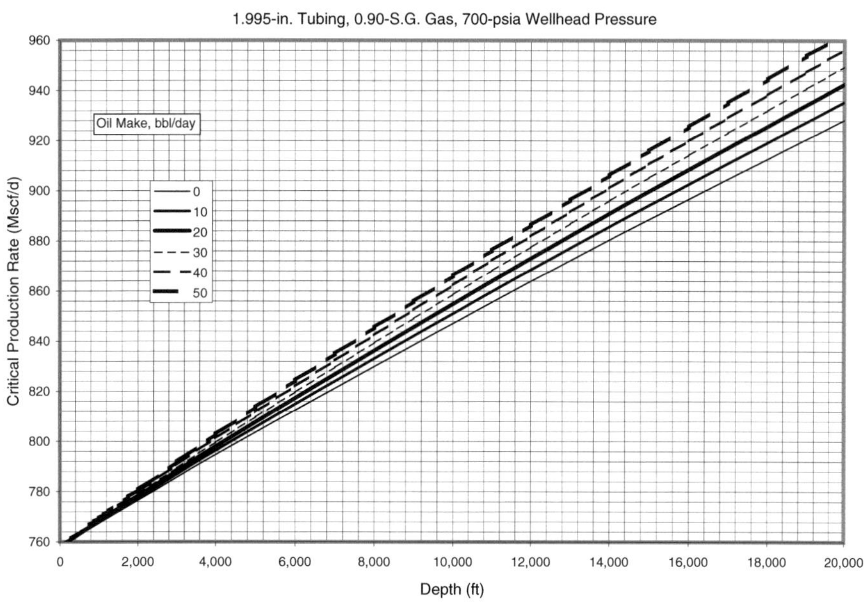

Figure E–84 **Critical gas production rate for condensate removal in 1.995-in tubing against 700 psia wellhead pressure, S.G. 0.90 gas.**

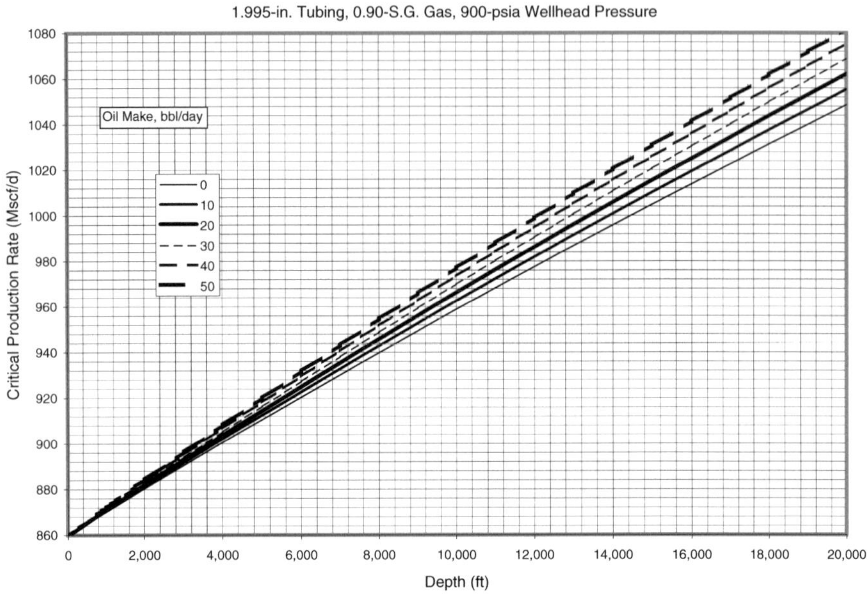

Figure E–85 **Critical gas production rate for condensate removal in 1.995-in tubing against 900 psia wellhead pressure, S.G. 0.90 gas.**

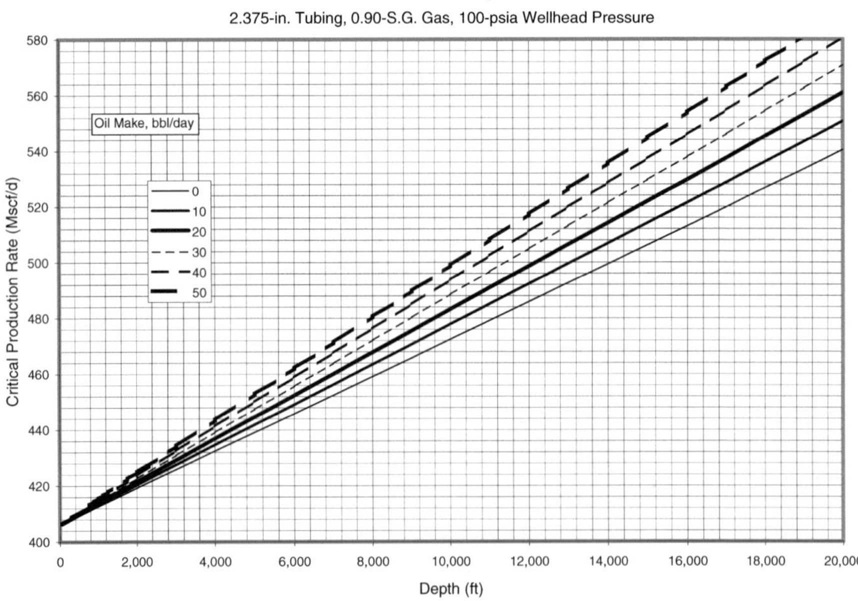

Figure E–86 **Critical gas production rate for condensate removal in 2.375-in tubing against 100 psia wellhead pressure, S.G. 0.90 gas.**

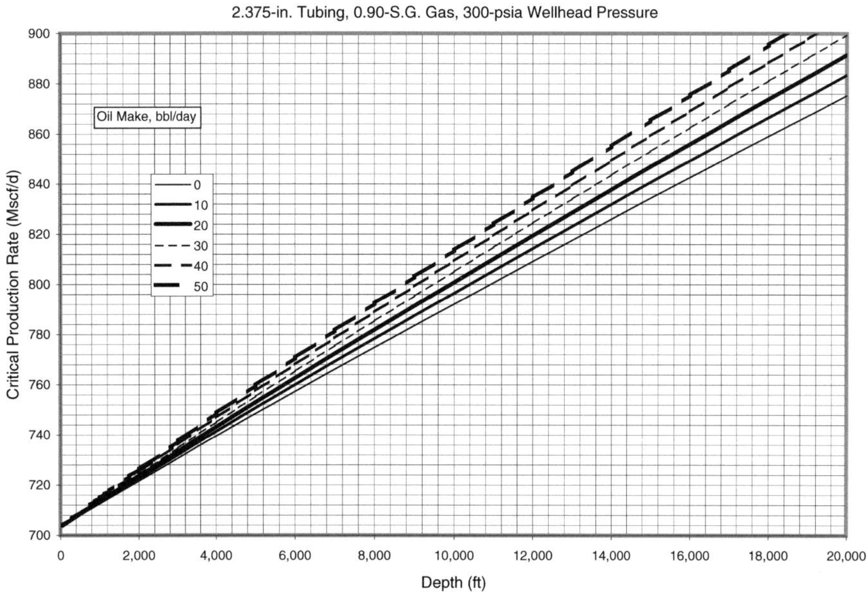

Figure E–87 **Critical gas production rate for condensate removal in 2.375-in tubing against 300 psia wellhead pressure, S.G. 0.90 gas.**

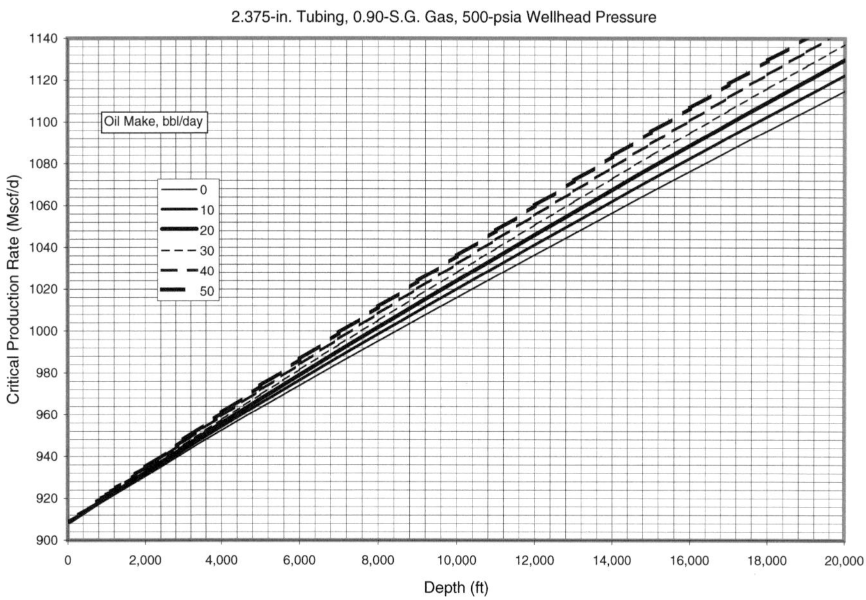

Figure E–88 **Critical gas production rate for condensate removal in 2.375-in tubing against 500 psia wellhead pressure, S.G. 0.90 gas.**

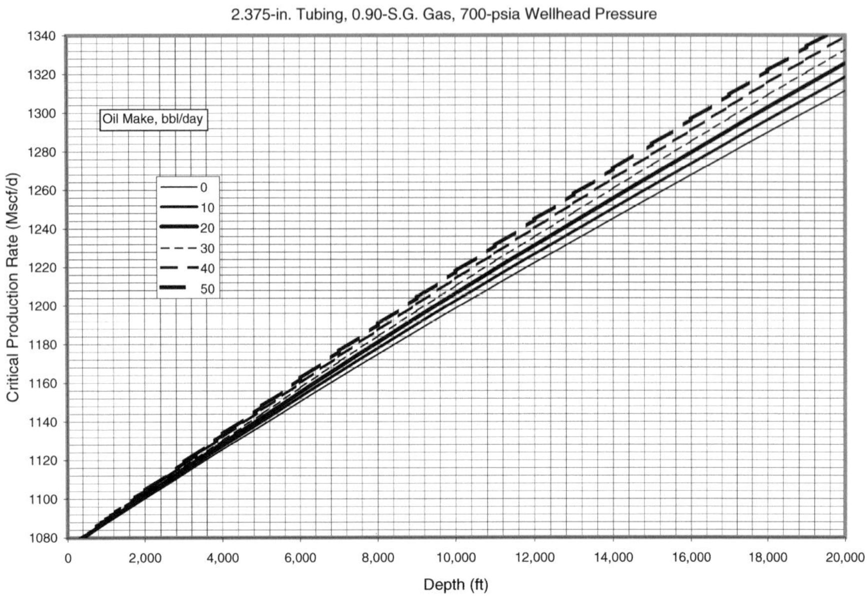

Figure E–89 Critical gas production rate for condensate removal in 2.375-in tubing against 700 psia wellhead pressure, S.G. 0.90 gas.

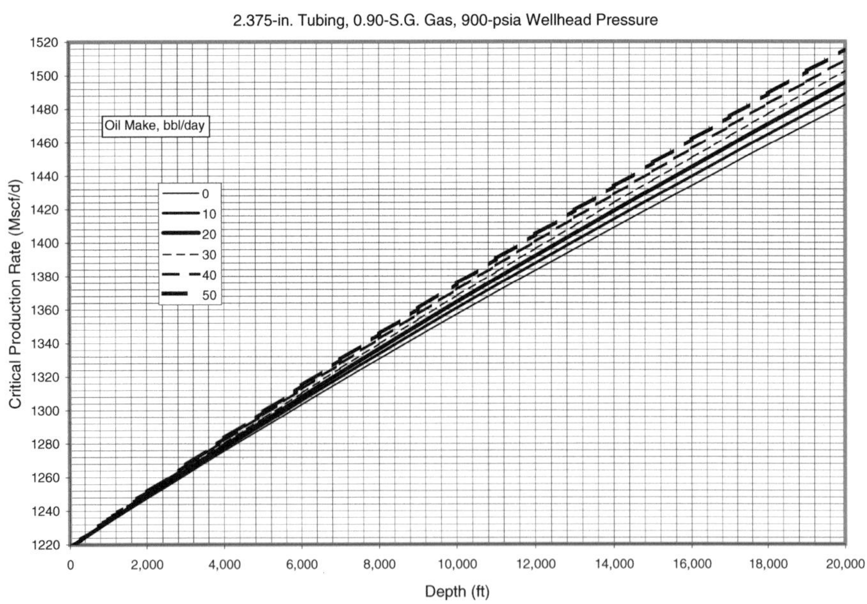

Figure E–90 Critical gas production rate for condensate removal in 2.375-in tubing against 900 psia wellhead pressure, S.G. 0.90 gas.

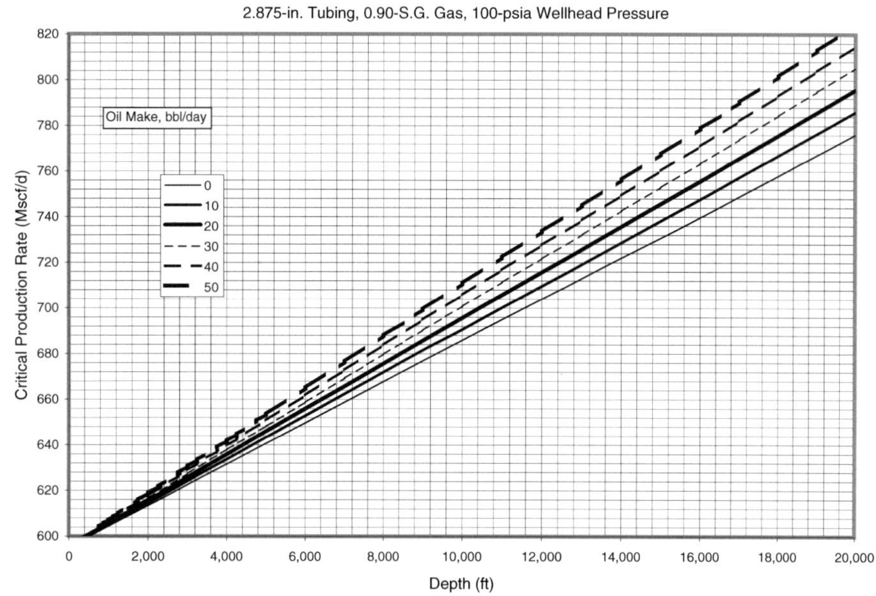

Figure E–91 Critical gas production rate for condensate removal in 2.875-in tubing against 100 psia wellhead pressure, S.G. 0.90 gas.

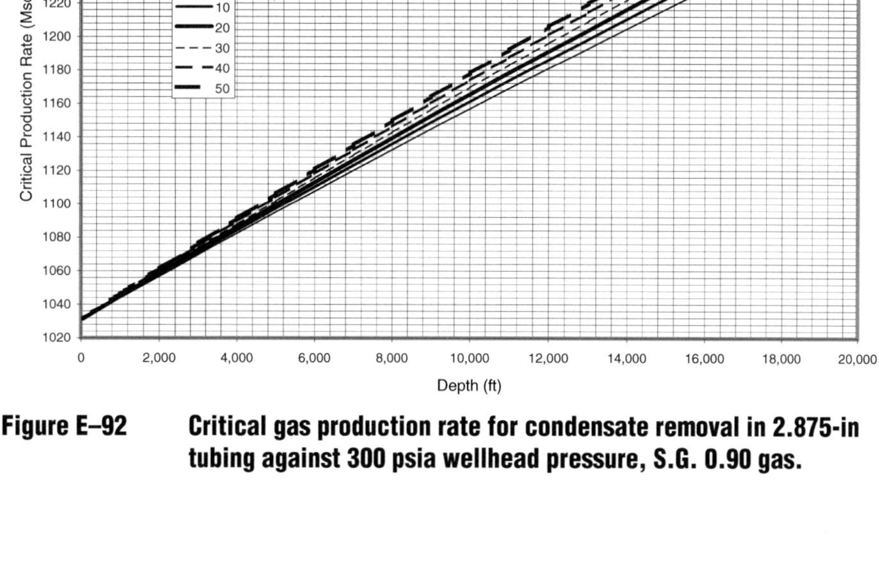

Figure E–92 Critical gas production rate for condensate removal in 2.875-in tubing against 300 psia wellhead pressure, S.G. 0.90 gas.

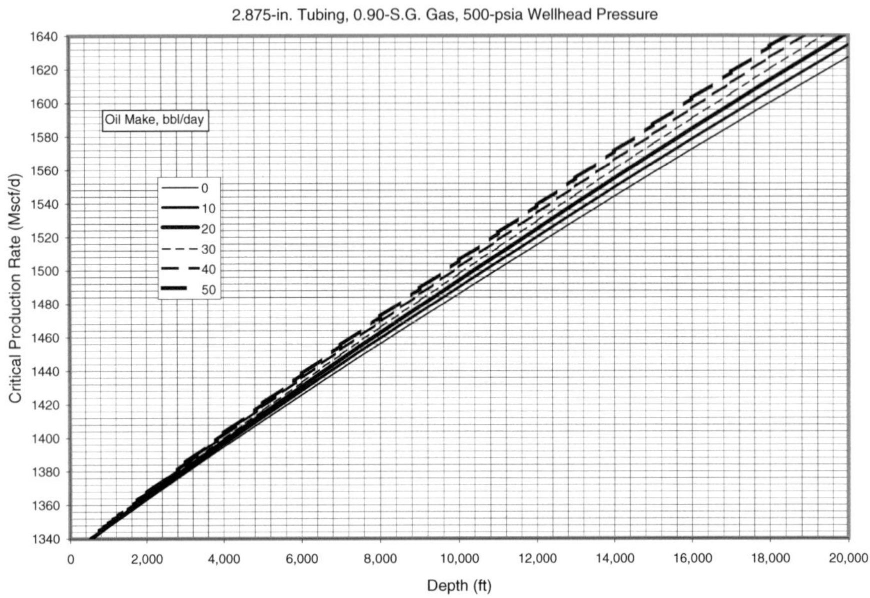

Figure E–93 Critical gas production rate for condensate removal in 2.875-in tubing against 500 psia wellhead pressure, S.G. 0.90 gas.

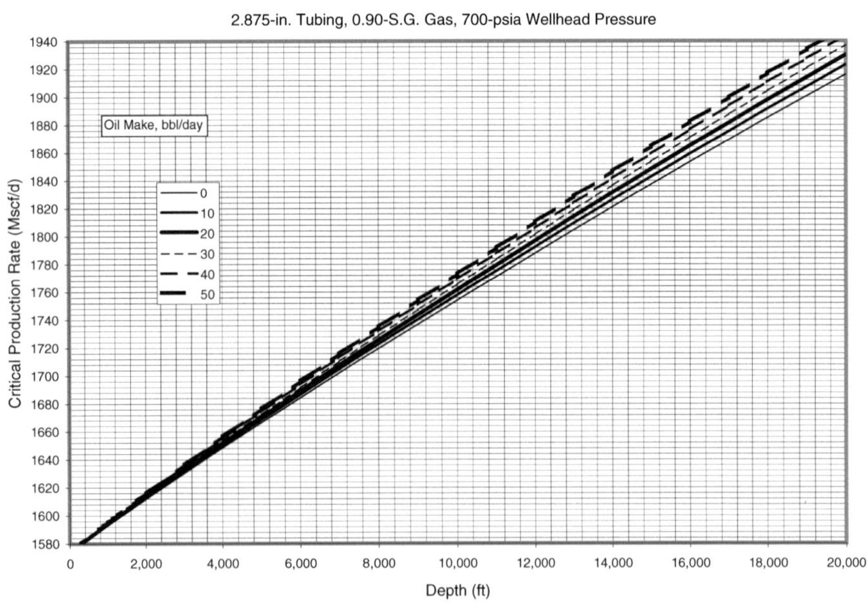

Figure E–94 Critical gas production rate for condensate removal in 2.875-in tubing against 700 psia wellhead pressure, S.G. 0.90 gas.

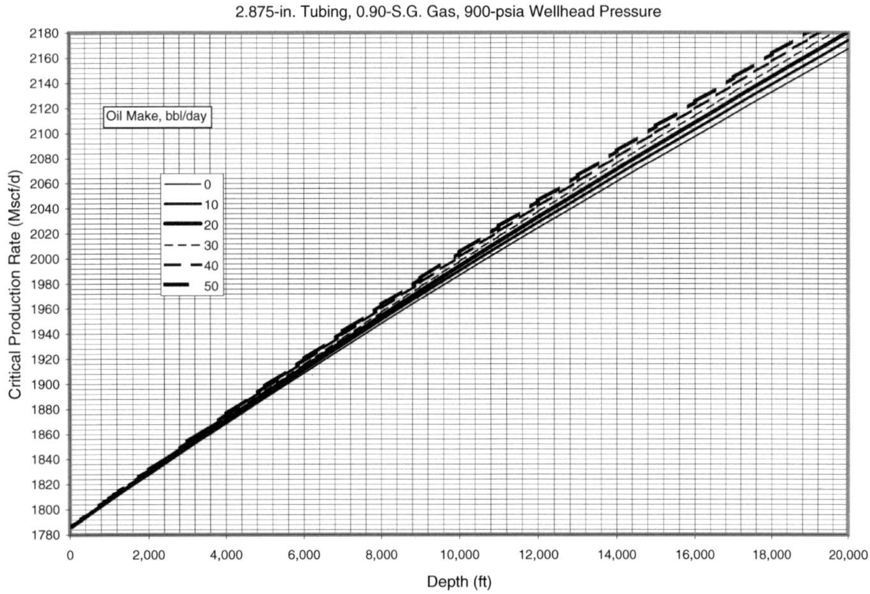

Figure E–95 **Critical gas production rate for condensate removal in 2.875-in tubing against 900 psia wellhead pressure, S.G. 0.90 gas.**

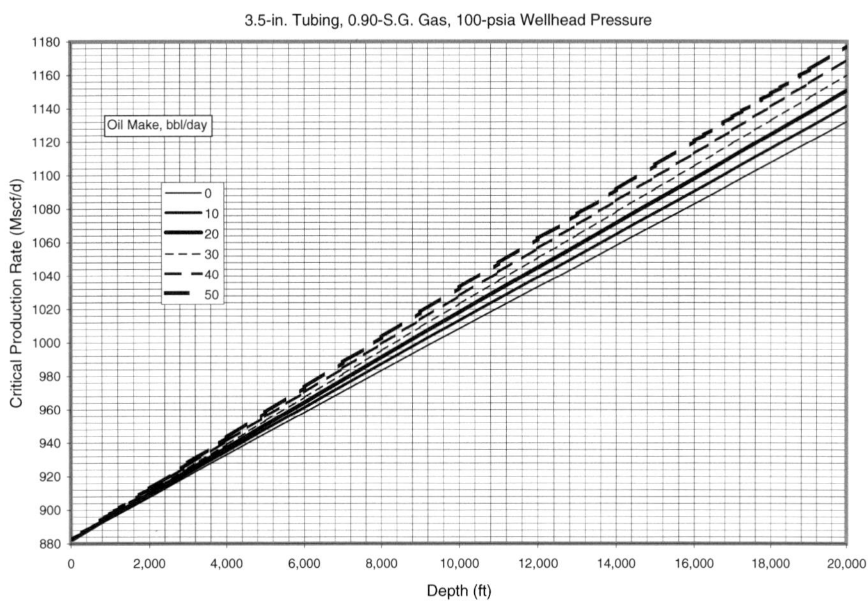

Figure E–96 **Critical gas production rate for condensate removal in 3.5-in tubing against 100 psia wellhead pressure, S.G. 0.90 gas.**

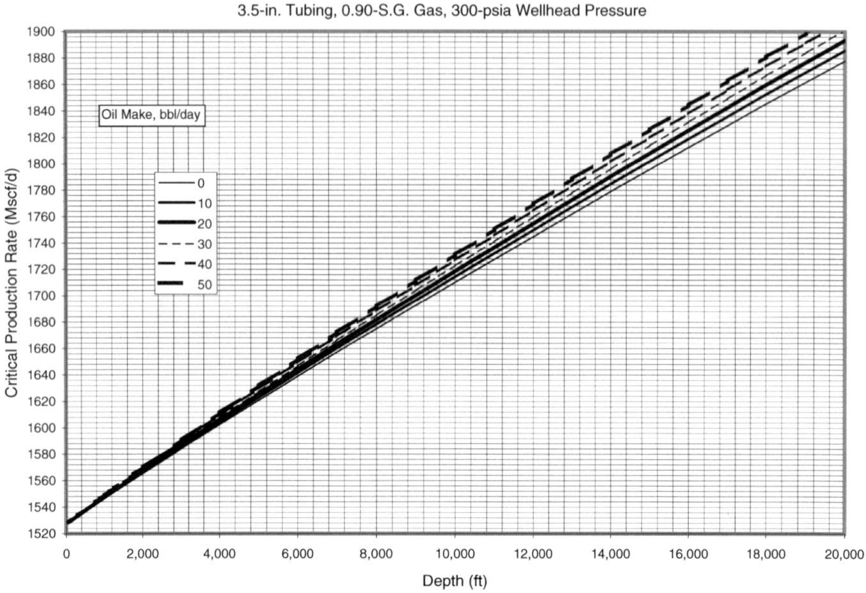

Figure E–97 **Critical gas production rate for condensate removal in 3.5-in tubing against 300 psia wellhead pressure, S.G. 0.90 gas.**

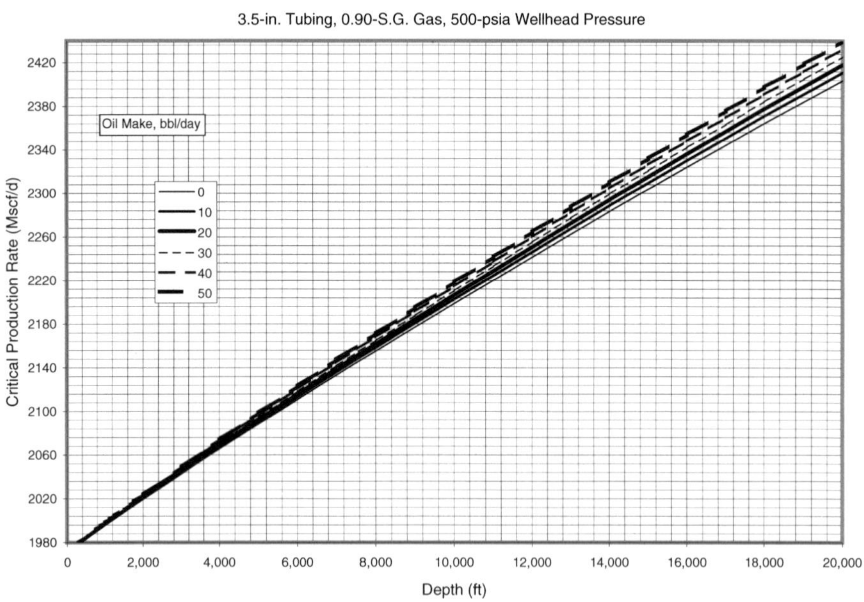

Figure E–98 **Critical gas production rate for condensate removal in 3.5-in tubing against 500 psia wellhead pressure, S.G. 0.90 gas.**

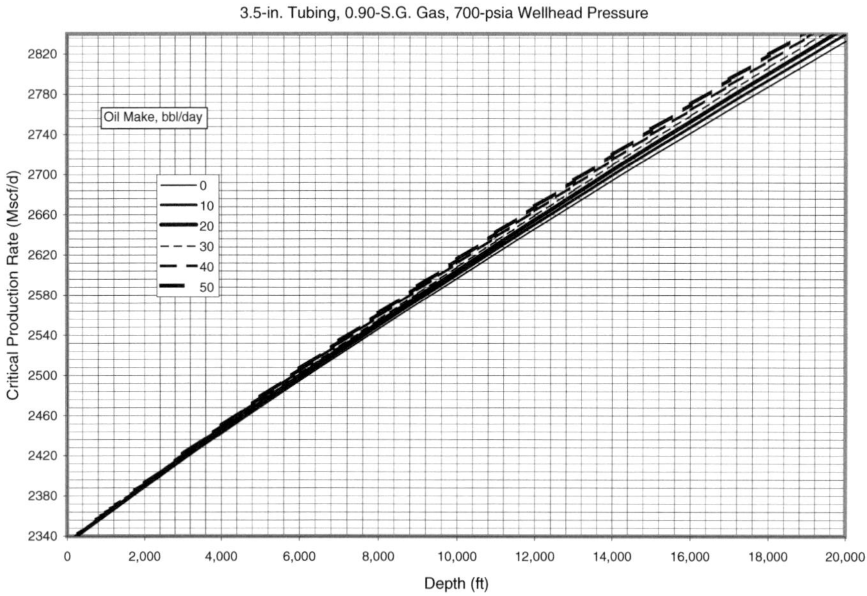

Figure E–99 **Critical gas production rate for condensate removal in 3.5-in tubing against 700 psia wellhead pressure, S.G. 0.90 gas.**

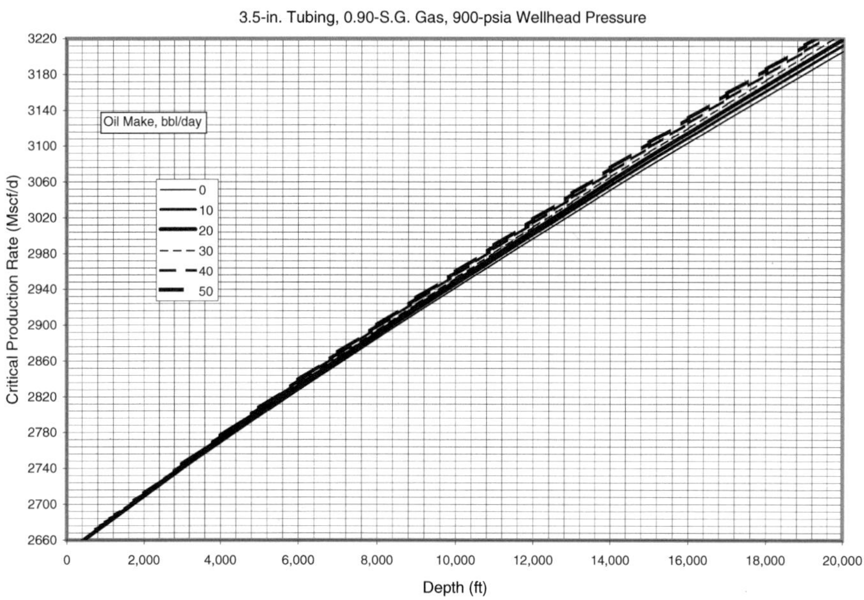

Figure E–100 **Critical gas production rate for condensate removal in 3.5-in tubing against 900 psia wellhead pressure, S.G. 0.90 gas.**

Mathematical Model for Obtaining Correction Factor F_g

The gas rate correction factor for wellbore friction is defined as:

$$F_g = \frac{Q_{gH,Friction}}{Q_{gH,No-friction}} \quad \textbf{(F.1)}$$

where $Q_{gH,Friction}$ and $Q_{gH,No-friction}$ are the gas production rates predicted by mathematical models with and without considering wellbore friction. The $Q_{gH,No-friction}$ can be estimated using the inflow model derived assuming a fully-penetrated box-shape reservoir:

$$Q_{gH,No-friction} = J_{sp,g} L\left(p_e^2 - p_{wf}^2\right) \quad \textbf{(F.2)}$$

where

$$J_{sp,g} = \frac{k_H}{1424\bar{\mu}_g \bar{z} T\left\{I_{ani} \ln\left[\frac{h}{r_w\left(I_{ani}+1\right)}\right] + \frac{\pi y_b}{h} - I_{ani}\left(1.224 - s\right)\right\}} \quad \textbf{(F.3)}$$

and

$$I_{ani} = \sqrt{\frac{k_H}{k_V}} \quad \textbf{(F.4)}$$

where

L = length of drain hole, ft

p_e = reservoir pressure, psi

p_{wf} = flowing bottom hole pressure psi

h = pay zone thickness, ft

k_H = horizontal permeability, md

k_V = vertical permeability, md

y_b = distance of boundary from drain hole, ft

s = skin face, dimensionless

T = reservoir temperature, ^{o}R

$\overline{z}$ = gas compressibility factor, dimensionless

$\overline{\mu}_g$ = gas viscosity, cp

The $Q_{gH,Friction}$ for a fully-penetrated box-shape reservoir is given by

$$Q_{gH,Friction} = \frac{3J_{sp}p_r}{\left(\frac{3}{C}\right)^{2/3}}\left\{2\left[F_1(z_0) - F_1(z)\right] - \left[F_2(z_0) - F_2(z)\right]\right\} \quad \textbf{(F.5)}$$

where

$$z = \frac{p_e}{3}\left[C_2 - \left(\frac{3}{C}\right)^{2/3} L\right] \quad \textbf{(F.6)}$$

$$z_0 = \frac{p_e C_2}{3} \quad \textbf{(F.7)}$$

$$F_1(z) = 3^{-1/3}\left\{\log\left(z + 3^{-1/3}\right) - \frac{1}{2}\log\left(z^2 - 3^{-1/3}z + 3^{-2/3}\right) + 3^{1/2}\arctan\left[\frac{3^{1/2}}{3}\left(2\times 3^{1/3}z - 1\right)\right]\right\} \quad \textbf{(F.8)}$$

$$F_1(z_0) = 3^{-1/3}\left\{\log\left(z_0 + 3^{-1/3}\right) - \frac{1}{2}\log\left({z_0}^2 - 3^{-1/3}z_0 + 3^{-2/3}\right) + 3^{1/2}\arctan\left[\frac{3^{1/2}}{3}\left(2\times 3^{1/3}z_0 - 1\right)\right]\right\} \quad \textbf{(F.9)}$$

$$F_2(z) = 2F_1(z) + \frac{3z}{3z^3+1} \quad \textbf{(F.10)}$$

$$F_2(z_0) = 2F_1(z_0) + \frac{3z_0}{3z_0^3+1} \quad \textbf{(F.11)}$$

$$C = \frac{140.86}{J_{sp}}\sqrt{\frac{p_{wH}d_h^5}{f_f\gamma_g T}} \quad \textbf{(F.12)}$$

$$C_2 = \left(\frac{3}{C}\right)^{2/3} L + \frac{3}{p_e}\left(\frac{p_e - \frac{1}{3}\left(p_e - p_{wH}\right)}{p_e - p_{wH}}\right)^{1/3} \quad \textbf{(F.13)}$$

where

p_{wH} = pressure at the heel of drain hole, psi

d_h = equivalent diameter of the drain hole, in

f_f = Fanning friction factor, dimensionless

g_c = gravitational conversion factor, 32.17 lbm-ft/lbf-s^2

γ_g = gas specific gravity, air = 1.

Index

H

I

J

L

M

N

O

P

R

S

Made in the USA
Monee, IL
15 September 2026

63317833R00295